Progress in Mathematics
Volume 298

Series Editors
Hyman Bass
Joseph Oesterlé
Yuri Tschinkel
Alan Weinstein

Jayce Getz • Mark Goresky

Hilbert Modular Forms with Coefficients in Intersection Homology and Quadratic Base Change

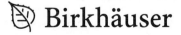

Jayce Getz
Department of Mathematics
McGill University
Montreal, Québec
Canada

Mark Goresky
School of Mathematics
Institute for Advanced Study
Princeton, N.J.
USA

1006822584

ISBN 978-3-0348-0350-2   ISBN 978-3-0348-0351-9 (eBook)
DOI 10.1007/978-3-0348-0351-9
Springer Basel Heidelberg New York Dordrecht London

Library of Congress Control Number: 2012936073

© Springer Basel 2012
This work is subject to copyright. All rights are reserved by the Publisher, whether the whole or part of the material is concerned, specifically the rights of translation, reprinting, reuse of illustrations, recitation, broadcasting, reproduction on microfilms or in any other physical way, and transmission or information storage and retrieval, electronic adaptation, computer software, or by similar or dissimilar methodology now known or hereafter developed. Exempted from this legal reservation are brief excerpts in connection with reviews or scholarly analysis or material supplied specifically for the purpose of being entered and executed on a computer system, for exclusive use by the purchaser of the work. Duplication of this publication or parts thereof is permitted only under the provisions of the Copyright Law of the Publisher's location, in its current version, and permission for use must always be obtained from Springer. Permissions for use may be obtained through RightsLink at the Copyright Clearance Center. Violations are liable to prosecution under the respective Copyright Law.
The use of general descriptive names, registered names, trademarks, service marks, etc. in this publication does not imply, even in the absence of a specific statement, that such names are exempt from the relevant protective laws and regulations and therefore free for general use.
While the advice and information in this book are believed to be true and accurate at the date of publication, neither the authors nor the editors nor the publisher can accept any legal responsibility for any errors or omissions that may be made. The publisher makes no warranty, express or implied, with respect to the material contained herein.

Printed on acid-free paper

Springer is part of Springer Science+Business Media (www.springer.com)

Ferran Sunyer i Balaguer (1912–1967) was a self-taught Catalan mathematician who, in spite of a serious physical disability, was very active in research in classical mathematical analysis, an area in which he acquired international recognition. His heirs created the Fundació Ferran Sunyer i Balaguer inside the Institut d'Estudis Catalans to honor the memory of Ferran Sunyer i Balaguer and to promote mathematical research.

Each year, the Fundació Ferran Sunyer i Balaguer and the Institut d'Estudis Catalans award an international research prize for a mathematical monograph of expository nature. The prize-winning monographs are published in this series. Details about the prize and the Fundació Ferran Sunyer i Balaguer can be found at

http://ffsb.iec.cat/EN/

This book has been awarded the Ferran Sunyer i Balaguer 2011 prize.

The members of the scientific commitee of the 2011 prize were:

Alejandro Adem
 University of British Columbia

Hyman Bass
 University of Michigan

Núria Fagella
 Universitat de Barcelona

Joseph Oesterlé
 Université de Paris VI

Joan Verdera
 Universitat Autònoma de Barcelona

Ferran Sunyer i Balaguer Prize winners since 2001:

2001   Martin Golubitsky and Ian Stewart
The Symmetry Perspective, PM 200

2002   André Unterberger
Automorphic Pseudodifferential Analysis and Higher Level Weyl Calculi, PM 209

Alexander Lubotzky and Dan Segal
Subgroup Growth, PM 212

2003   Fuensanta Andreu-Vaillo, Vincent Caselles and José M. Mazón
Parabolic Quasilinear Equations Minimizing Linear Growth Functionals, PM 223

2004   Guy David
Singular Sets of Minimizers for the Mumford-Shah Functional, PM 233

2005   Antonio Ambrosetti and Andrea Malchiodi
Perturbation Methods and Semilinear Elliptic Problems on R^n, PM 240

José Seade
On the Topology of Isolated Singularities in Analytic Spaces, PM 241

2006   Xiaonan Ma and George Marinescu
Holomorphic Morse Inequalities and Bergman Kernels, PM 254

2007   Rosa Miró-Roig
Determinantal Ideals, PM 264

2008   Luis Barreira
Dimension and Recurrence in Hyperbolic Dynamics, PM 272

2009   Timothy D. Browning
Quantitative Arithmetic of Projective Varieties, PM 277

2010   Carlo Mantegazza
Lecture Notes on Mean Curvature Flow, PM 290

To our families

Contents

1 Introduction
- 1.1 An observation of Serre 1
- 1.2 Notational conventions 3
- 1.3 The setting 4
- 1.4 First main theorem 6
- 1.5 Definition of $IH_n^{\chi_E}(X_0(\mathfrak{c}))$ and $\gamma_{\chi_E}(\mathfrak{m})$ 8
- 1.6 Second main theorem 9
- 1.7 Explicit cycles 11
- 1.8 Finding cycles dual to families of automorphic forms 13
- 1.9 Comments on related literature 14
- 1.10 Comparison with Zagier's formula 16
- 1.11 Outline of the book 17
- 1.12 Problematic primes 18
- Acknowledgement 18

2 Review of Chains and Cochains
- 2.1 Cell complexes and orientations 21
- 2.2 Subanalytic sets and stratifications 22
- 2.3 Sheaves and the derived category 23
- 2.4 The sheaf of chains 24
- 2.5 Homology manifolds 25
- 2.6 Cellular Borel-Moore chains 26
- 2.7 Algebraic cycles 28

3 Review of Intersection Homology and Cohomology
- 3.1 The sheaf of intersection chains 29
- 3.2 The sheaf of intersection cochains 30
- 3.3 Homological stratifications 32
- 3.4 Products in intersection homology and cohomology 35
- 3.5 Finite mappings 37
- 3.6 Correspondences 38

4 Review of Arithmetic Quotients

- 4.1 The setting 41
- 4.2 Baily-Borel compactification 42
- 4.3 L^2 differential forms 43
- 4.4 Invariant differential forms 46
- 4.5 Hecke correspondences for discrete groups 47
- 4.6 Mappings induced by a Hecke correspondence 48
- 4.7 The reductive Borel-Serre compactification 49
- 4.8 Saper's theorem 50
- 4.9 Modular cycles 51
- 4.10 Integration 53

5 Generalities on Hilbert Modular Forms and Varieties

- 5.1 Hilbert modular Shimura varieties 58
- 5.2 Hecke congruence groups 60
- 5.3 Weights 61
- 5.4 Hilbert modular forms 62
- 5.5 Cohomological normalization 65
- 5.6 Hecke operators 66
- 5.7 The Petersson inner product 68
- 5.8 Newforms 71
- 5.9 Fourier series 72
- 5.10 Killing Fourier coefficients 74
- 5.11 Twisting 76
- 5.12 L-functions 81
  - 5.12.1 The standard L-function 82
  - 5.12.2 Rankin-Selberg L-functions 83
  - 5.12.3 Adjoint L-functions 84
  - 5.12.4 Asai L-functions 85
- 5.13 Relationship with Hida's notation 89

6 Automorphic Vector Bundles and Local Systems

- 6.1 Generalities on local systems 92
- 6.2 Classical description of automorphic vector bundles 94
  - 6.2.1 Representations of Γ 94
  - 6.2.2 Representations of K_∞ 95
  - 6.2.3 Flat vector bundles 95
  - 6.2.4 Orbifold local systems 96
- 6.3 Classical description of automorphy factors 97
- 6.4 Adèlic automorphic vector bundles 98
  - 6.4.1 Definitions 98

	6.4.2	Flat bundles	99
	6.4.3	Orbifold bundles	100
6.5		Representations of $\mathbf{GL_2}$	100
6.6		Representations of $G = \operatorname{Res}_{L/\mathbb{Q}}\mathrm{GL}_2$	101
6.7		The section P_z	102
6.8		The local system $\mathcal{L}(\kappa,\chi_0)$	103
6.9		Adèlic geometric description of automorphic forms	105
6.10		Differential forms	107
6.11		Action of the component group	109

7 The Automorphic Description of Intersection Cohomology

7.1	The local system $\mathcal{L}(\kappa,\chi_0)$	112
7.2	The automorphic description of intersection cohomology	114
7.3	Complex conjugation	116
7.4	Atkin-Lehner operator	117
7.5	Pairings of vector bundles	122
7.6	Generalities on Hecke correspondences	126
7.7	Hecke correspondences in the Hilbert modular case	129
7.8	Atkin-Lehner-Hecke compatibility	131
7.9	Integral coefficients	132

8 Hilbert Modular Forms with Coefficients in a Hecke Module

8.1	Notation	136
8.2	Base change for the Hecke algebra	136
8.3	Hilbert modular forms with coefficients in a Hecke module	141
8.4	Hilbert modular forms with coefficients in intersection homology	143
8.5	Proof of Theorem 8.3	144
8.6	The Fourier coefficients of $[\gamma(\mathfrak{m}),\Phi_{\gamma,\chi_E}]_{IH_*}$	146

9 Explicit Construction of Cycles

9.1	Notation for the quadratic extension L/E	151
9.2	Canonical section over the diagonal	152
9.3	Homological properties of $Z_0(\mathfrak{c}_E)$	157
9.4	The twisting correspondence	160
9.5	Twisting the cycle $Z_0(\mathfrak{c}_E)$	165

10 The Full Version of Theorem 1.3

10.1	Statement of results	167
10.2	Rankin-Selberg integrals	170

11 Eisenstein Series with Coefficients in Intersection Homology
- 11.1 Eisenstein series 179
- 11.2 Invariant classes revisited 180
- 11.3 Definition of the $V_{\chi_E}(\mathfrak{m})$ 181
- 11.4 Statement and proof of Theorem 11.2 181

Appendices

A Proof of Proposition 2.4
- A.1 Cellular cosheaves 183
- A.2 Proof of Lemma 2.3 184
- A.3 Proof of Proposition 2.4 185

B Recollections on Orbifolds
- B.1 Effective actions 187
- B.2 Definitions 190
- B.3 Refinement 192
- B.4 Stratification 193
- B.5 Sheaves and cohomology 194
- B.6 Differential forms 196
- B.7 Groupoids 197

C Basic Adèlic Facts
- C.1 Adèles and idèles 199
- C.2 Characters of $L\backslash\mathbb{A}_L$ 200
- C.3 Characters of $\mathrm{GL}_1(L)\backslash\mathrm{GL}_1(\mathbb{A}_L)$ 201
- C.4 Haar measure on the adèles 203

D Fourier Expansions of Hilbert Modular Forms
- D.1 Statement of the theorem 205
- D.2 Fourier analysis on $\mathrm{GL}_2(L)\backslash\mathrm{GL}_2(\mathbb{A}_L)$ 206
- D.3 Whittaker models 207
- D.4 Decomposition of W_h 208
- D.5 Computing $W_{\phi\infty}$ and W_{h0} 209
- D.6 Final steps 212

E Review of Prime Degree Base Change for GL_2
- E.1 Automorphic forms and automorphic representations 214
- E.2 Hecke operators 218
- E.3 Agreement of L-functions 221
- E.4 Langlands functoriality 222
- E.5 Prime degree base change for GL_1 228
- E.6 Conductors of admissible representations of GL_2 229

E.7	The archimedean places	232
E.8	Global base change	234

Bibliography 237

Index of Notation 249

Index of Terminology 253

Chapter 1

Introduction

1.1 An observation of Serre

In their seminal paper [Hirz] on the intersection theory of Hilbert modular surfaces, F. Hirzebruch and D. Zagier mentioned that the motivation for their work was to explain an observation of J.-P. Serre. To describe his observation, let $p \equiv 1 \pmod 4$, let $\mathcal{O} = \mathcal{O}_{\mathbb{Q}(\sqrt{p})}$ be the ring of integers in the real quadratic field $\mathbb{Q}(\sqrt{p})$ and let $\mathfrak{h}$ denote the complex upper half-plane. Denote the associated (non-compact) Hilbert modular surface by

$$Y := \mathrm{SL}_2(\mathcal{O}) \backslash \mathfrak{h}^2.$$

In a letter dated December 8th, 1971, Serre observed that the integer $\lfloor \frac{p+19}{24} \rfloor$ came up in a computation of the arithmetic genus of a surface related to Y, and that

$$\dim_{\mathbb{C}} M^+ = \left\lfloor \frac{p+19}{24} \right\rfloor.$$

Here, $M^+ := M_2^+(\Gamma_0(p), (\frac{\cdot}{p}))$ is the "plus space" consisting of those holomorphic elliptic modular forms of (classical)[1] weight 2, level $\Gamma_0(p)$, and nebentypus $(\frac{\cdot}{p})$ whose ℓth Fourier coefficient is zero for all primes ℓ satisfying $(\frac{\ell}{p}) = 0$.

With the benefit of the modern theory of arithmetic quotients and automorphic forms, the work of [Hirz], and almost 40 years of hindsight, let us try to work out a possible explanation of Serre's observation. Let X be the Baily-Borel or minimal Satake compactification of Y. To explain Serre's observation, we could conjecture that there is a surjective Hecke-equivariant homomorphism

$$\Phi : IH_2(X) \to M^+ \tag{1.1.1}$$

---

[1] We say "classical weight" because, following Hida, we will later give an alternate normalization of the weight for an isomorphic space of automorphic forms.

where $IH_\bullet(X)$ is intersection homology (with respect to middle perversity, see Chapter 3 below).

In order to justify the introduction of intersection homology (as opposed to, say usual singular homology) we make two observations. First, the intersection pairing induces a canonical isomorphism

$$IH_2(X)^\wedge \cong IH_2(X).$$

Therefore Φ defines an element of

$$IH_2(X) \otimes M^+. \qquad (1.1.2)$$

We call an element of the vector space $IH_2(X) \otimes M^+$ an **elliptic modular form with coefficients in** $IH_2(X)$.

The second observation is that there is a canonical action of the Hecke algebra $\mathbb{T}$ attached to $\mathrm{SL}_2(\mathcal{O}_{\mathbb{Q}(\sqrt{p})})$ on $IH_2(X)$ via correspondences. This is explained in detail in Chapters 3.5 and 7 below. This gives a Hecke action on the left factor of (1.1.2). Things are a little more subtle on the right, since the plus space is not preserved by the action of the usual Hecke algebra $\mathbb{T}_\mathbb{Q}$ on spaces of modular forms. However, it is preserved by the image of the base change map

$$b : \mathbb{T}^p \longrightarrow \mathbb{T}_\mathbb{Q}.$$

Here $\mathbb{T}^p \subseteq \mathbb{T}$ is the subalgebra of Hecke operators whose "components at p" are trivial (see (8.2.1)). The map b sends

$$T(\mathfrak{P}') \mapsto \begin{cases} T(p') & \text{if } p' = \mathfrak{P}'\overline{\mathfrak{P}'} \text{ splits in } \mathbb{Q}(\sqrt{p}) \\ T(p'^2) - pT(p,p) & \text{if } p' \text{ is inert in } \mathbb{Q}(\sqrt{p}). \end{cases}$$

See Section 8.2.3 for more details. In Lemma E.5 it is shown that the map b is the homomorphism induced by the usual morphism of L-groups

$$^L\mathrm{GL}_{2/\mathbb{Q}} \to {}^L\mathrm{Res}_{\mathbb{Q}(\sqrt{p})/\mathbb{Q}}(\mathrm{GL}_2)$$

which defines the quadratic base change lifting (sending automorphic representations of $\mathrm{GL}_{2/\mathbb{Q}}$ to automorphic representations of $\mathrm{Res}_{\mathbb{Q}(\sqrt{p})/\mathbb{Q}}\mathrm{GL}_2$).

It is therefore possible to ask that the homomorphism defined by Φ satisfy

$$\Phi \circ T(\mathfrak{P}') = b(T(\mathfrak{P}')) \circ \Phi \qquad (1.1.3)$$

for all prime ideals $\mathfrak{P}' \subset \mathcal{O}_{\mathbb{Q}(\sqrt{p})}$ coprime to p.

The existence of a modular form $\Phi \in IH_2(X) \otimes M^+$ with coefficients in intersection homology satisfying (1.1.3) would give a reasonable explanation of Serre's observation. But there is more: one might ask if there is a family of cycles

$\{Z_n\}$ on X admitting classes $\{[Z_n]\}$ in intersection homology, such that Φ has a Fourier expansion of the form

$$\Phi := \sum_{n \geq 0} [Z_n] q^n. \tag{1.1.4}$$

Here $q := e^{2\pi i z}$, so that Φ can be regarded as a function of a complex parameter $z \in \mathfrak{h}$ with values in the intersection homology group $IH_2(X)$. In fact, such cycles Z_n were discovered by Hirzebruch and Zagier; for $n > 0$ they are "modular" cycles, and Z_0 is a hyperplane section. Consequently, the intersection numbers $\langle [Z_n], [Z_1] \rangle$ are the Fourier coefficients of an elliptic modular form.

Thus the paper [Hirz] may be viewed as having constructed a mapping $\Phi \in IH_2(X) \otimes M^+$ satisfying (1.1.3), admitting an expansion of the form (1.1.4), with a geometric interpretation in terms of modular curves on a Hilbert modular surface. The main result of this book is the existence of such a mapping Φ, not only in the classical setting described above, but also in the analogous setting when $\mathbb{Q}(\sqrt{p})/\mathbb{Q}$ is replaced by a general quadratic extension of totally real number fields L/E. We have attempted to place these results in a setting that might suggest possible generalizations to groups besides GL_2 and to field extensions other than L/E.

Since the work of Hirzebruch and Zagier, a host of generalizations of this sort of phenomenon have been discovered, mostly in the context of liftings connected to Shimura varieties of orthogonal type and/or involving arithmetic Chow groups in place of singular homology. We point out a few references with no claim to completeness: [Borc], [Bru], [Cog] [Go1], [Go2], [Gro], [Ku3], [Ku1], [KuM3], [KuM1], [KuM2], [KuR], [Mi], [Oda1], [Oda2], [Oda3], [Ton], [TonW1], [TonW2]. The Galois representation defined by the étale intersection (co)homology of Hilbert modular varieties was determined in [Bry]; see the further comments in Section 1.9 below.

1.2 Notational conventions

If L is an algebraic number field, let $\mathcal{O}_L$ denote its ring of integers and let $\Sigma(L)$ be the set of infinite places of L, which may be identified with the set of embeddings $\sigma : L \to \mathbb{C}$. Let $v(\mathfrak{p})$ be the finite place corresponding to a prime ideal $\mathfrak{p}$, let $\mathfrak{p}_v$ be the prime ideal corresponding to a finite place v; and if $\mathfrak{c}$ is an ideal, write $v|\mathfrak{c}$ for $\mathfrak{p}_v|\mathfrak{c}$. Denote by $\mathbb{A}_L = \mathbb{A}_{Lf} \mathbb{A}_{L\infty}$ the finite and infinite adèles of L, and by $\mathbb{A}_L^\times = GL_1(\mathbb{A}_L)$ the idèles of L, see Appendix C. (We set $\mathbb{A} = \mathbb{A}_\mathbb{Q}$.) For $b \in \mathbb{A}_L$ denote by b_∞ (resp. b_0) the projection to the subgroup $\mathbb{A}_{L\infty}$ (resp. $\mathbb{A}_{Lf}$). Let $\widehat{\mathbb{Z}} := \prod_{p<\infty} \mathbb{Z}_p$ and write $\widehat{\mathcal{O}}_L = \widehat{\mathbb{Z}} \otimes \mathcal{O}_L = \prod_{v<\infty} \mathcal{O}_v$. The *normalized absolute value* of $z \in \mathbb{A}_L$ is denoted $|z|_{\mathbb{A}_L}$, see Section C.3.

The standard additive character $e_L : \mathbb{A}_L \to \mathbb{C}^\times$ (see Section C.2) decomposes as the product $e_{Lf}(z_0) e_{L\infty}(z_\infty)$ with $e_{L\infty}(z_\infty) = \prod_{\sigma \in \Sigma(L)} \exp(2\pi i z_\sigma)$. Each finite idèle $x_0 \in \mathbb{A}_{Lf}^\times$ determines a fractional ideal

$$[x_0] = \prod_{\mathfrak{p}} \mathfrak{p}^{\mathrm{ord}_\mathfrak{p}(x_\mathfrak{p})}$$

where the product is over all prime ideals. A *Hecke character* or *quasicharacter* of L is a continuous homomorphism (see Section C.3),

$$\chi : L^\times \backslash \mathbb{A}_L^\times \to \mathbb{C}^\times.$$

If L/E is an abelian extension of number fields then, by class field theory, the group of characters $\mathrm{Gal}(L/E)^\wedge$ may be identified with those Hecke characters $\chi_E : E^\times \backslash \mathbb{A}_E^\times \to \mathbb{C}^\times$ that are trivial on the image of the norm $\mathrm{N} : \mathbb{A}_L^\times \to \mathbb{A}_E^\times$.

Throughout this book, with the exception of Appendix E, we fix a quadratic extension L/E of totally real number fields, with relative discriminant $d_{L/E}$ and relative different $\mathcal{D}_{L/E} \subset L$ (see Section C.2), so that $a \in \mathcal{D}_{L/E}^{-1}$ iff $\mathrm{Tr}_{L/E}(ax) \in \mathcal{O}_E$ for all $x \in \mathcal{O}_L$. Ideals, characters, etc. for E will often be denoted with a subscript E, while ideals, characters, etc. for L will often be denoted without a subscript. An element $\xi \in L$ is *totally positive* (written $\xi \gg 0$) if $\sigma(\xi) > 0$ for all $\sigma \in \Sigma(L)$. The group of characters $\mathrm{Gal}(L/E)^\wedge = \{1, \eta\}$ consists of two elements, of which $\eta = \eta_E$ may be considered to be a Hecke character that is trivial at the infinite places.

1.3 The setting

Consider the reductive algebraic $\mathbb{Q}$-rank one group $G_L := \mathrm{Res}_{L/\mathbb{Q}}(\mathrm{GL}_2)$. Here $\mathrm{Res}_{L/\mathbb{Q}}$ denotes the Weil restriction of scalars (see [PlaR] Section 2.1.2). Thus G_L is defined over $\mathbb{Q}$ and its group of rational points may be canonically identified as $G_L(\mathbb{Q}) = \mathrm{GL}_2(L)$. For any $\mathbb{Q}$ algebra A, the group of A-valued points of G_L is $G_L(A) = \mathrm{GL}_2(A \otimes_\mathbb{Q} L)$. For any rational prime p, we have: $G(\mathbb{Q}_p) = \prod_{\mathfrak{p}|p} \mathrm{GL}_2(L_\mathfrak{p})$, and $G(\mathbb{R}) = \prod_{\sigma \in \Sigma(L)} \mathrm{GL}_2(\mathbb{R})$. Let K_∞ be the normalizer of the standard $\mathbb{R}$-algebraic homomorphism (see 5.1.1)

$$\mathbb{C}^\times \cong \mathrm{Res}_{\mathbb{C}/\mathbb{R}}(\mathbb{G}_m)(\mathbb{R}) \longrightarrow G_L(\mathbb{R}).$$

The associated symmetric space is $G_L(\mathbb{R})/K_\infty \cong (\mathbb{C} - \mathbb{R})^{\Sigma(L)}$. If $\mathfrak{c} \subset \mathcal{O}_L$ is an ideal, there is a *Hilbert modular variety*

$$Y_0(\mathfrak{c}) := G_L(\mathbb{Q}) \backslash G_L(\mathbb{A}) / K_\infty K_0(\mathfrak{c}),$$

of (complex) dimension $n = [L : \mathbb{Q}]$. Here $\mathbb{A} = \mathbb{A}_\mathbb{Q}$ and as discussed in equation (1.3.1), (5.2.1),

$$K_0(\mathfrak{c}) := \left\{ u \in G_L(\widehat{\mathbb{Z}}) : u = \begin{pmatrix} a & b \\ c & d \end{pmatrix} \text{ with } [c] \subseteq \mathfrak{c} \right\} \tag{1.3.1}$$

1.3. The setting

is the standard compact open congruence subgroup $G_L(\mathbb{A}_f)$ of "Hecke type". We restrict to subgroups of this type in order to simplify the Fourier expansions, but more general compact open subgroups could be used. Since $K_0(\mathfrak{c})$ may contain torsion elements, the variety $Y_0(\mathfrak{c})$ is an orbifold, rather than a manifold; see Appendix B.

The variety $Y_0(\mathfrak{c})$ is a disjoint union of finitely many arithmetic quotients $\Gamma_j \backslash \mathfrak{h}^{\Sigma(D)}$ of a product of upper half-planes, see (5.1.7), on which $G_L(\mathbb{A}_\infty)$ acts transitively by fractional linear transformations: if $\mathbf{i} = (i, i, \ldots, i) \in \mathfrak{h}^{\Sigma(L)}$ is the base point, then
$$\begin{pmatrix} y_\infty & x_\infty \\ 0 & 1 \end{pmatrix} . \mathbf{i} = x_\infty + i y_\infty \in \mathfrak{h}^{\Sigma(L)}.$$

Let $\chi : L^\times \backslash \mathbb{A}_L^\times \to \mathbb{C}$ be a Hecke character. As explained in Section 5.4 there is a standard way to translate between Hilbert cusp forms (with *level* $K_0(\mathfrak{c})$ and *nebentypus*, or central character, χ), considered as holomorphic sections of certain vector bundles on $Y_0(\mathfrak{c})$, to functions $h : G_L(\mathbb{A}) \to \mathbb{C}$ satisfying certain equivariance, growth, and rigidity conditions, see Section 5.4. The collection of cusp forms of weight $(0,0)$ is denoted
$$S(K_0(\mathfrak{c}), \chi) = S_{(0,0)}(K_0(\mathfrak{c}), \chi).$$
(The subscript $(0,0)$, denoting the weight, will be omitted only in the introduction.) In the special case $L = \mathbb{Q}(\sqrt{d})$ for $d \in \mathbb{Z}_{>0}$, such a cusp form corresponds to a Hilbert modular form of weight $(2,2)$ on each connected component $\Gamma_j \backslash \mathfrak{h}^2$ of $Y_0(\mathfrak{c})$, where $\Gamma_i \subset \mathrm{SL}_2(\mathcal{O}_{\mathbb{Q}(\sqrt{d})})$ is a certain congruence subgroup.

The *Fourier expansion* of a cusp form $h \in S(K_0(\mathfrak{c}), \chi)$ is given by
$$h\left(\begin{pmatrix} y & x \\ 0 & 1 \end{pmatrix}\right) = |y|_{\mathbb{A}_L} \left(\sum_{\substack{\xi \in L^\times \\ \xi \gg 0}} b(\xi y_0) q(\xi x, \xi y) \right).$$

(See Theorem 5.8 and its explanation and proof in Appendix D for the general case.) Here,
$$q(x, y) = q_{(0,0)}(x, y) = e_f(x_0) \exp\left(-2\pi i \sum_{\sigma \in \Sigma(L)} x_\sigma \right) \exp\left(2\pi \sum_{\sigma \in \Sigma(L)} y_\sigma\right)$$
corresponds to the function q that occurs in the classical Fourier series (1.1.4); and $b(\xi y_0) \in \mathbb{C}$ is the Fourier coefficient of $q(\xi x, \xi y)$. (In the Introduction we suppress the subscript $(0,0)$, which denotes the weight.) In fact, $q(\xi x, \xi y)$ is a *Whittaker function*, normalized according to Hida's conventions in [Hid7] which we recall in Section 5.4. The Fourier coefficient $b(\xi y_0)$ of h depends only on the fractional ideal $[\xi y_0]$, hence also on the fractional ideal $\mathfrak{m} := [\xi y_0] \mathcal{D}_{L/\mathbb{Q}}$, and it vanishes unless this ideal is integral, in which case we refer to
$$a(\mathfrak{m}, h) := b(\xi y_0)$$
as the $\mathfrak{m}$th *Fourier coefficient* of h.

If $\mathfrak{c} \subset \mathcal{O}_L$ is an ideal, let $\mathbb{T}_\mathfrak{c}$ be the Hecke algebra. It is a subalgebra of the algebra of of compactly supported smooth functions on $K_0(\mathfrak{c})\backslash G_L(\mathbb{A}_f)/K_0(\mathfrak{c})$ with convolution as multiplication, and it is generated as a $\mathbb{Z}$-algebra by certain elements $T_\mathfrak{c}(\mathfrak{b})$ and $T_\mathfrak{c}(\mathfrak{b},\mathfrak{b})$ as $\mathfrak{b}$ ranges over the ideals of $\mathcal{O}_L$ (see Section 5.6). In Section 8.2 and Section E.4 below we recall the construction of the standard map of L groups ${}^L G_E \to {}^L G_L$ which gives rise to *base change*, a "lifting" from automorphic representations of G_E to automorphic representations of G_L. There is an associated base change map

$$b: \mathbb{T}^{\mathfrak{c}\mathcal{D}_{L/E}} \longrightarrow \mathbb{T}^{\mathfrak{c}_E d_{L/E}}$$

Here $\mathfrak{c}_E = \mathfrak{c} \cap \mathcal{O}_E$ and the superscripts denote the subalgebra of operators trivial at the places dividing $\mathfrak{c}\mathcal{D}_{L/E}$ (resp. $\mathfrak{c}_E d_{L/E}$) which is implicit in the theorems of Hirzebruch-Zagier and its generalizations.

Let $X_0(\mathfrak{c})$ be the Baily-Borel (Satake) compactification of $Y_0(\mathfrak{c})$. Each Hecke operator $T_\mathfrak{c}(\mathfrak{b})$ determines a correspondence and hence an endomorphism $T_\mathfrak{c}(\mathfrak{b})_*$ of the intersection homology groups $IH_*(X_0(\mathfrak{c})) = I^m H_*(X_0(\mathfrak{c}), \mathbb{C})$ (see Section 7.6, Section 4.5 and Section 5.3). In Section 1.5 and Section 8.3 below, we define a certain subgroup $IH_n^{\chi_E}(X_0(\mathfrak{c}))$ that is preserved by the Hecke algebra $\mathbb{T}_\mathfrak{c}$.

1.4 First main theorem

Hirzebruch and Zagier proved that the intersection numbers of (certain) modular curves on a Hilbert modular surface are the Fourier coefficients of an elliptic modular form. In Theorem 8.4 below, we show that results of this type are formal consequences of abelian base change (for certain cohomological automorphic representations of G_E), and as such, they hold for any intersection cohomology classes, not just those coming from modular cycles. For simplicity, we state a special case of Theorem 8.4; the notation is explained in the paragraphs following the statement of Theorem 1.1.

Set $n = [L : \mathbb{Q}] = \dim_\mathbb{C}(X_0(\mathfrak{c}))$. Let $\mathfrak{c} \subset \mathcal{O}_L$ be an ideal, set $\mathfrak{c}_E = \mathfrak{c} \cap \mathcal{O}_E$, and let $\chi_E \in \mathrm{Gal}(L/E)^\wedge = \{1, \eta\}$. For each $\gamma \in IH_n^{\chi_E}(X_0(\mathfrak{c}))$ and for each ideal $\mathfrak{m} \subset \mathcal{O}_E$, in Section 1.5 and Section 8.3 below, we define a certain Hecke translate $\gamma_{\chi_E}(\mathfrak{m})$ of γ. Theorem 1.1 below, a generalization of the Hirzebruch-Zagier theorem, says that the formal Fourier series constructed from these classes defines a Hilbert modular form on $\mathrm{R}_{E/\mathbb{Q}}\mathrm{GL}_2$ with coefficients in $IH_n^{\chi_E}(X_0(\mathfrak{c}))$, in the terminology of [Ku3], [KuM3]. This result is a formal consequence of the existence of base change for L/E. Theorem 1.2 below gives a way to explicitly compute some of the Fourier coefficients of the resulting Hilbert modular forms in terms of certain period integrals.

In order to state the theorem, we note that any

$$\Phi \in IH_n^{\chi_E}(X_0(\mathfrak{c})) \otimes S(K_0(\mathcal{N}(\mathfrak{c})), \chi_E)$$

1.4. First main theorem

can be multiplied with elements of $IH_n(X_0(\mathfrak{c}))$, yielding a linear map

$$\langle \cdot, \Phi_{\gamma,\chi_E} \rangle_{IH} : IH_n(X_0(\mathfrak{c})) \longrightarrow S(K_0(\mathcal{N}(\mathfrak{c})), \chi_E). \tag{1.4.1}$$

Theorem 1.1. *There is an ideal $\mathcal{N}(\mathfrak{c})$ and a unique*

$$\Phi_{\gamma,\chi_E} \in IH_n^{\chi_E}(X_0(\mathfrak{c})) \otimes S(K_0(\mathcal{N}(\mathfrak{c})), \chi_E)$$

such that

(1) *The map $\langle \cdot, \Phi_{\gamma,\chi_E} \rangle_{IH}$ is Hecke equivariant with respect to the base change map b, that is,*

$$\langle t_* \psi, \Phi_{\gamma,\chi_E} \rangle_{IH} = \langle \psi, \Phi_{\gamma,\chi_E} \rangle_{IH} | b(t)$$

for any $\psi \in IH_n(X_0(\mathfrak{c}))$ and $t \in \mathbb{T}^{\mathfrak{c}\mathcal{D}_{L/E}}$.

(2) *If $\mathfrak{m} \subseteq \mathcal{O}_E$ is a norm from $\mathcal{O}_L$, $\mathfrak{m} + (\mathfrak{c} \cap \mathcal{O}_E)d_{L/E} \ne \mathcal{O}_E$, or $\mathfrak{m} + d_{L/E} = \mathcal{O}_E$ and $\eta(\mathfrak{m}) = -1$ then the $\mathfrak{m}$th Fourier coefficient of Φ_{γ,χ_E} is $\gamma_{\chi_E}(\mathfrak{m})$.*

In Theorem 1.1, $\mathcal{D}_{E/\mathbb{Q}}$ is the absolute different (see Appendix C.2). The ideal $\mathcal{N}(\mathfrak{c})$ may be considered to be an "upper bound" for the level of the resulting cusp form. It is

$$\mathcal{N}(\mathfrak{c}) := \mathfrak{m}_2 \mathfrak{b}_{L/E} \mathfrak{c}_E \prod_{\mathfrak{p}} \mathfrak{p},$$

where the product is over those primes $\mathfrak{p}$ which divide $\mathfrak{c}_E = \mathfrak{c} \cap \mathcal{O}_E$, where $\mathfrak{m}_2 \subset \mathcal{O}_E$ is an ideal divisible only by dyadic primes that we may take to be $\mathcal{O}_E$ if $\mathfrak{c} + 2\mathcal{O}_L = \mathcal{O}_L$, and where $\mathfrak{b}_{L/E}$ is an ideal divisible only by those primes ramifying in L/E. Finally, the statement that the $\mathfrak{m}$th Fourier coefficient of Φ_{γ,χ_E} is $\gamma_{\chi_E}(\mathfrak{m})$ is (by definition) the statement that the $\mathfrak{m}$th Fourier coefficient of

$$\langle \psi, \Phi_{\gamma,\chi_E} \rangle_{IH} \tag{1.4.2}$$

is

$$\langle \psi, \gamma(\mathfrak{m}) \rangle$$

where $\langle \cdot, \cdot \rangle = \langle \cdot, \cdot \rangle_{IH}$ is the Poincaré dual pairing on intersection homology, and where $\psi \in IH_n(X_0(\mathfrak{c}))$. Thus, the intersection numbers $\langle \psi, \gamma(\mathfrak{m}) \rangle$ are the Fourier coefficients of a modular form if $\mathfrak{m}$ is a norm or $\eta(\mathfrak{m}) = -1$.

Remarks. The "full" version of Theorem 8.4 involves more general local coefficient systems on $X_0(\mathfrak{c})$ which in turn necessitates the introduction of higher weight Hilbert modular forms with nebentypus (see Chapter 8). The level of $\langle \Lambda, \Phi_{\gamma,\chi_E} \rangle$ for a given linear functional $\langle \Lambda, \cdot \rangle$ on $IH_n^E(X_0(\mathfrak{c}))$ could be an ideal strictly containing $\mathcal{N}(\mathfrak{c})$. The image of the mapping (1.4.1) is contained in a certain subspace, $S^+(\mathcal{N}(\mathfrak{c}_E), \chi_E)$, defined in equation (8.3.5), which is the analog of the "plus space" M^+ of Hirzebruch-Zagier. Theorem 1.1 remains true if we replace $IH_n(X_0(\mathfrak{c}))$ with any $\mathbb{T}_\mathfrak{c}$-module, as described in the "full" version, Theorem 8.4. This reflects the fact that the proof is a formal consequence of quadratic base change. Finally, we do not give an expression for the Fourier coefficient of Φ_{γ,χ_E} attached any ideal $\mathfrak{m}$ that is not a norm but still satisfies $\eta(\mathfrak{m}) = 1$. These coefficients seem slightly out of reach of our formal methods, but can probably be computed using the theta correspondence, the key tool in the approach of Kudla and Millson (see [KuM1], [KuM2]).

1.5 Definition of $IH_n^{\chi_E}(X_0(\mathfrak{c}))$ and $\gamma_{\chi_E}(\mathfrak{m})$

The "new" space $S^{\text{new}}(K_0(\mathfrak{c}))$ (see Section 5.8 and Section 1.7), has a basis consisting of *newforms*, each element of which is a simultaneous eigenform (of multiplicity one) for all Hecke operators. For such a newform f the eigenvalues are denoted $\lambda_f(\mathfrak{m})$, meaning that

$$f|T_{\mathfrak{c}}(\mathfrak{m}) = \lambda_f(\mathfrak{m})f \qquad (1.5.1)$$

for all ideals $\mathfrak{m} \subset \mathcal{O}_L$. Denote by

$$IH_n(X_0(\mathfrak{c}))(f)$$

the f-isotypical component of $IH_n(X_0(\mathfrak{c}))$ viewed as a Hecke module (see Section 7.2). Let $\chi_E \in \text{Gal}(L/E)^\vee$. Then

$$IH_n^{\chi_E}(X_0(\mathfrak{c})) := \bigoplus_g IH_n(X_0(\mathfrak{c}))(g)$$

where the sum is over those g such that $g \in S^{\text{new}}(K_0(\mathfrak{c}))$ is the base change of a Hilbert modular form of nebentypus χ_E (see Section 8)[2].

Fix $\gamma \in IH_n^{\chi_E}(X_0(\mathfrak{c}))$. For any ideal $\mathfrak{m} \subset \mathcal{O}_E$ define the Hecke translate

$$\gamma_{\chi_E}(\mathfrak{m}) := \begin{cases} \widehat{T}(\mathfrak{m})_*\gamma & \text{if } \mathfrak{m} + d_{L/E}\mathfrak{c}_E = \mathcal{O}_L \\ 0 & \text{otherwise} \end{cases} \qquad (1.5.2)$$

where the Hecke operator (cf. Section 8.2) $\widehat{T}(\mathfrak{m}) := \widehat{T}_{\mathfrak{c},\chi_E}(\mathfrak{m}) \in \mathbb{T}_{\mathfrak{c}} \otimes \mathbb{C}$ is defined as follows. If $\mathfrak{m} \subset \mathcal{O}_E$ is not a norm from $\mathcal{O}_L$ set $\widehat{T}(\mathfrak{m}) := 0$. Otherwise, the operator $\widehat{T}(\mathfrak{m})$ is defined multiplicatively, but involves a choice: if $\mathfrak{p} \subset \mathcal{O}_E$ is a prime ideal that splits in $\mathcal{O}_L$ (say, $\mathfrak{p}\mathcal{O}_L = \mathfrak{P}\overline{\mathfrak{P}}$) then choose a prime $\mathfrak{P}$ lying above $\mathfrak{p}$. Otherwise let $\mathfrak{P} = \mathfrak{p}\mathcal{O}_L$. Then

$$\widehat{T}(\mathfrak{p}^r) := \begin{cases} \text{Id} & \text{if } r = 0 \\ T_{\mathfrak{c}}(\mathfrak{P}^r) & \text{if } \mathfrak{p} \text{ splits in } L \\ T_{\mathfrak{c}}(\mathfrak{P}^{r/2}) + \chi_E(\mathfrak{p})N_{E/\mathbb{Q}}(\mathfrak{p})T_{\mathfrak{c}}(\mathfrak{P}^{r-2}) & \text{if } \mathfrak{p} \text{ is inert and } r \text{ is even} \\ 0 & \text{otherwise} \end{cases}$$

where "otherwise" means that either $\mathfrak{p}$ ramifies in L, or $\mathfrak{p}$ is inert and r is odd.

---

[2]In the introduction, we only consider Hilbert modular forms on L with trivial nebentypus χ, so that χ_E is just an element of $\text{Gal}(L/E)^\wedge = \{1, \eta\}$. In Chapter 8, we allow Hilbert modular forms on L with more general nebentypus χ, and hence more general χ_E.

Remarks. According to Theorem 1.1, the translate $\gamma_{\chi_E}(\mathfrak{m})$ is the "Fourier coefficient" of a modular form whose Fourier components have been killed at primes $\mathfrak{p}$ that divide $d_{L/E}\mathfrak{c}_E$.

The Hecke operator $\widehat{T}(\mathfrak{m})$ is constructed so as to have the following property. Suppose $f \in S^{\mathrm{new}}(K_0(\mathfrak{c}_E), \chi_E)$ is a newform (on E) which is a simultaneous eigenform for all Hecke operators, so that $f|T(\mathfrak{m}) = \lambda_f(\mathfrak{m})f$. Suppose its base change $\widehat{f}$ to L is an element of $S^{\mathrm{new}}(K_0(\mathfrak{c}))$. Then in Proposition 8.2 we show that

$$\widehat{f}|\widehat{T}(\mathfrak{m}) = \lambda_f(\mathfrak{m})\widehat{f} \tag{1.5.3}$$

if $\mathfrak{m}$ is a norm from $\mathcal{O}_L$ coprime to $d_{L/E}(\mathfrak{c} \cap \mathcal{O}_E)$. In fact, the linear map

$$\begin{aligned} \widehat{} : \mathbb{T}_{\mathfrak{c}_E} &\longrightarrow \mathbb{T}_{\mathfrak{c}} \\ T(\mathfrak{m}) &\longmapsto \widehat{T}(\mathfrak{m}) \end{aligned} \tag{1.5.4}$$

is roughly a section of the base change map

$$b : \mathbb{T}_{\mathfrak{c}} \longrightarrow \mathbb{T}_{\mathfrak{c}_E}$$

(where $\mathfrak{c}_E = \mathfrak{c} \cap \mathcal{O}_E$) associated to the usual map of L-groups ${}^LG_E \to {}^LG_L$. We will return to this point in Section 8.2 and Section E.4 below.

1.6 Second main theorem

In [Za] p. 159, equation (98), Zagier gave an integral formula for the Fourier coefficients of the modular forms that he and Hirzebruch had constructed from intersection numbers of modular cycles. The second main theorem in this book is a generalization of Zagier's result. It uses more of the structure of the Hecke module $IH_*(X_0(\mathfrak{c}))$ in order to compute the Fourier coefficients of $\langle \beta, \Phi_{\gamma, \chi_E} \rangle$ for certain β. In order to state it, let

$$S^{\mathrm{new}, E}(\mathfrak{c}, \chi_{\mathrm{triv}}) := \bigoplus_f \mathbb{C}f,$$

where the sum is over those (normalized) newforms $f \in S^{\mathrm{new}}(K_0(\mathfrak{c}), \chi_{\mathrm{triv}})$ (cf. Section 5.6), such that for almost all primes $\mathfrak{P} \subset \mathcal{O}_L$ we have $\lambda_f(\mathfrak{P}^\sigma) = \lambda_f(\mathfrak{P})$ for all $\sigma \in \mathrm{Gal}(L/E)$. The theory of abelian base change implies that $S^{\mathrm{new}, E}(\mathfrak{c}, \chi_{\mathrm{triv}})$ is precisely the subspace of $S^{\mathrm{new}}(K_0(\mathfrak{c}), \chi_{\mathrm{triv}})$ spanned by forms that are base changes from E.

For any subset $J \subset \Sigma(L)$ and for any $f \in S^{\mathrm{new}, E}(\mathfrak{c})$ there is a standard differential form $\omega_J(f^{-\iota})$ on $Y_0(\mathfrak{c})$ that is antiholomorphic on (the image of) the copies of $\mathrm{GL}_2(\mathbb{R})$ associated to the places in J (under the canonical projection $G_L(\mathbb{A}) \to Y_0(\mathfrak{c})$) and is holomorphic on the image of the copies of $\mathrm{GL}_2(\mathbb{R})$ associated to the places in $\Sigma(L) - J$ (see Section 7.2). (The $-\iota$ arises when translating between the two natural conventions for weights, see Section 5.4.)

Let $W_{\mathfrak{c}}^*$ be the Atkin-Lehner operator (see Section 7.5). It acts on the modular variety $Y_0(\mathfrak{c})$ and on the compactification $X_0(\mathfrak{c})$. The action preserves newforms and it has the effect of modifying a newform f by replacing its Fourier coefficients with their complex conjugates. To be precise, for a newform $f \in S^{\mathrm{new}}(K_0(\mathfrak{c}))$, there is a complex number $W(f)$ of norm 1 and a newform $f_{\mathfrak{c}} \in S^{\mathrm{new}}(K_0(\mathfrak{c}))$ such that
$$W_{\mathfrak{c}}^* f = W(f) f_{\mathfrak{c}}$$
and
$$\overline{a(\mathfrak{P}, f)} = a(\mathfrak{P}, f_{\mathfrak{c}})$$
for almost all primes $\mathfrak{P} \subset \mathcal{O}_L$. Set
$$[\cdot,\cdot]_{IH_*} : IH_n(X_0(\mathfrak{c})) \times IH_n(X_0(\mathfrak{c})) \longrightarrow \mathbb{C}$$
$$(a, b) \longmapsto \langle a, W_{\mathfrak{c}}^* b \rangle_{IH_*}$$
(see Section 7.5). We then have the following special case of Theorem 8.5:

Let $X_0(\mathfrak{c})$ be the Baily-Borel Satake compactification of the Hilbert modular variety $Y_0(\mathfrak{c})$ corresponding to the algebraic group G_L and ideal $\mathfrak{c} \subset \mathcal{O}_L$. Let $Z \subset X_0(\mathfrak{c})$ be an oriented subanalytic cycle of dimension $n = [L : \mathbb{Q}]$ (the middle dimension). Suppose its homology class $[Z] \in H_n(X_0(\mathfrak{c}))$ lifts to a class $[Z] \in IH_n^{\chi_E}(X_0(\mathfrak{c}))$ and let $\Phi_{W_{\mathfrak{c}}^{*-1}[Z], \chi_E}$ be the formal Fourier series provided by Theorem 1.1. Let $\mathfrak{m}, \mathfrak{n} \subset \mathcal{O}_E$ be ideals and let $W_{\mathfrak{c}}^{*-1}[Z](\mathfrak{n})$ be the Hecke translate defined by equation (1.5.2). Theorem 1.1 says that the intersection product
$$h := [W_{\mathfrak{c}}^{*-1}[Z](\mathfrak{n}), \Phi_{W_{\mathfrak{c}}^{*-1}[Z], \chi_E}]_{IH_*}$$
$$= \langle W_{\mathfrak{c}}^{*-1}[Z](\mathfrak{n}), W_{\mathfrak{c}}^* \Phi_{W_{\mathfrak{c}}^{*-1}[Z], \chi_E} \rangle_{IH_*} \in S(K_0(\mathcal{N}(\mathfrak{c})), \chi_E)$$
is a modular form on E whose $\mathfrak{m}$th Fourier coefficient is
$$a(\mathfrak{m}, h) := \begin{cases} \langle W_{\mathfrak{c}}^{*-1}[Z](\mathfrak{n}), \widehat{T}(\mathfrak{m})_*[Z]\rangle_{IH_*} & \text{if } \mathfrak{m} + d_{L/E}\mathfrak{c}_E = \mathcal{O}_L \\ 0 & \text{otherwise} \end{cases}$$

Theorem 1.2. *Suppose $\mathfrak{m}, \mathfrak{n} \subset \mathcal{O}_E$ are norms from $\mathcal{O}_L$ such that*
$$\mathfrak{m} + \mathrm{N}_{L/E}(\mathfrak{c})d_{L/E} = \mathfrak{n} + \mathrm{N}_{L/E}(\mathfrak{c})d_{L/E} = \mathcal{O}_L. \tag{1.6.1}$$

Then

(1) *The integral $\int_Z \omega_J(f^{-\iota})$ converges for all $f \in S(K_0(\mathfrak{c}), \chi_{\mathrm{triv}})$ and all $J \subset \Sigma(E)$.*

(2) *the $\mathfrak{m}$th Fourier coefficient $a(\mathfrak{m}, h)$ is equal to*
$$\frac{1}{4} \sum_{J \subset \Sigma(E)} \sum_{f} \frac{\int_Z \omega_J(\widehat{f}^{-\iota}) \int_Z \omega_{\Sigma(L)-J}(\widehat{f}^{-\iota})}{T(\widehat{f}, \Sigma(L) - J)(\widehat{f}, \widehat{f})_P} \lambda_f(\mathfrak{n})\overline{\lambda_f(\mathfrak{m})}$$

where the sum is over the normalized newforms f on E of nebentypus η_E whose base change $\widehat{f}$ to L is an element of $S^{\mathrm{new}, E}(K_0(\mathfrak{c}), \chi_{\mathrm{triv}})$.

(3) *If equation (1.6.1) does not hold then the $\mathfrak{m}$th Fourier coefficient is zero.*

Here $(\cdot, \cdot)_P$ denotes the Petersson inner product and $T(\widehat{f}, \Sigma(L) - J)$ is the nonzero constant defined in (7.5.10). (see Section 5.7). In the "full" version of Theorem 8.5, we also allow nontrivial local coefficient systems on $Y_0(\mathfrak{c})$, and we consider subanalytic cycles Z which define homology classes with local coefficients. This added generality results in a number of technical complications which we postpone addressing until Chapter 8.

1.7 Explicit cycles

In the "classical" case of Hilbert modular surfaces, Hirzebruch and Zagier considered the cycle Z (and its associated homology class $[Z]$) that arises as the image of the composition

$$\mathfrak{h} \longrightarrow \mathfrak{h}^2 \longrightarrow \mathrm{SL}_2(\mathcal{O}_{\mathbb{Q}(\sqrt{p})}) \backslash \mathfrak{h}^2.$$

They also considered Hecke translates of this cycle and cycles coming from compact Shimura curves. It is easy to show that these cycles lift to classes in intersection homology (although Hirzebruch and Zagier did not express their intersection numbers in this language). Theorem 1.2 raises the question of whether or not the diagonal embedding $G_E \hookrightarrow G_L$ similarly gives rise to (non-zero) classes in $IH_n^{\chi_E}(X_0(\mathfrak{c}))$, to which one might apply Theorem 1.2[3]. This can be done, although it is not as straightforward as in the Hirzebruch-Zagier case. The first problem is that the variety $X_0(\mathfrak{c})$ has isolated singularities at its cusps but the cycle $Z \subset X_0(\mathfrak{c})$ contains some of these cusps, so it fails the allowability conditions (3.1.1) of intersection homology.

Nevertheless, it turns out that the cycle $Z \subset X_0(\mathfrak{c})$ lifts canonically to an intersection homology class, for reasons involving subtle properties of torus weights in the local cohomology of $X_0(\mathfrak{c})$ at a cusp. The simplest route to this result involves a theorem of L. Saper and M. Stern [SaS2] and M. Rapoport [Rap], recently generalized in [Sa1], which is reviewed in Section 4.9 and applied to the Hilbert modular case in Theorem 4.6.

The second problem (which is not present in the Hirzebruch-Zagier case) is that $IH_n(X_0(\mathfrak{c}))$ does not have a basis of simultaneous eigenvectors for $\mathbb{T}_\mathfrak{c}$. Thus it is more convenient to consider the "new" part, $P_{\mathrm{new}}[Z]$ where

$$P_{\mathrm{new}} : IH_n(X_0(\mathfrak{c})) \longrightarrow IH_n^{\mathrm{new}}(X_0(\mathfrak{c}))$$

is the projection to the orthogonal complement of the subspace of "old classes". (A class ξ is "old" if there exists $\mathfrak{c}' \supset \mathfrak{c}$ such that ξ is the pullback of a class $\xi' \in IH_n(X_0(\mathfrak{c}'))$ under the canonical projection $X_0(\mathfrak{c}) \to X_0(\mathfrak{c}')$. See Section 10.)

Despite these potential difficulties, Theorem 1.2 is applicable to Z. Moreover, it is possible to express the resulting Fourier coefficients in terms of special values

---

[3]We do not treat the higher-dimensional analogues of the Shimura curves in this paper, as they are unnecessary in our approach.

of L functions as follows. For $g \in S(K_0(\mathfrak{c}), \chi_{\text{triv}})$ let

$$A(g, J) = \frac{T(g, J)(g, g)_P}{L^*(\text{Ad}(g), 1)}.$$

This nonzero constant depends only on L, the weight, and the root number of g (see (7.4.9)) by Theorems 5.16 and 7.11.

Theorem 1.3. *Let $[Z] \in IH_n(X_0(\mathfrak{c}))$ be the class represented by the cycle that comes from the diagonal embedding $G_E \to G_L$. Then $P_{\text{new}} W_\mathfrak{c}^{*-1}[Z]$ is a non-zero element of $IH_n^{\chi_E}(X_0(\mathfrak{c}))$ where $\chi_E = \eta$ is the nontrivial element of $\text{Gal}(L/E)^\wedge$. If $\mathfrak{m} + N_{L/E}(\mathfrak{c}) d_{L/E} = \mathfrak{n} + N_{L/E}(\mathfrak{c}) d_{L/E} = \mathcal{O}_E$ and $\mathfrak{m}, \mathfrak{n}$ are both norms from $\mathcal{O}_L$, then the $\mathfrak{m}$th Fourier coefficient $a(\mathfrak{m}, h)$ of*

$$h := [W_\mathfrak{c}^{*-1}[Z(\mathfrak{n})], \Phi_{[W_\mathfrak{c}^{*-1} P_{\text{new}}[Z]], \chi_E}]_{IH_*}$$

is given by the following formula,

$$\frac{c_1^2}{4}(-1)^{[E:\mathbb{Q}]} \sum_f \frac{\left(L^{*, d_{L/E} \mathfrak{f}(\theta) \cap \mathcal{O}_E}(\text{Ad}(f) \otimes \eta, 1) L_{\mathfrak{b}'}(\text{As}(\widehat{f} \otimes \theta^{-1}), 1)\right)^2}{A(\widehat{f}, \emptyset) L^*(\text{Ad}(\widehat{f}), 1)} \lambda_f(\mathfrak{n}) \overline{\lambda_f(\mathfrak{m})},$$

where the sum is over the normalized newforms f on E of nebentypus η whose base change $\widehat{f}$ to L is an element of $S^{\text{new}}(K_0(\mathfrak{c}), \chi_{\text{triv}})$. Here c_1 is an explicit nonzero scalar (see Theorem 10.2) and $\lambda_f(\mathfrak{n})$ is the $\mathfrak{n}$th Hecke eigenvalue of f. Moreover

$$\mathcal{N}(\mathfrak{c}) := \mathfrak{m}_2 \mathfrak{d}_{L/E}(\mathfrak{c} \cap \mathcal{O}_E) \prod_{\mathfrak{p} | (\mathfrak{c} \cap \mathcal{O}_E)} \mathfrak{p}^2$$

where $\mathfrak{m}_2 \subset \mathcal{O}_E$ is an ideal divisible only by dyadic primes, which we may take to be $\mathcal{O}_E$ if $\mathfrak{c} + 2\mathcal{O}_L = \mathcal{O}_L$ and $\mathfrak{d}_{L/E}$ is an ideal divisible only by primes ramifying in L/E.

In the statement of Theorem 1.3, the product

$$[\cdot, \cdot]_{IH_*} : IH_n(X_0(\mathfrak{c})) \times IH_n(X_0(\mathfrak{c})) \longrightarrow \mathbb{C}$$

denotes a twist of the canonical intersection pairing by the Atkin-Lehner operator $W_\mathfrak{c}$ (see (7.5.9)). The "standard" L-function $L(Ad(f), s)$ and the "Asai" L-function $L(As(\widehat{f}), s)$ will be defined in Section 5.12. The base change $\widehat{f}$ of f will be defined in Section 5.12.4. We will show how to deduce this theorem from Theorem 1.2 and a Rankin-Selberg integral computation in Chapter 9 below.

Remark. It is possible, by "twisting" the cycle Z, to obtain a nonzero class in $IH_n^{\chi_E}(X_0(\mathfrak{c}))$ where χ_E is the trivial character as opposed to the nontrivial character of Theorem 1.3. We refer the reader to Section 9.4 for more details. This twist is the geometric manifestation of the well-known operation of twisting automorphic forms by characters (see Section 5.11).

1.8 Finding cycles dual to families of automorphic forms

In the previous section we mentioned in passing that we used the theory of distinguished representations to prove that $P_{\text{new}}[Z] \in IH_n^{\chi_E}(X_0(\mathfrak{c}))$. In this section we pause to briefly describe the theory of distinguished representations and how it can be used in certain cases to predict the existence of cycles representing classes in subspaces of the intersection homology of locally symmetric varieties.

In this work we are interested in cycles on locally symmetric spaces that are defined by sub-symmetric spaces. We show that these cycles are nontrivial in intersection cohomology by showing that there are differential forms representing classes in L_2-cohomology that have nonzero integrals against them. Since that L_2-cohomology of locally symmetric spaces can be described in terms of automorphic representations [BoW, Section XIV.3], it is natural to seek a representation-theoretic analogue of the geometric statement that a differential form on a locally symmetric space has a nonzero integral over a sub-symmetric space. Such a representation-theoretic analogue is provided by the notion of distinction, introduced by Harder, Langlands and Rapoport [HarL].

Let $G' \subseteq G$ be a pair of reductive groups over $\mathbb{Q}$. One says that an automorphic representation π of $G(\mathbb{A}_\mathbb{Q})$ is G'-distinguished if

$$\int_{G'(F)\backslash G'(\mathbb{A}_F) \cap {}^0G(\mathbb{R})G(\mathbb{A}_{\mathbb{Q},f})} \phi(g) dg \neq 0 \tag{1.8.1}$$

is convergent and nonzero for some ϕ in the space of π. One says that (1.8.1) is the integral of ϕ over G'. Here dg is induced by a choice of Haar measure and ${}^0G(\mathbb{R})$ is defined as in Section 4.1 below.

Even if π is distinguished and cohomological (i.e., has nonzero $(\mathfrak{g}, K)$-cohomology with coefficients in some representation), this does not necessarily mean that there is a nonzero intersection homology class attached to G' in the π_f-isotypic component of the intersection homology of some locally symmetric space attached to G. One must prove that a "cohomological vector" in the space of π has nonzero period over G'. To see an example of what one must prove, see [AshG]. In particular, in order to understand cycles on locally symmetric spaces, one is forced at some point to work at the level of automorphic forms inside an automorphic representation, or at least its associated $(\mathfrak{g}, K)$-module. This is one justification for the introduction of explicit spaces of automorphic forms in Chapter 5 below.

Remark. The problem of finding cohomological vectors in distinguished representations dual to specific cycles is an important problem that, at the time of this writing, is unsolved in almost all nontrivial cases. We refer the reader to Section 6 of [Rag] and the references therein for examples and applications.

Suppose G' is the fixed point set of an involution. Jacquet has suggested that the automorphic representations of $G(\mathbb{A}_\mathbb{Q})$ that are distinguished by G' are precisely those automorphic representations that are functorial lifts from some other group H with (absolute) root datum determined by G and G' [JaLai]. In other words, there should be an L-map

$$^LH \longrightarrow {}^LG \qquad (1.8.2)$$

such that an L-packet of automorphic representations on $G(\mathbb{A}_\mathbb{Q})$ contains an automorphic representation distinguished by G' if and only if the L-packet is in the image of the putative Langlands transfer attached to (1.8.2) (see Section E.4).

In this language, the present manuscript is concerned with the case where $G = G_L$ and $G' = G_E$ is the fixed point set of the involution of G_L induced by $\text{Gal}(L/E)$. Jacquet's formalism predicts that H should be a form of G_E. In fact, Flicker and Rallis have suggested that it should be a unitary group attached to the extension L/E [Fl]. However, since the derived group of G_L has $\mathbb{Q}$-rank one, the set of cuspidal automorphic representations of $G_L(\mathbb{A}_\mathbb{Q})$ that are a lift from a unitary group attached to L/E is roughly the same as those that are a lift from $G_E(\mathbb{A}_\mathbb{Q})$, the point being that automorphic representations of G_L are all self-dual up to a twist (see Theorem E.11 and [Rog, Section 11.5]). This is why it is possible for us to relate the cycles we construct to automorphic forms on G_E as opposed to some unitary group.

Given Jacquet's conjectural formalism and the work in this manuscript, one is lead to the following rough conjectural generalization: Suppose that $G' \subseteq G$ is the fixed point set of an involution, and assume that one can find a group H satisfying Jacquet's desiderata above. Suppose that G' defines a cycle class Z in some cohomology group $H(\Gamma \backslash X)$ attached to a locally symmetric space $\Gamma \backslash X$ defined by G and an arithmetic subgroup $\Gamma \subseteq X$. If there is a good enough theory of models on H, then there should be an automorphic form $\Phi_{G,G'}$ on $H(\mathbb{A}_\mathbb{Q})$ with coefficients in $H(\Gamma \backslash X)$. The automorphic form on $H(\mathbb{A}_\mathbb{Q})$ defined by

$$\langle Z, \Phi_{G,G'} \rangle_H$$

should have coefficients given in terms of special values of L-functions. Here by a "good enough theory of models" on H we mean something that can take the place of the Whittaker models which provide the Fourier coefficients we use in this book to make sense out of the notion of the coefficient of an automorphic form. Incidentally, making this notion of coefficients precise in the $G = G_L$ case is another justification for introducing explicit spaces of automorphic forms in Chapter 5 below.

1.9 Comments on related literature

Several excellent introductory books are available including [Ge], [Fr], and [Oda2]. The Hirzebruch-Zagier results are described in detail in [Ge]. The little book [Gel] gives a good introduction to the adelic viewpoint, with further details available in [Ga]. We have reproduced some of the standard material on Hilbert modular forms, Fourier series, and Hecke operators from [Hid5] and [Hid7].

As a special case of their far-reaching study of of explicit subvarieties of locally symmetric spaces of orthogonal and unitary type, Kudla and Millson (generalizing previous work of Oda) produce a generating series with coefficients in $H_n(X_0(\mathfrak{c}))$ out of cycles whose irreducible components are birational to the components of (possibly non-compact) Shimura curves attached to quaternion algebras over E split by L ([KuM3]). They then go on to prove that this generating series is a modular form with coefficients in a certain cohomology group.

There are (at least) three fundamental differences between this special case of the theory in [KuM3] and our Theorem 8.4. First, as opposed to one Hilbert modular form (with coefficients in cohomology), we obtain a family of modular forms (with coefficients in intersection homology), one for each local system $\mathbf{E}$ on $X_0(\mathfrak{c})$ and each intersection homology class in $IH_n(X_0(\mathfrak{c}), \mathbf{E})$.

Second, whereas [KuM3] consider the intersection product of their generating series with a cycle that is compactly supported in $Y_0(\mathfrak{c})$, we consider naturally occuring compact and noncompact cycles (e.g., noncompact subvarieties birational to Hilbert modular varieties associated to E), which are lifted to intersection homology (see Theorem 4.6). Our construction, however, does not require the sophisticated intersection theory [Tol] used by Tong in his study [Ton] of weighted intersection numbers on Hilbert modular surfaces.

Finally, the method of proof in [KuM3] differs from ours, in that we use quadratic base change as a tool to produce our results, whereas the cases of quadratic base change that Kudla and Millson require are incorporated into their arguments using theta liftings. It would be interesting to see if the formal arguments we use to prove Theorem 1.1 in this manuscript could be modified and extended to establish similar theorems in the context of other liftings of automorphic forms, especially when the lifting is not a theta lifting. A relatively simple case of such an extension is provided in the remark after Proposition 8.2, when the automorphic lifting is the GL_2 base change associated to a prime-degree Galois extension of totally real fields.

In [Bry], J.L. Brylinski and J.P. Labesse determine the Hasse-Weil L-functions of the Hilbert modular varieties that are considered in this book. Although the focus of their article differs from that of ours, their L-functions are related to the Asai L-functions considered in Section 5.12.4. Period integrals on Hilbert modular surfaces are also considered in [Oda2]. Hilbert modular varieties are, in a natural way, moduli spaces of abelian varieties with real multiplication. This point of view, and its relation to Hilbert modular forms, is developed in [Goren].

1.10 Comparison with Zagier's formula

Theorem 1.2 provides a generalization of a formula for a generating series Φ_{HZ} given by Zagier [Za], equation (98). In order to state his formula, denote by $\widehat{f}$ the Naganuma lift of an $f \in S_2(\Gamma_0(p), (\frac{p}{\cdot}))$ (here we have used the classical normalization of the weight). Assume for simplicity that the narrow class number of $\mathbb{Q}(\sqrt{p})$ is 1. Then, denoting by $[Z_m]_{m \geq 0}$ the family of classes introduced by Hirzebruch and Zagier in [Hirz], Zagier proved that the intersection product $\langle [Z_m], \Phi_{HZ} \rangle_H$ is given by:

$$\pi_+ \left(t(m) E_{2,p}(z) - \sum_{n=1}^{\infty} r' \left(\sum' \frac{\left(\int_{Z_1} \eta(\widehat{f}) \right)^2}{(\widehat{f}, \widehat{f})} a_f(m) a_f(n) \right) q^n \right).$$

Here the prime indicates summation over a basis of normalized newforms (i.e., eigenforms for all the Hecke operators)

$$f(z) := \sum_{n=1}^{\infty} a_f(n) q^n$$

in $S_2(\Gamma_0(p), (\frac{p}{\cdot}))$, π_+ is the canonical projection to $S_2^+(\Gamma_0(p), (\frac{p}{\cdot}))$, the rational number $t(m)$ depends only on m, the nonzero complex number r' is a certain explicit constant, and $E_{2,p}$ is a weight two Eisenstein series in $S_2^+(\Gamma_0(p), (\frac{p}{\cdot}))$ (see [Za, (98–99)]). Moreover $\eta(\widehat{f})$ is a certain differential $(1,1)$-form attached to $\widehat{f}$, and we used the proof of Oda's period relation for the Naganuma lifting (see [Ge, p. 154 (7.9)], [Oda3] and [Oda4]) to modify Zagier's expression.

Let Z^0 be the (open) modular subvariety of $Y_0(\mathcal{O}_{\mathbb{Q}(\sqrt{p})})$ given by the image of the diagonal embedding:

$$G_E(\mathbb{A}) \hookrightarrow G_L(\mathbb{A}) \longrightarrow Y_0(\mathcal{O}_{\mathbb{Q}(\sqrt{p})}).$$

Here the first map is the diagonal embedding and the second map is the canonical projection. Denote by Z the closure of Z^0 in $X_0(\mathcal{O}_{\mathbb{Q}(\sqrt{p})})$. Let

$$\pi : IH_2(X_0(\mathcal{O}_{\mathbb{Q}(\sqrt{p})})) \longrightarrow IH_2(X_0(\mathcal{O}_{\mathbb{Q}(\sqrt{p})}))$$

be the projection onto the orthogonal complement of the invariant forms (see Section 4.4 for generalities on invariant forms). Choose an embedding $\sigma : \mathbb{Q}(\sqrt{p}) \hookrightarrow \mathbb{R}$. If m is a norm from $\mathbb{Q}(\sqrt{p})$, then

$$\langle [Z]_{\chi_{\mathbb{Q}}}(m\mathbb{Z}), \Phi_{\pi[Z], \chi_{\mathbb{Q}}} \rangle_{IH_*}$$

$$= |y|_{\mathbb{A}_{\mathbb{Q}}} \sum_{\substack{n=1 \\ (n,p)=1,\ \chi_{\mathbb{Q}}(n)=1}}^{\infty} r'' \left(\sum' \frac{\left(\int_Z \omega_{\{\sigma\}}(\widehat{f}) \right)^2}{(\widehat{f}, \widehat{f})} a(m\mathbb{Z}, f) a(n\mathbb{Z}, f) \right) q(nx, ny)$$

for some explicit constant $r'' \in \mathbb{C}^\times$, where the sum is over a basis of normalized newforms $f \in S(K_0(p\mathbb{Z}), \chi_\mathbb{Q})$. (See equation (5.9.5) for the equality between the Hecke eigenvalue $\lambda_f(n\mathbb{Z})$ and the Fourier coefficient $a(m\mathbb{Z}, f)$.) In order to obtain this expression, we used the fact that $W^*_{\mathcal{O}_L}$ is the identity for any totally real field L, along with Corollary 4.10 and Proposition 7.9 of [Ge, Chapter VI].

1.11 Outline of the book

The goal of Chapters 2 to 4 is to review the construction and basic properties of the integral of a differential form ω over a cycle ξ. This "standard" material is known to experts, but, to our knowledge, it is not recorded anywhere in the form that we require. In particular, we will need to make use of this formalism in the following context:

- The differential form ω is defined on an orbifold Y, rather than on a smooth manifold.
- The differential form ω and the chain ξ take values in local coefficient systems $\mathbf{E_1}$ and $\mathbf{E_2}$.
- The orbifold Y may be the largest stratum of a stratified space X, and the local systems $\mathbf{E_1}, \mathbf{E_2}$ may fail to extend to all of X.
- The cycle ξ represents a class in the intersection homology of X, rather than in ordinary homology.

We will also need to know that such an integral can be interpreted using various homologically defined products such as the "Kronecker" pairing between (intersection) homology and (intersection) cohomology, the cup product on cohomology, and the intersection product on intersection homology; and that these various products are compatible with each other whenever there are natural identifications among the different homology and cohomology groups.

Although there are no essential difficulties in constructing such an integral, and in describing it homologically, a complete proof of any statement to this effect necessarily involves chain-level operations. These in turn involve geometric properties of the cycle ξ (such as a sub-analytic structure, or triangulability) and analytic properties (such as growth rates) of the differential form. We do not know of any published literature that specifically deals with these (relatively straight forward) issues, so we have included the relevant details in the early chapters: in Chapters 2 (chains and cochains), 3 (intersection homology and cohomology) and 4 (arithmetic quotients). We have also included Appendix B on the definition and basic properties of orbifolds, which fills in some of the technical details that are not easily extracted from the standard references.

Then in Chapter 5 we review relevant facts from the theory of Hilbert modular varieties and Hilbert modular forms. This is in preparation for Chapter 7, where we recall the well-known description of the intersection cohomology of the Hilbert modular varieties $X_0(\mathfrak{c})$ in terms of Hilbert modular forms.

After this preparatory material, we move on to the core of the second half of this material, namely the proofs of Theorems 1.1, 1.2, and 1.3 (see Sections 8.3, 8.6, and 10, respectively). Theorem 1.1 is actually implied by the more general Theorem 8.3 where intersection homology is replaced by an arbitrary Hecke module (see Section 8.3 for details). As indicated above, Theorem 8.3 relies crucially on the theory of quadratic base change for GL_2; thus we have included a synopsis of prime degree base change for GL_2 in Appendix E. Theorem 1.3 relies on Theorem 1.2 and a Rankin-Selberg computation that is contained in Chapter 9.

Finally, in Chapter 11, we prove an analogue of Theorems 1.1 and 1.2 with cuspidal classes replaced by invariant classes (see Section 7.2 for the definition of a cuspidal and invariant class). In particular, we prove that suitable generating series created out of invariant classes are Eisenstein series with coefficients in intersection homology.

As the discussion above indicates, this paper touches on a wide range of topics from the topology and number theory of $\mathbb{Q}$-rank one hermitian symmetric spaces. However, Chapters 2 to 4 can be read independently of the rest of this work (though their content and structure reflect the requirements of the later chapters). Similarly, Chapters 5 through 11 only use results from Chapters 2 through 4 that one can reasonably take to be a "black box." Appendices B and E are also self-contained.

1.12 Problematic primes

It is a subtle problem to determine the minimal level of a Hilbert modular form whose base change has a given level. The $\mathfrak{m}$th Fourier coefficient of the Hilbert modular form with coefficients Φ_{γ,χ_E} that was constructed in Theorem 1.1 is zero if $\mathfrak{m} + d_{L/E}(\mathfrak{c} \cap \mathcal{O}_E) \neq \mathcal{O}_E$. By placing more assumptions on the level and the character, it is possible to produce an analogue of Φ_{γ,χ_E} (satisfying an analogue of Theorem 1.1) that may have nonzero $\mathfrak{m}$th Fourier coefficients when $\mathfrak{m} + d_{L/E}(\mathfrak{c} \cap \mathcal{O}_E) \neq \mathcal{O}_E$. Proving a theorem along these lines would either require substantial hypotheses on the local admissible representations involved, or a substantial digression on local representation theory. For brevity, we have not attempted to do either. However, we hope that Appendix E might be a useful starting point for further investigations in this direction.

Acknowledgement

The first author would like to thank Tonghai Yang for answering many questions on automorphic forms and representations and Ken Ono for his constant support. L. Saper, J.-P. Wintenberger and several referees made valuable suggestions for

1.12. Problematic primes

improvements in the manuscript. The authors are grateful to the Institute for Advanced Study for providing an ideal research environment where much of this work was done. The research of Getz was partially supported by the Amy Research Office through the National Defense Science and Engineering Graduate Fellowship program and the National Science Foundation through their Mathematical Sciences Postdoctoral Research program. The research of Goresky was partially funded by the Ambrose Monell Foundation and the Association of Members of the Institute for Advanced Study. Goresky is grateful to the Defense Advanced Research Projects Agency for its support by way of grant numbers HR0011-04-1-0031 and HR0011-09-1-0010. The authors are grateful to the Ferran Sunyer i Balaguer Foundation, and to its director Manuel Castellet, for their gracious hospitality during our visit to Barcelona. Commutative diagrams were typeset using Paul Taylor's diagrams package. The body of the book was typeset with LaTeX2ε.

Chapter 2
Review of Chains and Cochains

2.1 Cell complexes and orientations

Recall (e.g., [Hud]) that a *closed convex linear cell* is the convex hull of finitely many points in Euclidean space. A *convex linear cell complex* K is a finite collection of closed convex linear cells in some $\mathbb{R}^N$ such that if $\sigma \in K$ then every face of σ is in K, and if $\sigma, \tau \in K$ then the intersection $\sigma \cap \tau$ is in K. The underlying closed subset of Euclidean space is denoted $|K|$. Such a complex is a *regular* cell complex, meaning that each (closed) cell is homeomorphic to a closed ball: no identifications occur on its boundary. If $\tau \in K$ is a face of $\sigma \in K$ we write $\tau < \sigma$. A *finite simplicial complex* is a convex linear cell complex, all of whose cells are simplices. Every cell complex admits a simplicial refinement with no extra vertices.

An *orientation* of a finite-dimensional real vector space V is a choice of ordered basis, two being considered equivalent if one can be continuously deformed to the other, through ordered bases. Every (finite-dimensional real) vector space has two orientations. An orientation of a convex linear cell is an orientation of the real affine space that it spans. An orientation of a smooth manifold is a continuously varying choice of orientation of each of its tangent spaces.

Let K be a convex linear cell complex and let L be a (closed) subcomplex. Let $X = |K|$ and let $Y = |K| - |L|$. Although Y is not a union of cells, it is a union of interiors of cells. We refer to this decomposition of Y as a *pseudo cell decomposition* (or a *pseudo-triangulation* if K is a simplicial complex).

Every cell $\sigma \in K$ has two orientations. A choice of orientation for σ determines a unique orientation for each codimension 1 face $\tau < \sigma$ such that the orientation of τ followed by the inward pointing vector $\overrightarrow{\tau\sigma}$ agrees with the orientation of σ. The complex K is *purely d-dimensional* if every cell is the face of some d-dimensional cell and there are no cells of dimension greater than d.

A *oriented cellular pseudomanifold* is a convex linear cell complex K, purely of some dimension d, such that every $d-1$-dimensional cell is a face of exactly two d-dimensional cells; together with a choice of orientation of each d-dimensional cell such that the induced orientations cancel on every $d-1$-dimensional cell.

2.2 Subanalytic sets and stratifications

Let F = semi-algebraic, semi-analytic, or subanalytic. Any finite union, intersection, or difference of F-subsets of $\mathbb{R}^N$ is again an F-subset of $\mathbb{R}^N$. The closure and the interior of any F-subset of $\mathbb{R}^N$ is again an F-subset of $\mathbb{R}^N$. The image of an F-subset $X \subset \mathbb{R}^N$ by an F-mapping $f : \mathbb{R}^N \to \mathbb{R}^k$ is again an F-set for F= semi-algebraic or subanalytic (but this last statement is false for F= semi-analytic).

Let $X \subset \mathbb{R}^N$ be a set of type F. A *F-Whitney stratification* of X is a locally finite decomposition $X = \bigcup_\alpha X_\alpha$ into disjoint real analytic manifolds or *strata*, such that

- the closure $\overline{X_\alpha}$ of X_α is an F-subset of $\mathbb{R}^N$
- if $X_\alpha \cap \overline{X_\beta} \neq \phi$ then $X_\alpha \subset \overline{X_\beta}$. and the pair (X_α, X_β) satisfies Whitney's conditions A and B.

An F-stratified space is such a set X together with an F-Whitney stratification. Every closed subset $X \subset \mathbb{R}^N$ of type F admits an F-Whitney stratification.

A Whitney stratification of a closed F-set X implies that the local topological type of X is locally constant along each stratum S in the following sense. Without loss of generality we may assume that S is connected. R. Thom [Th] and J. Mather [Ma1] proved the following:

Theorem 2.1. *There exists a compact F-Whitney stratified space ℓ (the link of the stratum S) such that every point $x \in S$ has a neighborhood basis in X consisting of neighborhoods N_x homeomorphic to $\mathbb{R}^s \times \mathrm{cone}(\ell)$ by a stratum-preserving homeomorphism that is smooth on each stratum and takes $\mathbb{R}^s \times \{pt\}$ to $N_x \cap S$.* □

(Here, $s = \dim(S)$ and $\{pt\}$ denotes the cone point.) Such a neighborhood is called a *basic neighborhood*. The Thom-Mather theorem implies, in particular, that the local homology $H_i(X, X - x; \mathbb{Z})$ of X is finitely generated at every point $x \in X$. An *F-triangulation* of a closed set $X \subset \mathbb{R}^N$ of type F is a locally finite simplicial complex K with $|K| \subset \mathbb{R}^N$ together with an F-isomorphism $f : \mathbb{R}^N \to \mathbb{R}^N$ such that $f(|K|) = X$ and

- For each simplex $\sigma \in K$ the restriction $f|\sigma^o$ of f to its interior is a real analytic isomorphism $\sigma^o \to f(\sigma^o)$.
- Each $f(\sigma)$ is a (closed) set of type F in $\mathbb{R}^N$.

Any two F-triangulations of X have a common refinement. An F-triangulation f of a closed F-set $X \subset \mathbb{R}^N$ is *compatible* with an F-Whitney stratification $X = \bigcup_\alpha X_\alpha$ if, for each α the set $f^{-1}(\overline{X_\alpha})$ is a (closed) subcomplex of K. If X is F-Whitney stratified and F-triangulated by a compatible triangulation, and if x is a point in some stratum $S \subset X$ then the *link* L_x of x (in the sense of P.L. topology) is homeomorphic

$$L_x \cong \Sigma^s \ell \qquad (2.2.1)$$

to the $s = \dim(S)$-fold suspension of the link ℓ of the stratum S.

Theorem 2.2. *Let $Z \subset X \subset \mathbb{R}^N$ be closed subsets of type F. Then X admits an F-Whitney stratification such that Z is a union of strata. Given any F-Whitney stratification of X there exists an F-triangulation that is subordinate to the stratification.*

The proof of this result has a long history. We list here a few of the important references: [Lo1], [Lo2], [Hardt7], [Hardt4], [Hardt5], [Hardt6], [Hardt3], [Hardt2], [Hardt1], [Hir5], [Hir4], [Hir3], [Hir1], [Hir2], [Gre1], [Joh].

A closed F-subset $X \subset \mathbb{R}^N$ is *purely d-dimensional* if there exists an F-stratification of X such that X is the closure of the union of all of its d-dimensional strata. Then X is an *oriented pseudomanifold* if there exists a (simplicial) oriented pseudomanifold K of pure dimension d and an F-triangulation $f : |K| \to X$.

For Whitney stratified sets, an oriented pseudomanifold structure may be described without reference to a triangulation. Let X be a purely d-dimensional F-set. Then X is a pseudomanifold if it can be Whitney stratified with no strata of dimension $d - 1$. In this case an orientation of X is determined by a choice of orientation of each of the d-dimensional strata.

2.3 Sheaves and the derived category

Let X be a real or complex algebraic, analytic, semi-analytic or sub-analytic set. Then X is locally compact, Hausdorff, and is homeomorphic to a (locally finite) simplicial complex. Throughout this section we fix a regular, commutative, Noetherian ring R (with unit) of finite cohomological dimension. (A principal ideal domain, for example, is such a ring.) Recall that a *complex of sheaves* of R-modules $\mathbf{S}^\bullet$ on X is a collection of sheaves $\mathbf{S}^i$ and differentials $d_i : \mathbf{S}^i \to \mathbf{S}^{i+1}$ with $d^2 = 0$. The associated *cohomology sheaf* of degree i is $\mathbf{H}^i(\mathbf{S}^\bullet) = \ker d_i / \operatorname{Im} d_{i-1}$. If each $\mathbf{S}^i$ is fine, flabby, soft, or injective, then the cohomology $H^*(X, \mathbf{S}^\bullet)$ (resp. cohomology with compact supports $H^*_c(X, \mathbf{S}^\bullet)$) is given by the cohomology of the complex of global sections (resp. global sections with compact supports).

It is customary to denote by $\mathbf{S}^\bullet[n]$ the shift of $\mathbf{S}^\bullet$ by n, that is, $(\mathbf{S}^\bullet[n])^k = \mathbf{S}^{n+k}$. A morphism $\mathbf{S}^\bullet \to \mathbf{T}^\bullet$ is a *quasi-isomorphism* if it induces an isomorphism on the associated cohomology sheaves. In this case, the complex $T^\bullet$ is called an *injective resolution* of $\mathbf{S}^\bullet$ if each $\mathbf{T}^j$ is injective (in the category of sheaves of R modules).

A complex of sheaves $\mathbf{S}^\bullet$ is *cohomologically locally constant* (CLC) if each of the cohomology sheaves $\mathbf{H}^i(\mathbf{S}^\bullet)$ is locally constant. The complex $\mathbf{S}^\bullet$ is *cohomologically constructible* with respect to a given stratification of X if each of the cohomology sheaves $\mathbf{H}^i(\mathbf{S}^\bullet)$ is locally constant on each stratum. Let $D^b_c(X)$ denote the bounded constructible derived category: its objects consist of complexes of sheaves that are bounded from below and are cohomologically constructible with respect to some Whitney F-stratification of X. In this category, every quasi-isomorphism is invertible. See, for example, [Iv], [GelM], [Gre5]. Many functors F

defined on the category $\mathrm{Sh}(X)$ of sheaves on X pass to derived functors RF. In particular, we shall use the standard notations Rf_*, $Rf_!$, f^*, $f^!$ for the derived push-forward, derived push-forward with proper supports, the pull-back and the extraordinary pull-back on sheaves. If $\mathbf{S}^\bullet, \mathbf{T}^\bullet$ are complexes of sheaves on X then $\mathbf{RHom}^\bullet(\mathbf{S}^\bullet, \mathbf{T}^\bullet)$ denotes the complex of sheaves that is obtained from the double complex of pre-sheaves which associates to any open subset $j : U \subset X$ the R module $\mathrm{Hom}(j^*\mathbf{S}^p, j^*\mathbf{I}^q)$ where $\mathbf{T}^\bullet \to \mathbf{I}^\bullet$ is an injective resolution of $\mathbf{T}^\bullet$. In this case
$$\mathrm{Hom}_{D^b_c(X)}(\mathbf{S}^\bullet, \mathbf{T}^\bullet) = H^0(X, \mathbf{RHom}^\bullet(\mathbf{S}^\bullet, \mathbf{T}^\bullet)).$$

If $\mathbf{S}^\bullet$ is cohomologically constructible then it follows from the Thom-Mather theorem (Section 2.2) that the stalk cohomology (or "local cohomology")
$$H^i(j_x^*\mathbf{S}^\bullet) = H^i_x(\mathbf{S}^\bullet) = \mathbf{H^i}(\mathbf{S}^\bullet)_x$$
(of $\mathbf{S}^\bullet$ at the point $x \in X$) coincides with the cohomology $H^i(U_x, \mathbf{S}^\bullet)$ of any basic neighborhood U_x of x in X. Here $j_x : \{x\} \to X$ denotes the inclusion. Similarly
$$H^i_c(U_x, \mathbf{S}^\bullet) \cong H^i(j_x^!(\mathbf{S}^\bullet))$$
is the stalk cohomology with compact supports.

2.4 The sheaf of chains

Let $\mathbf{E}$ be a local coefficient system (= locally constant sheaf) of R modules on a set $X \subset \mathbb{R}^n$ of type F. There are many quasi-isomorphic versions of the sheaf $\mathbf{C}^\bullet(X, \mathbf{E})$ of chains on X. We briefly recall the construction of the sheaf of F-chains.

Let T be a (locally finite) F-triangulation of X. For each simplex σ of T the restriction of $\mathbf{E}$ to σ has a canonical trivialisation, so we may unambiguously refer to the *fiber* $\mathbf{E}_\sigma$. An i-dimensional (T-simplicial) *Borel-Moore chain* with coefficients in $\mathbf{E}$ is a (locally finite) linear combination of oriented simplices $\xi = \sum_t e_t \sigma_t$ with $e_t \in \mathbf{E}_{\sigma_t}$ whose support $|\xi|$ is closed in X; we identify $e_t\sigma_t$ with $-e_t\sigma'_t$ where σ' is the same simplex as σ but with the opposite orientation. The collection of all Borel-Moore i-chains with respect to the F-triangulation T forms an R module $C_i^{BM,T}(X, \mathbf{E})$ and the usual boundary map gives a homomorphism $\partial_i : C_i^{BM,T}(X, \mathbf{E}) \to C_{i-1}^{BM,T}(X, \mathbf{E})$. If T' is a refinement of the triangulation T then the natural homomorphism $C_i^{BM,T}(X, \mathbf{E}) \to C_i^{BM,T'}(X, \mathbf{E})$ induces an isomorphism
$$H_i^{BM,T}(X, \mathbf{E}) \cong H_i^{BM,T'}(X, \mathbf{E})$$
on homology. Define the complex of (Borel-Moore) F-chains
$$C_i^{BM}(X, \mathbf{E}) = \varinjlim_T C_i^{BM,T}(X, \mathbf{E})$$

to be the direct limit over all F-triangulations of X. The Borel-Moore chains then form a pre-sheaf (with respect to the open subsets of type F). For if $U \subset V$ are open F-subsets of X and if T is a triangulation of V then it is possible to find a triangulation T' of U such that each simplex of T' is contained in a unique simplex of T. This procedure gives a homomorphism $C_i^{BM}(U, \mathbf{E}) \to C_i^{BM}(V, \mathbf{E})$. The *sheaf of F-chains* $\mathbf{C}^\bullet(X, \mathbf{E})$ on X is the complex of sheaves whose R module of sections over an open set $U \subset X$ is

$$\Gamma\left(U, \mathbf{C}^{-\mathbf{i}}(X, \mathbf{E})\right) = C_i^{BM}(U, \mathbf{E})$$

with $d_{-i} = \partial_i$ (for $i \geq 0$). (It is placed in negative degrees so that the differentials raise degree.) The sheaf $\mathbf{C}^\bullet(X, \mathbf{E})$ is soft ([Hab] Section II.5), so the sheaf cohomology over any open set $U \subset X$ can be obtained as the cohomology of the complex of sections over U. With this in mind, the *Borel-Moore* homology is defined by

$$H_i^{BM}(U, \mathbf{E}) := H^{-i}(U, \mathbf{C}^\bullet(X, \mathbf{E})).$$

The complex of (compact) F-chains on $U \subset X$ is the complex

$$C_i(U, \mathbf{E}) = \Gamma_c(U, \mathbf{C}^{-\mathbf{i}}(X, \mathbf{E}))$$

of sections with compact support. The (local) *homology sheaf*

$$\mathbf{H}^{-i}(\mathbf{C}^\bullet(\mathbf{X}, \mathbf{E}))$$

is the cohomology sheaf of the sheaf of chains. It is a topological invariant and its stalk cohomology is the *local homology*, that is, $H_x^{-i}(\mathbf{C}^\bullet(X, \mathbf{E})) = H_i(X, X - x; \mathbf{E})$. The cohomology with compact support $H_c^{-i}(X, \mathbf{C}^\bullet(X, \mathbf{E}))$ is the ordinary homology $H_i(X, \mathbf{E})$.

A similar construction [Bre] may be made with singular chains, and the resulting complex of sheaves (which is a topological invariant and does not depend on a choice of piecewise linear structure) is canonically quasi-isomorphic to the sheaf of F-chains. We will sometimes refer to "the" sheaf of chains $\mathbf{C}^\bullet(X, E)$ without reference to a particular PL or analytic structure on X.

If R is a field and if $\mathbf{E} = \mathbf{R}$ is the constant local system then the sheaf of chains on X is called the *dualizing sheaf* (with coefficients in $\mathbf{R}$) and it is denoted $\mathbb{D}_X^\bullet$. (For an arbitrary regular Noetherian ring R of finite cohomological dimension, the dualizing sheaf is obtained from the sheaf of chains by tensoring with an injective resolution of R [Bo84, Section 7.A].)

We remark that if $\xi = \sum_t a_t \sigma_t \in C_i^{BM}(U, R)$ is a chain with constant coefficients, and if $s \in \Gamma(|\xi|, \mathbf{E})$ is a section of $\mathbf{E}$ over the support of ξ then we obtain, in a natural way a chain $s\xi = \sum_t a_t s(\sigma_t) \sigma_t \in C_i^{BM}(U, \mathbf{E})$.

2.5 Homology manifolds

As in the previous section, we assume the coefficient ring R is a regular Noetherian ring of finite cohomological dimension, and we let F refer to semi-algebraic, semi-analytic, or subanalytic. Let Y be a *purely n-dimensional* set of type F (so Y is contained in some Euclidean space and its closure is also an n-dimensional set of type F). The set Y is an R-*homology manifold* if

$$H_j(Y, Y-y; R) = \begin{cases} 0 & \text{if } j \neq n \\ R & \text{if } j = n \end{cases}$$

or, equivalently, if the local homology sheaf $\mathbf{H}^{-j}(\mathbf{C}^\bullet(Y, R))$ is a local system of rank 1 for $j = n$ and vanishes for $j \neq n$. Assume Y is an R-homology manifold. The R-*orientation sheaf*

$$\mathcal{O}_Y = \mathbf{H}^{-n}(\mathbf{C}^\bullet(Y, R)).$$

is the local system whose fiber at each point $y \in Y$ is $H_n(Y, Y-y; R)$.

If an orientation of Y exists (Section 2.1) then it determines an isomorphism between $\mathcal{O}_Y$ and the trivial local system $\mathbf{R}$. If Y is also connected (but not necessarily compact) then $H_n^{BM}(Y, \mathcal{O}_Y) \cong R$. A choice of generator $[Y]$ of this group is called a *fundamental class*. It can be represented by a (Borel-Moore) chain $\xi \in C_n^{BM}(Y, \mathcal{O}_Y)$ whose support is the union of all the n-dimensional simplices in a triangulation of Y, that is, $|\xi| = Y$. There is a canonical quasi-isomorphism $\mathcal{P} : \mathcal{O}_Y \to \mathbf{C}^\bullet(Y, R)[-n]$ which assigns to each sufficiently small open F-ball $U \subset Y$ the chain $[U] \in \Gamma(U, \mathbf{C}^{-n}(Y, \mathcal{O}_Y))$. It induces a quasi-isomorphism

$$\mathcal{O}_Y \otimes \mathbf{E} \to \mathbf{C}^\bullet(Y, \mathbf{E})[-n] \tag{2.5.1}$$

for any finite-dimensional local system $\mathbf{E}$ of R modules on Y. The resulting isomorphisms of cohomology groups are often referred to as *Poincaré duality* isomorphisms,

$$H^i(Y, \mathcal{O}_Y \otimes \mathbf{E}) \cong H_{n-i}^{BM}(Y, \mathbf{E})$$
$$H_c^i(Y, \mathcal{O}_Y \otimes \mathbf{E}) \cong H_{n-i}(Y, \mathbf{E}).$$

The morphism $\mathcal{P}$ may also be viewed as a quasi-isomorphism $\mathcal{P} : \mathbf{E} \to \mathbf{C}^\bullet(Y, \mathcal{O}_Y \otimes \mathbf{E})[-n]$ with resulting Poincaré duality isomorphisms

$$H^i(Y, \mathbf{E}) \cong H_{n-i}^{BM}(Y, \mathcal{O}_Y \otimes \mathbf{E})$$
$$H_c^i(Y, \mathbf{E}) \cong H_{n-i}(Y, \mathcal{O}_X \otimes \mathbf{E}).$$

2.6 Cellular Borel-Moore chains

In Chapters 8 and 9 we will need to integrate differential forms (defined on the non-compact top stratum Y of a modular variety X) over chains (which are themselves non-compact), with coefficients in local systems on Y that may not extend over its compactification X. Integration of non-compact chains on non-compact

2.6. Cellular Borel-Moore chains

manifolds leads to a host of potential pathological difficulties, none of which (fortunately) occur in the setting of modular cycles on modular varieties. The purpose of this section is to provide a few standard but not previously easily referenceable technical tools which will be used to guarantee that the integrals we will eventually consider are well behaved. The main point is that the manifold Y and the modular cycles in Y are *compactifiable*.

Let X be a set of type F (= semi-algebraic, semi-analytic, or subanalytic) and let $f : |K| \to X$ be an F-triangulation of X, where K is a locally finite simplicial complex. Let $L \subset K$ be a closed subcomplex such that the open set $Y = f(|K| - |L|)$ is dense in X. The resulting decomposition of Y is a *pseudo cell decomposition* in the sense of Section 2.1.

Let $\mathbf{E}$ be a local coefficient system of R modules on Y. If σ is a cell of K whose interior σ^o is contained in Y, then the fibers E_x, E_y of E over any two points $x, y \in \sigma \cap Y$ are canonically isomorphic. Therefore we may refer unambiguously to the fiber E_σ. If $\tau < \sigma$ and $\tau^o \subset Y$ then there is a canonical isomorphism $\Phi_{\sigma\tau} : E_\sigma \to E_\tau$.

An r-dimensional *elementary Borel-Moore cellular chain* (on Y with coefficients in $\mathbf{E}$) is an equivalence class of formal products $a_\sigma \sigma$ where σ is an oriented r-dimensional cell of K such that $\sigma^o \subset Y$ and where $a_\sigma \in \mathbf{E}_\sigma$; modulo the identification

$$a_\sigma \sigma \sim (-a_\sigma)\sigma'$$

where σ' is the same cell but with the opposite orientation.

The boundary $\partial a_\sigma \sigma$ of an elementary r-dimensional chain is defined to be

$$\partial a_\sigma \sigma = \sum_\tau \Phi_{\sigma\tau}(a_\sigma)\tau$$

where the sum is taken over those $r-1$-dimensional faces $\tau < \sigma$ such that $\tau^o \subset Y$.

The R-module of *cellular Borel-Moore chains* $\widehat{C}_r^K(Y, \mathbf{E})$ (with respect to the pseudo-cell decomposition K) is the module of finite formal linear combinations of elementary r chains. Let $\xi = \sum_i a_i \sigma_i \in \widehat{C}_r^K(Y, \mathbf{E})$ be a cellular Borel-Moore chain. Its *support* $|\xi|$ is the intersection of Y with the union of those cells σ_i such that $a_i \neq 0$. If K' is a (finite) refinement of K (and we write $K' < K$) there is a canonical injection $\widehat{C}_r^K(Y, \mathbf{E}) \to \widehat{C}_r^{K'}(Y, \mathbf{E})$ which preserves supports. The proof of the following Lemma will appear in Appendix A below.

Lemma 2.3. *If K' is a finite refinement of K then the induced mapping on homology*

$$\widehat{H}_r^K(Y, \mathbf{E}) \to \widehat{H}_r^{K'}(Y, \mathbf{E})$$

is an isomorphism.

Now let T be a (piecewise linear) triangulation of Y that is subordinate to K, that is, a triangulation such that every (closed) simplex in T is contained in a (closed) cell of K as a convex linear subset. (If Y is not compact then T will consist of infinitely many simplices.) Then we obtain a canonical injection

$$\widehat{C}_r^K(Y, \mathbf{E}) \to C_r^{BM}(Y, \mathbf{E}). \tag{2.6.1}$$

Proposition 2.4. *The mapping* (2.6.1) *induces an isomorphism on homology,*

$$\widehat{H}_r^K(Y, \mathbf{E}) \to H_r^{BM,T}(Y, \mathbf{E}) \cong H_r^{BM}(Y, \mathbf{E}).$$

In summary, *any pseudo cell decomposition of Y may be used to compute its Borel-Moore homology.* The proof, which is standard but surprisingly messy, is in Appendix A below.

2.7 Algebraic cycles

Let X be a nonsingular complex algebraic variety with a local coefficient system $\mathbf{E}$ and let $Y \subset X$ be a complex algebraic closed subvariety of complex dimension n. Suppose the local system E has the underlying structure of a (real or complex) vector bundle with a flat connection. (See Section 6.1.) When does Y determine a homology class in $H_{2n}^{BM}(X, \mathbf{E})$?

Proposition 2.5. *Let $Z \subset Y$ be a proper complex algebraic subvariety containing the singularities of Y and let S be a flat section of $\mathbf{E}|(Y-Z)$. Then the pair (Y, S) determines a Borel-Moore homology class $[Y, S] \in H_{2n}^{BM}(X, \mathbf{E})$.*

Proof. According to the above, in order to make (Y, S) into a (Borel-Moore) cycle, one needs a triangulation or a cell decomposition or a pseudo-cell decomposition of X so that Y is a union of cells; plus the data of a cellular chain on Y whose boundary is zero. In other words, one needs an assignment, to each $2n$-dimensional simplex $\sigma \subset Y$, of an element $a_\sigma \in \mathbf{E}_\sigma$ such that the boundary cancels on every $2n-1$-dimensional simplex. Here, E_σ is the fiber of $\mathbf{E}$ over any point in σ. Since $\mathbf{E}$ has a flat connection, it is possible to find a trivialization of $\mathbf{E}|\sigma$ such that the constant sections are flat, so we may interpret the element a_σ as a flat section over σ. Since $Y - Z$ is a manifold, each $2n-1$-dimensional simplex τ is a face of exactly two n-dimensional simplices, say, σ, σ'. The requirement that the boundaries cancel is the same as saying that, in a flat local trivialization of $\mathbf{E}$ on a neighborhood of τ, the two flat (= constant) sections a_σ and $a_{\sigma'}$ agree.

But this is exactly what is given in the proposition. It is possible (Theorem 2.2) to find a (smooth) triangulation of X so that Z and Y are unions of simplices, in which case Z consists entirely of simplices with dimension $2n - 2$ or less. So all of the $2n$-dimensional and $2n - 1$-dimensional simplices of Y are contained in $Y - Z$ on which the flat section S therefore defines a chain whose boundary vanishes. $\square$

Chapter 3
Review of Intersection Homology and Cohomology

In this chapter we recall the relation between intersection homology, constructed using (p, i)-allowable chains as in [Gre4], and intersection cohomology, constructed via sheaf theory.

3.1 The sheaf of intersection chains

In this section we suppose X is a (piecewise linear or subanalytic) purely n-dimensional stratified pseudomanifold ([Gre4, Gre5, Gre6]), but see also Section 3.3. As in Chapter 2, we denote by R a regular, commutative, Noetherian ring (with unit) of finite cohomological dimension; for example, any principal ideal domain.

Let $\mathsf{p} : \mathbb{N} \to \mathbb{N}$ be a perversity, that is, a mapping such that $\mathsf{p}(0) = \mathsf{p}(1) = \mathsf{p}(2) = 0$ and $\mathsf{p}(c) \leq \mathsf{p}(c+1) \leq \mathsf{p}(c) + 1$ for all c. We will make use of the following special perversities:

$$\text{zero: } \mathsf{0}(c) = 0$$
$$\text{lower middle: } \mathsf{m}(c) = \lfloor c/2 \rfloor - 1$$
$$\text{upper middle: } \mathsf{n}(c) = \lceil c/2 \rceil - 1$$
$$\text{top: } \mathsf{t}(c) = c - 2.$$

Notation. Intersection cohomology with perversity p and coefficients in a local system $\mathbf{E}$ will be denoted $I^{\mathsf{p}}H^*(X, \mathbf{E})$ and it is defined below. We will usually be concerned with pseudomanifolds X for which the intersection cohomology with perversities m and n coincide, in which case we refer to it as *the middle* intersection cohomology, and we write

$$IH^*(X, \mathbf{E}) = I^{\mathsf{m}}H^*(X, \mathbf{E}) = I^{\mathsf{n}}H^*(X, \mathbf{E}).$$

Let **E** be a local coefficient system (see Section 6.1) of R modules on the nonsingular part $Y \subset X$ of X. Let $i \geq 0$ be an integer. The pre-sheaf $I^pC^{-i}(\mathbf{E})$ of (subanalytic or piecewise linear) intersection chains assigns to each open set $U \subset X$ the subgroup $I^pC_i^{BM}(U, \mathbf{E})$ of all (subanalytic or piecewise linear) Borel-Moore chains ξ in U (= chains with closed support in U) with coefficients in **E** which satisfy the following (p, i)-*allowability condition*:

$$\dim(\xi \cap S) \leq i - \text{cod}(S) + \mathsf{p}(\text{cod}(S)), \tag{3.1.1}$$
$$\dim(\partial \xi \cap S) \leq i - 1 - \text{cod}(S) + \mathsf{p}(\text{cod}(S)) \tag{3.1.2}$$

for each singular stratum $S \subset X$. The allowability condition guarantees that all the i-dimensional and all the $i-1$-dimensional simplices in ξ are contained in the nonsingular part Y, where the local system **E** is defined. This pre-sheaf is in fact a soft sheaf ([Hab] Section II.5) so for any open set $U \subset X$ the sheaf cohomology of U may be obtained as the cohomology of the complex of sections over U. It is denoted $I^pH_i^{BM}(U,\mathbf{E})$ or $I^pH_{\text{closed}}^{-i}(U,\mathbf{E})$ and it is referred to as the *intersection homology with closed supports* of U with coefficients in **E**. The *intersection homology with compact supports* of U is the homology of the complex $I^pC_*(U,\mathbf{E})$ of sections with compact support, that is, the chains ξ in U with coefficients in **E** satisfying (3.1.1) and (3.1.2) such that $|\xi|$ is compact. It is denoted $I^pH_i(U,\mathbf{E})$. Clearly $I^pH_i(X,\mathbf{E}) = I^pH_i^{BM}(X,\mathbf{E})$ if X is compact.

The stalk cohomology of the sheaf $I^p\mathbf{C}^\bullet(\mathbf{E})$ at a point x in some stratum S of X is given by

$$H_x^{-i}(I^p\mathbf{C}^\bullet(\mathbf{E})) = I^pH_i(X, X-x, \mathbf{E}) = \begin{cases} 0 & \text{if } -i \geq \mathsf{p}(c) - n + 1 \\ I^pH_{i-s-1}(\ell_x, \mathbf{E}) & \text{if } -i \leq \mathsf{p}(c) - n. \end{cases} \tag{3.1.3}$$

Here, $s = \dim(S)$, $c = n - s$ is the codimension of S and ℓ_x is the link of the stratum S at the point x. The intersection homology is a topological invariant of $(X, \mathbf{E})$ and it does not depend on a choice of subanalytic or PL structure, or the choice of stratification.

3.2 The sheaf of intersection cochains

As in the previous section we suppose X is a (real) n-dimensional (PL or subanalytic, not necessarily compact) pseudomanifold. If **E** is a local system of R modules on the nonsingular part Y of X then Deligne [Gre5] has given an alternate construction of the sheaf of intersection cochains as an element of the constructible derived category $D_c^b(X)$ of sheaves on X. Given a stratification of X let X_k denote the closed subset of X consisting of all strata of dimension less than or equal to k. We follow the indexing scheme of [Bo84], starting with the local system **E** in degree 0 and setting

$$I^p\mathbf{S}^\bullet(\mathbf{E}) = \tau_{\leq \mathsf{p}(n)} Ri_{n*} \ldots \tau_{\leq \mathsf{p}(3)} Ri_{3*} \tau_{\leq \mathsf{p}(2)} Ri_{2*}(\mathbf{E})$$

3.2. The sheaf of intersection cochains

where the *truncation functor* $\tau_{\leq k}$ kills all stalk cohomology in degrees greater than k and where i_k is the inclusion of $U_k = X - X_{n-k}$ into U_{k+1}. The cohomology of $\mathbf{I^pS^\bullet(E)}$ is the *intersection cohomology* of X,

$$H^i(X, \mathbf{I^pS^\bullet(E)}) = I^pH^i(X, \mathbf{E}).$$

If $S \subset X$ is a stratum of codimension c and if $x \in S$ then the stalk cohomology $H^i_x(\mathbf{I^pS^\bullet(E)})$ at x of this intersection cohomology sheaf is

$$H^i(j_x^* \mathbf{I^pS^\bullet(E)}) = \begin{cases} 0 & \text{if } i > p(c) \\ I^pH^i(\ell_x, \mathbf{E}) & \text{if } i \leq p(c) \end{cases} \quad (3.2.1)$$

where ℓ_x is the *link of the stratum* S at the point x, cf. Section 2.2 and where $j_x : \{x\} \to X$ denotes the inclusion. The stalk cohomology with compact support at x is

$$H^i(j_x^! \mathbf{I^pS^\bullet(E)}) = \begin{cases} I^pH^{i-1-n+c}(\ell_x, \mathbf{E}) & \text{if } i > \mathsf{p}(c) + 1 + n - c \\ 0 & \text{if } i \leq \mathsf{p}(c) + 1 + n - c. \end{cases} \quad (3.2.2)$$

Let $\mathbf{T^\bullet}$ be a cohomologically constructible complex of sheaves on X. This means that for each k, the (local) cohomology sheaves of the restriction $\mathbf{T^\bullet_k} = \mathbf{T^\bullet}|(X_{n-k} - X_{n-k-1})$ are finite-dimensional local systems (i.e., locally constant sheaves). Fix a perversity p. Consider the following possible conditions:

(a) $H^m(j_x^* \mathbf{T^\bullet}) = 0$ for all $m > \mathsf{p}(k)$, all $x \in X_{n-k} - X_{n-k-1}$, and all k.
(b) $H^m(j_x^! \mathbf{T^\bullet}) = 0$ for all $m < \mathsf{p}(k) + n - k$, all $x \in X_{n-k} - X_{n-k-1}$, and all k.

The following lemma may be proven using the same argument as in [Gre5] Section 3.5, or using Section 1.3.4 of [Bei]:

Lemma 3.1. *Let $\mathbf{E}$ be a local system on Y. If $\mathbf{T^\bullet}$ satisfies condition (a) above then any quasi-isomorphism $\mathbf{T^\bullet}|Y \to \mathbf{E}$ has a unique extension in $D^b_c(X)$ to a morphism $\mathbf{T^\bullet} \to \mathbf{I^pS^\bullet}(X, \mathbf{E})$. If $\mathbf{T^\bullet}$ satisfies condition (b) above then any quasi-isomorphism $\mathbf{E} \to \mathbf{T^\bullet}|Y$ has a unique extension in $D^b_c(X)$ to a morphism $\mathbf{I^pS^\bullet}(X, \mathbf{E}) \to \mathbf{T^\bullet}$.* □

Consequently the quasi-isomorphism of equation (2.5.1) extends (uniquely) to Poincaré duality quasi-isomorphisms

$$\mathcal{P} : \mathbf{I^pS^\bullet}(\mathbf{E} \otimes \mathcal{O}_Y) \longrightarrow \mathbf{I^pC^\bullet}(\mathbf{E})[-n] \quad (3.2.3)$$
$$\mathcal{P} : \mathbf{I^pS^\bullet}(\mathbf{E}) \longrightarrow \mathbf{I^pC^\bullet}(\mathcal{O}_Y \otimes \mathbf{E})[-n] \quad (3.2.4)$$

which determine isomorphisms

$$I^pH^i(X, \mathbf{E} \otimes \mathcal{O}_Y) \cong I^pH^{BM}_{n-i}(X, \mathbf{E})$$
$$I^pH^i(X, \mathbf{E}) \cong I^pH^{BM}_{n-i}(X, \mathbf{E} \otimes \mathcal{O}_Y)$$
$$I^pH^i_c(X, \mathbf{E} \otimes \mathcal{O}_Y) \cong I^pH_{n-i}(X, \mathbf{E})$$
$$I^pH^i_c(X, \mathbf{E}) \cong I^pH_{n-i}(X, \mathbf{E} \otimes \mathcal{O}_Y).$$

3.3 Homological stratifications

We will eventually be concerned with the Baily-Borel compactification X of a modular variety Y that is obtained as an arithmetic quotient $Y = \Gamma\backslash D$ of a symmetric domain D by an arithmetic group Γ that is not necessarily torsion free. Such a space has a canonical "stratification" by boundary strata (cf. [BaiB]), but the strata are not necessarily smooth manifolds; rather they are orbifolds (see Appendix B), and hence they are rational homology manifolds. Thus one is led to consider orbifold stratifications and more generally, homological stratifications, in which each stratum is a homology manifold, along which the space is locally homologically a product. Homological stratifications were defined and shown to exist in [Gre5] Section 4.2 where they were used to prove that intersection homology is a topological invariant. See also [Rou]. The requirements on a stratification vary with the application, and there are many variants on the notion of a homological stratification.

Let X be an n-dimensional pseudomanifold of type F ($=$ semi-algebraic, semi-analytic or subanalytic). Let X_{n-2} be a closed subset of dimension $\leq n-2$ such that $X - X_{n-2}$ is an R-homology manifold. Let $\mathbf{E}$ be a local coefficient system of R modules on $X - X_{n-2}$. By assumption, there exists a stratification of X such that the largest stratum $Y \subset X$ is contained in $X - X_{n-2}$. Therefore the intersection complex $\mathbf{I}^\mathbf{p}\mathbf{S}^\bullet(\mathbf{E})$ is well defined on X.

Definition 3.3.1. A homological stratification of X (depending on R, p, and $\mathbf{E}$) is a filtration by closed F-subsets

$$X = X_n \supset X_{n-2} \supset X_{n-3} \supset \cdots \supset X_{-1} = \phi \qquad (3.3.1)$$

such that

(1) each $X_j - X_{j-1}$ is a j-dimensional R-homology manifold,

(2) the intersection complex $\mathbf{I}^\mathbf{p}\mathbf{S}^\bullet(\mathbf{E})$ is cohomologically locally constant (CLC) on each stratum $X_j - X_{j-1}$ (meaning that each of its cohomology sheaves are locally constant on each stratum),

(3) for each $j \geq 2$ the complex $Rh_{j*}\mathbf{I}^\mathbf{p}\mathbf{S}^\bullet(\mathbf{E})$ is cohomologically locally constant on the stratum $X_j - X_{j-1}$.

Here, $h_j : X - X_{j-1} \to X - X_j$ denotes the inclusion. Similarly let us say that a filtration (3.3.1) is an *R-orbifold stratification* if each $X_j - X_{j-1}$ is a j-dimensional R-orbifold along which X is locally topologically trivial in the sense of the Thom-Mather theorem (2.2) (where ℓ is assumed to have an R-orbifold stratification rather than a Whitney stratification).

Proposition 3.2. *Let X be an F-set with a filtration (3.3.1) by closed F-subsets. If this filtration is a Whitney stratification then it is also a $\mathbb{Q}$-orbifold stratification*

3.3. Homological stratifications

(*with the trivial orbifold structure on each stratum*). *If it is a R-orbifold stratification then it is also a R-homological stratification for any local system* **E** *of R-modules and any perversity* p. *If it is an R-homological stratification then there exists a refinement that is a Whitney stratification.* □

In other words,

$$\text{Whitney} \implies R\text{-orbifold} \implies R\text{-homological} \implies \text{Whitney refinement.}$$

Proposition 3.3. *Let* $\mathcal{S} = \{X = X_n \supset X_{n-2} \supset \cdots \supset X_{-1} = \phi\}$ *be a homological stratification (for R, p, and* **E**) *as defined above. Let*

$$\mathbf{I}_{\mathcal{S}}^{\mathsf{p}} \mathbf{S}^\bullet(\mathbf{E}) = \tau_{\leq \mathsf{p}(n)} R i_{n*} \ldots \tau_{\leq \mathsf{p}(3)} R i_{3*} \tau_{\leq \mathsf{p}(2)} R i_{2*}(\mathbf{E})$$

be the complex of sheaves obtained from Deligne's construction with respect to the homological stratification $\mathcal{S}$. Let $\mathbf{I}_{\mathcal{S}}^{\mathsf{p}} \mathbf{C}^\bullet(\mathbf{E})$ be the complex of sheaves of (Borel-Moore) chains that satisfy the allowability conditions (3.1.1) with respect to the homological stratification $\mathcal{S}$. Then the identity mapping $\mathbf{E} \to \mathbf{E}$ *extends, uniquely in $D^b(X)$, to quasi-isomorphisms*

(1) $\mathbf{I}^{\mathsf{p}} \mathbf{S}^\bullet(\mathbf{E}) \to \mathbf{I}_{\mathcal{S}}^{\mathsf{p}} \mathbf{S}^\bullet(\mathbf{E})$ *and*
(2) $\mathbf{I}^{\mathsf{p}} \mathbf{C}^\bullet(\mathbf{E}) \to \mathbf{I}_{\mathcal{S}}^{\mathsf{p}} \mathbf{C}^\bullet(\mathbf{E})$.

In other words, a homological stratification may be used in place of an honest stratification for either of these constructions of intersection (co)homology.

Proof. Statement (1) is proven in [Gre5] Section 4.2. Now let us prove statement (2). Let $\mathcal{S}'$ be an honest stratification that refines the homological stratification $\mathcal{S}$. Then every stratum $A' \in \mathcal{S}'$ is contained in a unique stratum $A \in \mathcal{S}$. Let p^+ be the stratum-dependent perversity that assigns to any such stratum $A' \in \mathcal{S}'$ the number $\mathsf{p}^+(A') = \mathsf{p}(A) + \dim(A) - \dim(A')$. Let p^- be the stratum-dependent perversity that assigns to any such stratum $A' \in \mathcal{S}'$ the number $\mathsf{p}^-(A') = \mathsf{p}(A)$. (Here we have written $\mathsf{p}(A)$ rather than $\mathsf{p}(\mathrm{cod}(A))$ for simplicity.) The sheaves $\mathbf{I}_{\mathcal{S}}^{\mathsf{p}} \mathbf{C}^\bullet(\mathbf{E})$ and $\mathbf{I}_{\mathcal{S}'}^{\mathsf{p}^+} \mathbf{C}^\bullet(\mathbf{E})$ are identical, because the p^+-allowability restrictions (3.1.1) with respect to strata $A' \in \mathcal{S}'$ coincide with the p-allowability restrictions with respect to the corresponding strata $A \in \mathcal{S}$. The argument of [Gre5] Section 3.6 implies that there are unique quasi-isomorphisms

$$\mathbf{I}_{\mathcal{S}'}^{\mathsf{p}} \mathbf{S}^\bullet(\mathbf{E} \otimes \mathcal{O}_X)[n] \cong \mathbf{I}_{\mathcal{S}'}^{\mathsf{p}} \mathbf{C}^\bullet(\mathbf{E}) \text{ and } \mathbf{I}_{\mathcal{S}'}^{\mathsf{p}^+} \mathbf{S}^\bullet(\mathbf{E} \otimes \mathcal{O}_X)[n] \cong \mathbf{I}_{\mathcal{S}'}^{\mathsf{p}^+} \mathbf{C}^\bullet(\mathbf{E}).$$

(Even though the perversity p^+ is stratum-dependent, the same proof works.) So it remains to show that $\mathbf{I}_{\mathcal{S}'}^{\mathsf{p}^+} \mathbf{S}^\bullet(\mathbf{E} \otimes \mathcal{O}_X)$ and $\mathbf{I}_{\mathcal{S}'}^{\mathsf{p}} \mathbf{S}^\bullet(\mathbf{E} \otimes \mathcal{O}_X)$ are canonically quasi-isomorphic. For notational simplicity we now drop explicit mention of the local system $\mathbf{E} \otimes \mathcal{O}_X$.

For any stratum $A' \in \mathcal{S}'$ we have $\mathsf{p}^+(A') \geq \mathsf{p}(A') \geq \mathsf{p}^-(A')$. Therefore there are canonical morphisms

$$\mathbf{I}_{\mathcal{S}'}^{\mathsf{p}^-}\mathbf{S}^\bullet \to \mathbf{I}_{\mathcal{S}'}^{\mathsf{p}}\mathbf{S}^\bullet \to \mathbf{I}_{\mathcal{S}'}^{\mathsf{p}^+}\mathbf{S}^\bullet. \tag{3.3.2}$$

We claim these are quasi-isomorphisms. Suppose by induction that we have proven that these are quasi-isomorphisms over the open set $X - X_{n-c}$ that consists of all homological strata of codimension $< c$. We wish to conclude that these are also quasi-isomorphisms over the homological strata of codimension c. Let $A \supset A'$ be strata of $\mathcal{S}$ and $\mathcal{S}'$ respectively, with codimensions c and $c+r$ respectively. Let $x' \in A'$ and let ℓ' denote the link of the stratum A' at the point x'. Since the homological stratum A is refined by the stratification $\mathcal{S}'$ there exists an honest stratum $A^0 \in \mathcal{S}$ which is open in A. Let $x \in A^0$ be a point, sufficiently close to x', and let ℓ denote the link of the stratum A^0 at the point x.

Since the homological stratification satisfies hypothesis (3) there is a canonical isomorphism between the stalk cohomology (3.2.1) of the $\mathbf{I}^\mathsf{p}\mathbf{S}^\bullet$ sheaf at the points x and x', that is, between

$$I^\mathsf{p} H_x^i = \begin{cases} 0 & \text{if } i > \mathsf{p}(c) \\ I^\mathsf{p} H^i(\ell) & \text{if } i \leq \mathsf{p}(c) \end{cases} \quad \text{and} \quad I^\mathsf{p} H_{x'}^i = \begin{cases} 0 & \text{if } i > \mathsf{p}(c+r) \\ I^\mathsf{p} H^i(\ell') & \text{if } i \leq \mathsf{p}(c+r) \end{cases}$$

Therefore $I^\mathsf{p} H^i(\ell') = 0$ for $\mathsf{p}(c) < i \leq \mathsf{p}(c+r)$. Moreover, a similar isomorphism holds for the stalk cohomology with compact supports (3.2.2). This is because, by hypothesis (3) again, each term in the following isomorphic triangles of sheaves on $X - X_{n-c-1}$ is CLC:

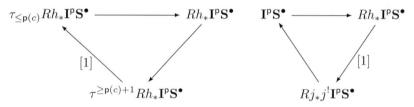

Here, $h : X - X_{n-c} \to X - X_{n-c-1}$ denotes the inclusion and $\tau^{\geq \mathsf{p}(c)+1}$ is the functor that kills cohomology in all degrees $\leq \mathsf{p}(c)$, cf. [Gre5] Section 1.14. (The upper left-hand corners of the two triangles are isomorphic by statement (1) of Proposition 3.3.) Since the stratum A is also a homology manifold, $j_x^! = i_x^* j^! [n-c]$ where j_x denotes the composition of inclusions

$$\{x\} \xrightarrow{i_x} A \xrightarrow{j} X - X_{n-c}.$$

Hence the cohomology of $j_x^! \mathbf{I}^\mathsf{p}\mathbf{S}^\bullet$ is locally constant as x varies in A. This gives a canonical isomorphism between (cf. equation (3.2.2)),

$$H^i\left(j_x^! \mathbf{I}^\mathsf{p}\mathbf{S}^\bullet\right) = \begin{cases} I^\mathsf{p} H^{i-1-n+c}(\ell) & \text{if } i > \mathsf{p}(c) + 1 + n - c \\ 0 & \text{if } i \leq \mathsf{p}(c) + 1 + n - c \end{cases}$$

and

$$H^i\left(j_{x'}^!\mathbf{I^p S^\bullet}\right) = \begin{cases} I^p H^{i-1-n+c+r}(\ell') & \text{if } i > \mathsf{p}(c+r)+1+n-c-r \\ 0 & \text{if } i \leq \mathsf{p}(c+r)+1+n-c-r \end{cases}$$

Therefore

$$I^p H^{i-1-n+c+r}(\ell') = 0 \quad \text{for} \quad \mathsf{p}(c+r)+1+n-c-r < i \leq \mathsf{p}(c)+1+n-c.$$

In summary we conclude that

$$I^p H^i(\ell') = 0 \quad \text{for} \quad \mathsf{p}(c)+1 \leq i \leq \mathsf{p}(c)+r. \tag{3.3.3}$$

(If $\mathcal{S}$ were also an honest stratification, the link ℓ' would be the r-fold suspension of ℓ, and this statement would follow directly.) On the other hand, the stalk cohomology at the point $x' \in A'$ of the sheaves (3.3.2) is given in the following table,

Stalk cohomology $H^i_{x'}(\mathbf{IS^\bullet})$					
p^+		p		p^-	
0	$(i > \mathsf{p}(c)+r)$	0	$(i > \mathsf{p}(c+r))$	0	$(i > \mathsf{p}(c))$
$I^{p^+}H^i(\ell')$	$(i \leq \mathsf{p}(c)+r)$	$I^p H^i(\ell')$	$(i \leq \mathsf{p}(c+r))$	$I^{p^-}H(\ell')$	$(i \leq \mathsf{p}(c))$

We have assumed by induction that $I^{p^-}H^i(\ell') \cong I^p H^i(\ell') \cong I^{p^+}H^i(\ell')$. Then equation (3.3.3) implies that the morphisms (3.3.2) induce isomorphisms on the stalk cohomology at the point $x' \in A'$. Since x' was arbitrary we conclude that the morphisms (3.3.2) are quasi-isomorphisms. This completes the proof of Proposition 3.3. □

3.4 Products in intersection homology and cohomology

As in Section 3.2, let R denote a regular Noetherian ring of finite cohomological dimension. We suppose that X is a (not necessarily compact) subanalytic or piecewise linear n-dimensional pseudomanifold with singular set Σ and let $Y = X - \Sigma$. Let

$$\mathbf{E_1} \times \mathbf{E_2} \to \mathbf{E_3} \tag{3.4.1}$$

be a bilinear pairing of local systems of R-modules on Y. Suppose p, q are perversities such that $\mathsf{p}+\mathsf{q}$ is also a perversity. The pairing (3.4.1) of local systems extends (uniquely in $D^b_c(X)$) to a morphism

$$\mathbf{I^p S^\bullet}(\mathbf{E_1}) \otimes \mathbf{I^q S^\bullet}(\mathbf{E_2}) \to \mathbf{I^{p+q} S^\bullet}(\mathbf{E_3}) \tag{3.4.2}$$

giving a product $I^p H^i(X, \mathbf{E_1}) \otimes I^q H^j(X, \mathbf{E_2}) \to I^{p+q} H^{i+j}(X, \mathbf{E_3})$.

By an *orientation* of X we mean an orientation of its nonsingular part, Y. Suppose X is orientable and oriented. Then the transverse intersection of chains ([Gre4]) induces, for any open set $U \subset X$, an *intersection pairing*

$$I^{\mathsf{p}} H_i^{BM}(U, \mathbf{E_1}) \times I^{\mathsf{q}} H_j^{BM}(U, \mathbf{E_2}) \to I^{\mathsf{p+q}} H_{i+j-n}^{BM}(U, \mathbf{E_3})$$

using the fact that orientations of transverse chains ξ, η determines an orientation of the intersection $|\xi| \cap |\eta|$ in the presence of an orientation of Y. The orientation of X also determines a Poincaré duality morphism $\mathcal{P}$, cf. (2.5.1) and (3.2.3), and the following diagram commutes,

$$\begin{array}{ccc} I^{\mathsf{p}} H^i(X, \mathbf{E_1}) \otimes I^{\mathsf{q}} H^j(X, \mathbf{E_2}) & \longrightarrow & I^{\mathsf{p+q}} H^{i+j}(X, \mathbf{E_3}) \\ {\scriptstyle \mathcal{P} \otimes \mathcal{P}} \downarrow & & \downarrow {\scriptstyle \mathcal{P}} \\ I^{\mathsf{p}} H_{n-i}(X, \mathbf{E_1}) \otimes I^{\mathsf{q}} H_{n-j}(X, \mathbf{E_2}) & \longrightarrow & I^{\mathsf{p+q}} H_{n-i-j}(X, \mathbf{E_3}) \end{array} \qquad (3.4.3)$$

where the bottom row is the intersection pairing. These pairings and mappings are independent of the stratification ([Gre5]).

For any perversity r let

$$\epsilon : I^{\mathsf{r}} H_0(X, R) \to H_0(X, R) \to R$$

denote the augmentation. Suppose $\mathbf{E_1} \times \mathbf{E_2} \to \mathbf{R}$ is a pairing of local systems of R-modules (where $\mathbf{R}$ denotes the constant local system). If p, q are perversities such that $\mathsf{p} + \mathsf{q}$ is a perversity and if X is oriented, then the above products give rise to pairings on intersection homology and cohomology, which are essentially three different ways to express the same product:

$$\langle, \rangle_{IH^*} : I^{\mathsf{p}} H^i(X, \mathbf{E_1}) \times I^{\mathsf{q}} H^{n-i}(X, \mathbf{E_2}) \to R \qquad (3.4.4)$$

$$\langle, \rangle_{IH_*} : I^{\mathsf{p}} H_j(X, \mathbf{E_1}) \times I^{\mathsf{q}} H_{n-j}(X, \mathbf{E_2}) \to R \qquad (3.4.5)$$

$$\langle, \rangle_K : I^{\mathsf{p}} H^k(X, \mathbf{E_1}) \times I^{\mathsf{q}} H_k(X, \mathbf{E_2}) \to R \qquad (3.4.6)$$

defined by $\langle a, b \rangle_{IH^*} = \epsilon(\mathcal{P}(a \cdot b))$ in (3.4.4), $\langle a, b \rangle_{IH_*} = \epsilon(a \cdot b)$ in (3.4.5), and $\langle a, b \rangle_K = \epsilon(\mathcal{P}(a) \cdot b)$ in (3.4.6), where $\cdot$ denotes the intersection product. These pairings will be referred to as the *cup product pairing*, the *intersection pairing*, and the *Kronecker pairing* respectively. They are compatible with the corresponding products in the (ordinary) homology and cohomology of Y. For example, in the following diagram,

$$\begin{array}{ccc} H_c^k(Y; \mathbf{E_1}) \times H_k^{BM}(Y; \mathbf{E_2}) & \xrightarrow{\langle, \rangle} & R \\ {\scriptstyle \alpha_{1*}} \downarrow \quad\quad \uparrow {\scriptstyle \beta_{1*}} & & \\ I^{\mathsf{p}} H^k(X, \mathbf{E_1}) \times I^{\mathsf{q}} H_k(X, \mathbf{E_2}) & \xrightarrow{\langle, \rangle_K} & R \\ {\scriptstyle \alpha_{2*}} \downarrow \quad\quad \uparrow {\scriptstyle \beta_{2*}} & & \\ H^k(Y, \mathbf{E_1}) \times H_k(Y, \mathbf{E_2}) & \xrightarrow{\langle, \rangle} & R \end{array} \qquad (3.4.7)$$

these "Kronecker products" satisfy

$$\langle \alpha_{1*}(x), y \rangle_K = \langle x, \beta_{1*}(y) \rangle \text{ and } \langle \alpha_{2*}(x'), y' \rangle = \langle x', \beta_{2*}(y') \rangle_K. \tag{3.4.8}$$

Let $\mathbb{D}_X^\bullet$ denote the *dualizing complex* on X. The *Poincaré duality* theorem ([Gre4, Gre5]) states:

Theorem 3.4. *Assume R is a field. Assume also*

(1) p *and* q *are complementary perversities (that is, $\mathsf{p}(c) + \mathsf{q}(c) = c - 2$ for all c),*

(2) *the pairing* $\mathbf{E_1} \times \mathbf{E_2} \to \mathbf{R}$ *is nondegenerate, and*

(3) X *is compact.*

Then (3.4.2) becomes a dual pairing, that is, a morphism

$$\mathbf{I^p S^\bullet(E_1)} \otimes \mathbf{I^q S^\bullet(E_2)} \to \mathbb{D}_X^\bullet[-n] \tag{3.4.9}$$

such that the induced mapping

$$\Psi : \mathbf{I^p S^\bullet(E_1)} \to \mathbf{RHom}^\bullet(\mathbf{I^q S^\bullet(E_2)}, \mathbb{D}_X^\bullet[-n]) \tag{3.4.10}$$

is a quasi-isomorphism. Therefore the resulting bilinear forms (3.4.4), (3.4.5), and (3.4.6) are nondegenerate. $\square$

3.5 Finite mappings

Let X, X' be purely n-dimensional oriented subanalytic pseudomanifolds and let R be a regular Noetherian ring of finite cohomological dimension. Let $f : X' \to X$ be a finite proper surjective subanalytic mapping. Then X, X' can be Whitney stratified so that f takes strata to strata, and f is a covering on each stratum. Suppose such stratifications have been chosen. (For the purposes of this section, R-orbifold stratifications of X, X' would also suffice. However one could always reduce to the case of Whitney stratifications because the intersection cohomology groups are independent of the stratification.) Let $\mathbf{E}$ be a local coefficient system of R-modules on the nonsingular part Y of X. Fix a perversity p. It is easy to see that the pullback $f^{-1}(|\xi|)$ of a (p, i)-allowable (subanalytic) chain $\xi \in I^\mathsf{p} C_i^{BM}(X, \mathbf{E})$ is again (p, i)-allowable. A section of $\mathbf{E}$ over the i-dimensional simplices (in some triangulation) of ξ gives a section of $f^*(\mathbf{E})$ over the i-dimensional simplices of $f^{-1}(|\xi|)$ so we obtain a chain $f^*(\xi) \in I^\mathsf{p} C_i(X', f^*(\mathbf{E}))$. Similarly, the image $f(|\eta|)$ of a (p, i)-allowable subanalytic chain $\eta \in I^\mathsf{p} C_j(X', f^*(\mathbf{E}))$ is (p, i)-allowable. Adding the values of a section over the fibers of f gives a chain $f(\eta) \in I^\mathsf{p} C_j(X, \mathbf{E})$. Therefore f induces mappings

$$I^\mathsf{p} H_i(X, \mathbf{E}) \xrightarrow{f^*} I^\mathsf{p} H_i(X', f^*(\mathbf{E})) \xrightarrow{f_*} I^\mathsf{p} H_i(X, \mathbf{E}). \tag{3.5.1}$$

If $\mathbf{E_1} \times \mathbf{E_2} \to \mathbf{E_3}$ is a bilinear pairing between local systems on X, and if p, q are perversities such that $\mathsf{p} + \mathsf{q}$ is a perversity, then these mappings are compatible with the intersection product, via the *projection formula*:

$$f_*(f^*(a) \cdot b) = a \cdot f_*(b) \in I^{\mathsf{p+q}} H_{i+j-n}(X, \mathbf{E_3}) \qquad (3.5.2)$$

for all $a \in I^{\mathsf{p}} H_i(X', f^*(\mathbf{E_1}))$ and all $b \in I^{\mathsf{p}} H_j(X, \mathbf{E_2})$.

3.6 Correspondences

A *correspondence* $(c_1, c_2) : X' \rightrightarrows X$ is a pair of finite proper surjective mappings. Given such a correspondence, the pseudomanifolds X, X' can be stratified with nonsingular strata Y, Y' respectively, so that the mappings $c_i : Y' \to Y$ are (unramified) coverings. If $\mathbf{E}$ is a local system on Y then a *lift of* $\mathbf{E}$ to the correspondence is an isomorphism of local systems, $c_1^*(\mathbf{E}) \cong c_2^*(\mathbf{E})$. Using (3.5.1), such a lift determines endomorphisms

$$h = (c_2)_* (c_1)^* \quad \text{and} \quad {}^t h = (c_1)_* (c_2)^*$$

on $I^{\mathsf{p}} H_*(X, \mathbf{E})$. Notice that if we only assume that $c_1^*(\mathbf{E}) \to c_2^*(\mathbf{E})$ is a morphism of local systems, we still obtain the endomorphism h, but not necessarily ${}^t h$.

Lemma 3.5. *Let* $(c_1, c_2) : X' \rightrightarrows X$ *be a correspondence with* $h, {}^t h$ *as above. Let* $\mathbf{E_1} \times \mathbf{E_2} \to \mathbb{Q}$ *be a pairing of local systems (of rational vector spaces) on the nonsingular part* $Y \subset X$. *Suppose we are given lifts of* $\mathbf{E_1}$ *and* $\mathbf{E_2}$ *to the correspondence. Let* p, q *be perversities such that* $\mathsf{p} + \mathsf{q}$ *is a perversity. Then the pairing* (3.4.4) *satisfies:*

$$\langle h(a), b \rangle = \langle a, {}^t h(b) \rangle \in \mathbb{Q}$$

for all $a \in I^{\mathsf{p}} H_i(X, \mathbf{E_1})$ *and all* $b \in I^{\mathsf{q}} H_{n-i}(X, \mathbf{E_2})$.

Proof. Using the projection formula (3.5.2), calculate

$$\begin{aligned}
\langle h(a), b \rangle &= \langle c_{2*} c_1^* a, b \rangle \\
&= \epsilon((c_{2*} c_1^* a) \cdot b)) = \epsilon(c_{2*}(c_1^* a \cdot c_2^* b)) \\
&= \epsilon(c_1^* a \cdot c_2^* b) = \epsilon(c_{1*}(c_1^* a \cdot c_2^* b)) \\
&= \epsilon(a \cdot c_{1*} c_2^* b) = \langle a, {}^t h(b) \rangle \qquad \square
\end{aligned}$$

The mappings h and ${}^t h$ may also be constructed on intersection cohomology using the sheaf theoretic approach. Suppose $\mathbf{E}$ is a local system on $Y \subset X$ and $c_1^*(\mathbf{E}) \to c_2^*(\mathbf{E})$ is a morphism of local systems. The complex of sheaves $c_1^*\left(\mathbf{I^p S^\bullet}(X, \mathbf{E})\right)$ satisfies the support condition (a), and $c_2^!\left(\mathbf{I^p S^\bullet}(X, \mathbf{E})\right)$ satisfies the cosupport condition (b) of Lemma 3.1, which therefore provides canonical morphisms

$$c_1^* \mathbf{I^p S^\bullet}(\mathbf{E}) \to \mathbf{I^p S^\bullet}(c_1^* \mathbf{E}) \to \mathbf{I^p S^\bullet}(c_2^*(\mathbf{E})) \to c_2^! \mathbf{I^p S^\bullet}(\mathbf{E}).$$

3.6. Correspondences

Let $s: X \to \{pt\}$ be the map to a point. Then $s_*(c_1)_* = s_*(c_2)_!$ since X is compact. We obtain a morphism

$$Rs_*\mathbf{I^pS^\bullet(E)} \to Rs_*Rc_{1*}c_1^*\mathbf{I^pS^\bullet(E)} \to Rs_*Rc_{1*}c_2^!\mathbf{I^pS^\bullet(E)}$$
$$= Rs_*Rc_{2!}c_2^!\mathbf{I^pS^\bullet(E)} \to Rs_*\mathbf{I^pS^\bullet(E)}$$

using the adjunction morphisms

$$\mathbf{A^\bullet} \to Rf_*f^*\mathbf{A^\bullet} \quad \text{and} \quad Rf_!f^!\mathbf{A^\bullet} \to \mathbf{A^\bullet}$$

(which exist for any complex of sheaves $\mathbf{A^\bullet}$ on X). This gives the desired mapping $h = (c_2)_*(c_1)^*$ on intersection cohomology. It is compatible with the isomorphism $\mathcal{P}$, meaning that (for all p, i) the following diagram commutes up to a sign which is 1 if c_1 and c_2 are both orientation preserving:

$$\begin{array}{ccc} I^{\mathsf{p}}H^i(X,\mathbf{E}) & \xrightarrow{h} & I^{\mathsf{p}}H^i(X,\mathbf{E}) \\ {\scriptstyle \mathcal{P}}\downarrow & & \downarrow{\scriptstyle \mathcal{P}} \\ I^{\mathsf{p}}H_{n-i}(X,\mathbf{E}) & \xrightarrow{h} & I^{\mathsf{p}}H_{n-i}(X,\mathbf{E}). \end{array} \quad (3.6.1)$$

If $c_1^*(\mathbf{E}_1) \to c_2^*(\mathbf{E}_2)$ is an isomorphism, then similar remarks apply to ${}^t h$. Moreover, in this case the pairing (3.4.5) satisfies $\langle h(a), b \rangle = \langle a, {}^t h(b) \rangle$ for all $a \in I^{\mathsf{p}}H^i(X, \mathbf{E}_1)$ and all $b \in I^{\mathsf{q}}H^{n-i}(X, \mathbf{E}_2)$.

Chapter 4

Review of Arithmetic Quotients

4.1 The setting

Possible references for the geometry described in this section include [BoS, Sa2, Gre6, Gre3, Gre2]. Let G be a connected reductive algebraic group over $\mathbb{Q}$, let S_G be the greatest $\mathbb{Q}$-split torus in the center of G and let

$$^0G = \cap_\chi \ker(\chi^2)$$

be the intersection of the kernels of the squares of the rationally defined characters χ of G. Then the group of real points $G(\mathbb{R})$ splits as a direct product

$$G(\mathbb{R}) = {}^0G(\mathbb{R}) \times A_G$$

where $A_G = S_G(\mathbb{R})^0$ is the topologically connected component of the group of real points of S_G. Borel and Serre [BoS] consider the "symmetric space" $G(\mathbb{R})/K_\infty^1 A_G$ where K_∞^1 is a maximal compact subgroup of $G(\mathbb{R})$. In the case that we are most interested in, see Section 5.1, there is the annoying problem that the center of G contains an $\mathbb{R}$-split torus S that is not $\mathbb{Q}$-split, and we need to divide by it in order to get a symmetric space that is a product of copies of $\mathbb{C} - \mathbb{R}$.

For this purpose, let S be a central subgroup of G defined over $\mathbb{Q}$ that contains S_G. We denote by

$$A := S(\mathbb{R})^0$$

the identity (or neutral) component of $S(\mathbb{R})$. We are most interested in the case where S is the center of G as in Section 5.1. Let $K_\infty = K_\infty^1 A$. where $K_\infty^1 \subset G(\mathbb{R})$ is a maximal compact subgroup corresponding to a Cartan involution θ of G. We refer to $D = G(\mathbb{R})/K_\infty$ as the "symmetric space" associated to G. The choice of K_∞^1 corresponds to a choice of basepoint $x_0 \in D$. Then K_∞^1 acts on the tangent space $T_{x_0}D$ and a choice of K_∞^1-invariant inner product on $T_{x_0}D$ gives rise to a Riemannian metric on D. Such a metric is referred to as an *invariant metric* and in this metric, the manifold D is complete.

A *locally symmetric space* or *arithmetic quotient* is the quotient $\Gamma\backslash D$ by an arithmetic subgroup $\Gamma \subset G(\mathbb{Q})$. It is *Hermitian* if the symmetric space D is Hermitian, that is, if it carries a $G(\mathbb{R})$-invariant complex structure. If Γ is torsion-free then $\Gamma\backslash D$ is a smooth manifold, otherwise it is an orbifold. By a theorem [BoHC] of Borel and Harish-Chandra, the volume (in the invariant metric) of $\Gamma\backslash D$ is finite. If $K_0 \subset G(\mathbb{A}_f)$ is a compact open subgroup of the finite adèles, the space

$$Y = G(\mathbb{Q})\backslash G(\mathbb{A})/K_0 K_\infty$$

is the disjoint union of finitely many locally symmetric spaces. To describe this, let $Z \subseteq G$ be the center of G and let $G(\mathbb{R})^+ \subseteq G(\mathbb{R})$ be the subgroup whose elements map to the identity component of $G/Z(\mathbb{R})$. Assume that $S = Z$. Finally let $G(\mathbb{Q})^+ := G(\mathbb{Q}) \cap G(\mathbb{R})^+$. Using the strong approximation theorem ([PlaR] Thm. 7.12) it can be shown that the set $G(\mathbb{Q})^+\backslash G(\mathbb{A}_f)/K_0$ is finite. It is possible to choose finitely many elements $x_i \in G(\mathbb{A}_f)$ (say, $1 \leq i \leq r$) which form a complete set of representatives for the set $G(\mathbb{Q})^+\backslash G(\mathbb{A}_f)/K_0$. Then the space Y is the disjoint union of the locally symmetric spaces $\Gamma_i\backslash D^+$ where

$$\Gamma_i = G(\mathbb{Q})^+ \cap x_i K_0 x_i^{-1}, \quad 1 \leq i \leq r,$$

(compare Section 5 of [Mil]). Here D^+ is the component of D containing the identity.

For the remainder of this chapter we fix an arithmetic group $\Gamma \subset G(\mathbb{R})$ and set $X = \Gamma\backslash D$. We will be concerned with three compactifications of X: the Borel-Serre compactification $\overline{X}^{BS}$ (which is a manifold with corners), the reductive Borel-Serre compactification, $\overline{X}^{RBS}$ (which is a stratified singular space), and the Baily-Borel (Satake) compactification $\overline{X}^{BB}$ (which is a complex projective algebraic variety, usually singular, and is only defined when X is Hermitian). These compactifications are related by canonical stratified mappings

$$\overline{X}^{BS} \longrightarrow \overline{X}^{RBS} \xrightarrow{\mu} \overline{X}^{BB}.$$

4.2 Baily-Borel compactification

Standard references for this section include [AshM] Chapt. III, [Sat] Chapt. II and [BaiB]. Assume that the "symmetric space" $D = G(\mathbb{R})/K_\infty$ is Hermitian (meaning that it admits a $G(\mathbb{R})$-invariant complex structure. It may be holomorphically embedded in Euclidean space $\mathbb{C}^m$ as a bounded (open) domain, by the Harish Chandra embedding ([AshM] p. 170, [Sat] Section II.4). The action of $G(\mathbb{R})$ extends to the closure $\overline{D}$. The boundary $\partial D = \overline{D} - D$ is a smooth manifold which decomposes into a (continuous) disjoint union of *boundary components*, each of which is the intersection

$$F = \overline{D} \cap H$$

4.3. L^2 differential forms

with a supporting hyperplane H. Alternatively, it is possible ([Sat] III.8.13) to characterize each boundary component as a single holomorphic path component of ∂D: two points $x, y, \in \partial D$ lie in a single boundary component F iff they are both in the image of a holomorphic "path" $\alpha : \Delta \to \partial D$ (where Δ denotes the open unit disk). In this case $\alpha(\Delta)$ is completely contained in F.

The boundary component F is a bounded symmetric domain in the smallest affine space L that contains F. The normalizer $N_{G(\mathbb{R})}F$ (consisting of those group elements which preserve the boundary component F) is a proper parabolic subgroup of $G(\mathbb{R})$, and the boundary component F is *rational* if this subgroup is rationally defined.

If we decompose G into its $\mathbb{Q}$ simple factors, $G = G_1 \times \cdots \times G_k$ then the symmetric space D decomposes similarly, $D = D_1 \times \cdots \times D_k$. Each (rational) boundary component F of D is then the product $F = F_1 \times \cdots \times F_k$ where either $F_i = D_i$ or F_i is a proper (rational) boundary component of D_i. The normalizer of F is the product $N_G(F) = N_{G_1}(F_1) \times \cdots \times N_{G_k}(F_k)$ (writing $N_{G_i}(D_i) = G_i$ whenever necessary). If G is $\mathbb{Q}$-simple then the normalizer $N_G(F)$ is a *maximal* (rational) proper parabolic subgroup of G.

Definition 4.2.1. The Baily-Borel Satake partial compactification $\overline{D}^{BB}$ is the union of D together with all its rational boundary components, with the *Satake topology*.

Theorem 4.1 ([BaiB]). *The closure $\overline{F}$ of each rational boundary component $F \subset \overline{D}^{BB}$ is the Baily-Borel-Satake partial compactification $\overline{F}^{BB}$ of F. The group $G(\mathbb{Q})$ acts continuously, by homeomorphisms on the partial compactification $\overline{D}^{BB}$. The quotient $\overline{X}^{BB} = \Gamma\backslash\overline{D}^{BB}$ is compact. Moreover, it admits the structure of a complex projective algebraic variety. It has a canonical stratification, with one stratum $X_F = \Gamma \cap (N_{G(\mathbb{R})}F)\backslash F$ for each Γ-equivalence class of rational boundary components F. The closure of X_F is the Baily-Borel compactification of X_F.*

4.3 L^2 differential forms

As in Section 4.2, assume the "symmetric space" $D = G(\mathbb{R})/K_\infty$ is Hermitian, so that $X = \Gamma\backslash D$ is a complex algebraic variety. A finite-dimensional representation of G on a complex vector space V determines a local system $\mathbf{E} = V \times_\Gamma D$ on X. A choice of K_∞^1 invariant Hermitian form on V gives rise to a G invariant Hermitian metric on the local system $\mathbf{E}$. Let $\Omega^\bullet_{(2)}(X, \mathbf{E})$ denote the complex of smooth L^2 differential forms on X with coefficients in $\mathbf{E}$, that is, the collection of all smooth differential forms ω on the orbifold X such that $\int_X \|\omega\|^2 \, d\text{vol} < \infty$ and $\int_X \|d\omega\|^2 \, d\text{vol} < \infty$ (where $\|\|$ denotes the pointwise norm determined by the Hermitian metric, and $d\text{vol}$ denotes the invariant volume form on X). The cohomology $H^i_{(2)}(X, \mathbf{E})$ of this complex is called the L^2 cohomology of X.

Now consider the sheaf (or rather, the complex of sheaves) $\mathbf{\Omega}^\bullet_{(2)}(\mathbf{E})$ of smooth L^2 differential forms on the Baily-Borel compactification $\overline{X}^{BB}$. In degree j it is the sheafification of the presheaf whose sections over an open set $U \subset \overline{X}^{BB}$ consist of all smooth differential forms $\omega \in \Omega^j(U \cap X, \mathbf{E})$ such that

$$\int_{U \cap X} \|\omega\|^2 \, d\text{vol} < \infty \text{ and } \int_{U \cap X} \|d\omega\|^2 \, d\text{vol} < \infty.$$

This sheaf is known ([Zu1]) to be fine and (cohomologically) constructible, meaning that its stalk cohomology is finite-dimensional and is locally constant along each stratum of the canonical stratification of $\overline{X}^{BB}$. Its restriction to the "nonsingular" (or orbifold) part, X, is precisely the complex of sheaves $\mathbf{\Omega}^\bullet(X, \mathbf{E})$ of all smooth differential forms on X. Its vector space of global sections

$$\Gamma(\overline{X}^{BB}, \mathbf{\Omega}^{\mathbf{j}}_{(2)}(\mathbf{E})) = \Omega^j_{(2)}(X, \mathbf{E})$$

coincides with the L^2 differential forms on X. Hence, its cohomology is the L^2 cohomology, that is,

$$H^j(\overline{X}^{BB}, \mathbf{\Omega}^\bullet_{(2)}(\mathbf{E})) \cong H^j_{(2)}(X, \mathbf{E}).$$

In [Loo] and [SaS1] the following conjecture of S. Zucker was established, in the case that Γ acts freely on D.

Theorem 4.2. *The identity mapping* $\mathbf{E} \to \mathbf{E}$ *has a unique extension in* $D^b(\overline{X}^{BB})$ *to a (quasi-) isomorphism*

$$\mathcal{Z} : \mathbf{\Omega}^\bullet_{(2)}(\mathbf{E}) \to \mathbf{I}^{\mathsf{m}}\mathbf{S}^\bullet(\overline{X}^{BB}, \mathbf{E})$$

between the sheaf of L^2 differential forms and the intersection complex with middle perversity m *on* $\overline{X}^{BB}$. *It induces a natural isomorphism*

$$\mathcal{Z} : H^j_{(2)}(X, \mathbf{E}) \cong I^{\mathsf{m}} H^j(\overline{X}^{BB}, \mathbf{E})$$

between the L^2 cohomology of X and the (middle) intersection cohomology of $\overline{X}^{BB}$.

Proof. The general case of Theorem 4.2 can be reduced to the case when Γ acts freely. For, if the action of Γ on D is not free then there exists a subgroup $\Gamma' \subset \Gamma$ of finite index which acts freely. Then the mapping $\pi : X' = \Gamma' \backslash D \to X$ is a ramified covering of degree $[\Gamma : \Gamma']$ which also serves as an orbifold chart (see Appendix B). It extends to a mapping $\bar{\pi} : \overline{X'}^{BB} \to \overline{X}^{BB}$ which induces quasi-isomorphisms $R\bar{\pi}_*(\mathbf{\Omega}^\bullet_{(2)}(\mathbf{E'})) \to \mathbf{\Omega}^\bullet_{(2)}(\mathbf{E})$ and $R\bar{\pi}_*(\mathbf{I}^{\mathsf{m}}\mathbf{S}^\bullet(\overline{X'}^{BB}, \mathbf{E'})) \to \mathbf{I}^{\mathsf{m}}\mathbf{S}^\bullet(\overline{X}^{BB}, \mathbf{E})$, and the composition $\bar{\pi}_*\bar{\pi}^*$ is multiplication by $[\Gamma : \Gamma']$. $\square$

4.3. L^2 differential forms

The inclusion of complexes of smooth differential forms on the orbifold X

$$\Omega^\bullet_c(X, \mathbf{E}) \to \Omega^\bullet_{(2)}(X, \mathbf{E}) \to \Omega^\bullet(X, \mathbf{E})$$

induces mappings on the corresponding (de Rham) cohomology groups, and it is easy to verify that the following diagram commutes:

$$
\begin{array}{ccccc}
H^j_{dR,c}(X, \mathbf{E}) & \longrightarrow & H^j_{(2)}(X, \mathbf{E}) & \longrightarrow & H^j_{dR}(X, \mathbf{E}) \\
\downarrow & & {\scriptstyle \mathcal{Z}} \downarrow & & \downarrow \\
H^j_c(X, \mathbf{E}) & \longrightarrow & I^m H^j(X, \mathbf{E}) & \longrightarrow & H^j(X, \mathbf{E})
\end{array}
\quad (4.3.1)
$$

Proposition 4.3. *Let*

$$\langle \cdot, \cdot \rangle : \mathbf{E_1} \times \mathbf{E_2} \to \mathbb{C} \quad (4.3.2)$$

be a nondegenerate pairing of local systems on X that arise from finite-dimensional representations of G. Let $\omega_1 \in \Omega^j_{(2)}(X, \mathbf{E_1})$ and $\omega_2 \in \Omega^{n-j}_{(2)}(X, \mathbf{E_2})$ be smooth, closed, L^2 differential forms on X, representing cohomology classes

$$[\omega_1] \in H^j_{(2)}(X, \mathbf{E_1}) \quad \text{and} \quad [\omega_2] \in H^{n-j}_{(2)}(X, \mathbf{E_2})$$

respectively. Then

$$\int_X \omega_1 \wedge \omega_2 = \langle \mathcal{Z}([\omega_1]), \mathcal{Z}([\omega_2]) \rangle_{IH^*} \quad (4.3.3)$$

where $\langle \cdot, \cdot \rangle_{IH^*}$ denotes the (intersection cohomology) pairing of (3.4.4) and where $\int_X \omega_1 \wedge \omega_2$ is understood to mean that the values of the differential forms ω_1, ω_2 are multiplied using the pairing $\langle \cdot, \cdot \rangle$.

Proof. If $\mathbf{S}^\bullet$ is a (cohomologically) constructible complex of *fine* sheaves of complex vector spaces on $\overline{X}^{BB}$ then its Borel-Moore-Verdier dual is the complex of sheaves $\mathbf{T}^\bullet$ whose sections over an open set $U \subset \overline{X}^{BB}$ are defined as

$$\Gamma(U, \mathbf{T}^{-j}) = \mathrm{Hom}_\mathbb{C}(\Gamma_c(U, \mathbf{S}^j), \mathbb{C}). \quad (4.3.4)$$

(cf. [Iv] Section IV.1 or [GelM] Chapt. 4 Section 5.16.) Here Γ_c denotes sections with compact support. The resulting morphism $\mathbf{S}^\bullet \otimes \mathbf{T}^\bullet \to \mathbb{D}^\bullet_{\overline{X}^{BB}}$ (where $\mathbb{D}^\bullet_{\overline{X}^{BB}}$ denotes the dualizing complex on $\overline{X}^{BB}$) induces a quasi-isomorphism

$$\mathbf{T}^\bullet \to \mathbf{RHom}(\mathbf{S}^\bullet, \mathbb{D}^\bullet_{\overline{X}^{BB}}).$$

Let us apply this to the case when $\mathbf{S}^\bullet = \mathbf{\Omega}^\bullet_{(2)}(\mathbf{E_1})$ is the sheaf of L^2 differential forms on $\overline{X}^{BB}$. Let $\mathbf{T}^\bullet$ be the resulting (4.3.4) dual sheaf. The pairing (4.3.2) gives rise to a morphism of sheaves, $\Phi : \mathbf{\Omega}^\bullet_{(2)}(\mathbf{E_2}) \to \mathbf{T}^\bullet$ which is given on sections over $U \subset \overline{X}^{BB}$,

$$\Phi_U : \Gamma(U, \mathbf{\Omega}^\bullet_{(2)}(\mathbf{E_2})) \to \mathrm{Hom}_\mathbb{C}(\Gamma_c(U, \mathbf{\Omega}^\bullet_{(2)}(\mathbf{E_1})), \mathbb{C})$$

by $\Phi_U(\omega)(\eta) = \int_{U \cap Y} \omega \wedge \eta \in \mathbb{C}$. It suffices to show that the following diagram (of quasi-isomorphisms) commutes in $D^b \overline{X}^{BB}$,

$$\begin{array}{ccc}
\Omega^\bullet_{(2)}(\mathbf{E_2}) & \xrightarrow{\Phi} & \mathbf{RHom}(\Omega^\bullet_{(2)}(\mathbf{E_1}), \mathbb{D}^\bullet_{\overline{X}^{BB}}[-n]) \\
{\scriptstyle z} \downarrow & & \uparrow {\scriptstyle z} \\
\mathbf{I}^m \mathbf{S}^\bullet(\mathbf{E_2}) & \xrightarrow{\Psi} & \mathbf{RHom}(\mathbf{I}^m \mathbf{S}^\bullet(\mathbf{E_1}), \mathbb{D}^\bullet_{\overline{X}^{BB}}[-n])
\end{array}$$

where Ψ is given by (3.4.10).

Theorem 4.2 says that the vertical arrows are isomorphisms, so Lemma 3.1 may be applied, giving that the horizontal morphisms are uniquely determined by their restrictions to $X \subset \overline{X}^{BB}$. Thus, in order to show this diagram commutes, it suffices to check that it commutes when all these sheaves are restricted to X. This reduces the question to the classical situation: each of these complexes, when restricted to X is quasi-isomorphic to a local system. For any $y \in X$ let U_y be a neighborhood of y which is diffeomorphic to an open n-ball. Then the result follows from the commutativity of the following diagram.

$$\begin{array}{ccc}
H^0(U_y, \mathbf{E_1}) & \longrightarrow & \mathrm{Hom}(H^n_c(U_y, \mathbf{E_2}), \mathbb{C}) \\
{\scriptstyle \cong} \downarrow & & \uparrow {\scriptstyle \cong} \\
(\mathbf{E_1})_y & \longrightarrow & \mathrm{Hom}((\mathbf{E_2})_y, \mathbb{C})
\end{array} \qquad \square$$

4.4 Invariant differential forms

As in the preceding section assume that $X = \Gamma \backslash G(\mathbb{R}) / K_\infty$ is a Hermitian locally symmetric space. Let $\mathbf{E}$ be a local system on X corresponding to a finite-dimensional representation $G_\mathbb{R} \to GL(V)$ on some finite-dimensional complex vector space V. Let us say that a smooth differential form $\omega \in \Omega^j(X, \mathbf{E})$ is *invariant* if it pulls up to a (left) $G_\mathbb{R}$-invariant differential form on D. Let $\Omega^\bullet_{\mathrm{inv}}(X, \mathbf{E})$ denote the complex of smooth invariant differential forms with coefficients in $\mathbf{E}$, and let $H^j_{\mathrm{inv}}(X, \mathbf{E})$ denote its cohomology. If ω is such an invariant form then $\|\omega\|$ is constant on X so $\int_X \|\omega\|^2 \, d\mathrm{vol} < \infty$ and the same holds for $d\omega$. Hence, *the inclusion of the invariant differential forms into the complex of L^2 differential forms induces a canonical homomorphism*

$$H^j_{\mathrm{inv}}(X, \mathbf{E}) \to I^m H^j(\overline{X}^{BB}, \mathbf{E}). \qquad (4.4.1)$$

Each invariant differential form on X is determined by its value at the basepoint in D, hence as an element of $\bigwedge^j(\mathfrak{g}_\mathbb{R}/\mathfrak{k}, V)$. So the invariant cohomology $H^j_{\mathrm{inv}}(X, \mathbf{E})$ is just the relative Lie algebra cohomology $H^j(\mathfrak{g}_\mathbb{R}, K^1_\infty; V)$, which is known to be a direct summand of the L^2 cohomology ([Schwe]). It follows that the homomorphism (4.4.1) is injective.

4.5 Hecke correspondences for discrete groups

Hecke correspondences are treated from the adèlic point of view in Section 7.6. In this section we do not necessarily assume that $D = G(\mathbb{R})/K_\infty$ is Hermitian. Let $\Gamma_1, \Gamma_2 \subset G(\mathbb{Q})$ be commensurable arithmetic subgroups with corresponding locally symmetric spaces $X_i = \Gamma_i \backslash D$. Each $g \in G(\mathbb{Q})$ determines a *Hecke correspondence*

$$X_{12} \xrightarrow{(c_1, c_2)} X_1 \times X_2 \qquad (4.5.1)$$

where $c_i : X_{12} \to X_i$ ($i = 1, 2$) are finite coverings and are defined as follows. Set

$$\Gamma_{12} = \Gamma_1 \cap g^{-1} \Gamma_2 g$$

and $X_{12} = \Gamma_{12} \backslash D$. Then

$$c_1(\Gamma_{12} x K_\infty) = \Gamma_1 x K_\infty \quad \text{and} \quad c_2(\Gamma_{12} x K_\infty) = \Gamma_2 g x K_\infty$$

which are easily seen to be well defined for any $x \in G(\mathbb{R})$. The degree of c_1 is $[\Gamma_1 : \Gamma_{12}]$ which is finite, by [Bo69] Section 7.13. In fact, a decomposition into right cosets $\Gamma_1 = \coprod_{i=1}^m \Gamma_{12} h_i$ of elements $h_i \in \Gamma_1$ gives rise to a decomposition into right cosets $\Gamma_2 g \Gamma_1 = \coprod_{i=1}^m \Gamma_2 g h_i$ and conversely. Similarly, $\deg(c_2) = [g^{-1} \Gamma_2 g : \Gamma_{12}] = [\Gamma_2 : g \Gamma_{12} g^{-1}]$, because a decomposition $g^{-1} \Gamma_2 g = \coprod_{j=1}^r k_j \Gamma_{12}$ is equivalent to the decomposition $\Gamma_2 g \Gamma_1 = \coprod_{j=1}^r g k_j \Gamma_1$.

Up to isomorphism of correspondences, the Hecke correspondence (4.5.1) depends only on the double coset $\Gamma_2 g \Gamma_1 \subset G(\mathbb{Q})$. If $(s_1, t_1) : Y_1 \rightrightarrows X$ and $(s_2, t_2) : Y_2 \rightrightarrows X$ are two correspondences, their composition is the correspondence Y_3 defined to be the fiber product,

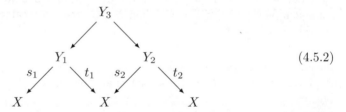

$$(4.5.2)$$

that is, $Y_3 = \{(y_1, y_2) \in Y_1 \times Y_2 |\ t_1(y_1) = s_2(y_2)\}$. As in [Shim1] Section 3.1 the multiplication law, which can be identified with composition of correspondences, determines an associative product

$$R(\Gamma_1, \Gamma_2) \times R(\Gamma_2, \Gamma_3) \to R(\Gamma_1, \Gamma_3)$$

where

$$R(\Gamma_a, \Gamma_b) = \mathbb{Z}\left[\Gamma_a \backslash G(\mathbb{Q}) / \Gamma_b\right]$$

is the group of finite formal linear combinations of double cosets. In particular, the group $R(\Gamma, \Gamma)$ obtains the structure of an algebra, the *Hecke algebra* of Γ.

It is easy to check that the mapping $\mathcal{Z}$ of Theorem 4.2 is Hecke-equivariant in the following sense: If $(c_1, c_2) : X_{12} \to X_1 \times X_2$ is a Hecke correspondence and if $\omega \in \Omega^j_{(2)}(X_2, \mathbf{E})$ then $c_2^*(\omega)$ is an L^2 form on X_{12} and $(c_1)_* c_2^*(\omega)$ (obtained by summing over the finitely many points in each fiber of c_1) is an L^2 form on X_1 whose associated cohomology class $[(c_1)_* c_2^* \omega] \in H^j_{(2)}(X_1, \mathbf{E})$ satisfies

$$\mathcal{Z}\left([(c_1)_* c_2^* \omega]\right) = (c_1)_* c_2^*(\mathcal{Z}([\omega])). \tag{4.5.3}$$

and

$$\int_{c_{2*} c_1^*(Z)} \omega = \int_Z (c_1)_* c_2^* \omega \tag{4.5.4}$$

whenever $Z \subset X$ is a subanalytic cycle such that $\int_Z \omega < \infty$.

4.6 Mappings induced by a Hecke correspondence

The Hecke correspondence (4.5.1) has unique continuous extensions

$$\overline{X}_{12}^{BB} \xrightarrow{(\bar{c}_1, \bar{c}_2)} \overline{X}_1^{BB} \times \overline{X}_2^{BB} \quad \text{and} \quad \overline{X}_{12}^{RBS} \xrightarrow{(\bar{c}_1, \bar{c}_2)} \overline{X}_1^{RBS} \times \overline{X}_2^{RBS}$$

(the first of which is only defined when D is Hermitian), and in each case (BB or RBS) the resulting mappings $\bar{c}_i : \overline{X}_{12} \to \overline{X}_i$ are finite. In the Hermitian case, these mappings commute with the projection $\mu : \overline{X}^{RBS} \to \overline{X}^{BB}$ of equation (4.8.1).

Let $\lambda : G \to \mathrm{GL}(V)$ be a finite-dimensional irreducible representation on some complex vector space V. Then λ gives rise to local systems $\mathbf{E}_1, \mathbf{E}_2, \mathbf{E}_{12}$ on X_1, X_2, X_{12} respectively.

It is easy to see that the action of $G(\mathbb{R})$ on V determines canonical isomorphisms of local systems, $c_2^*(\mathbf{E}_2) \to \mathbf{E}_{12} \to c_1^*(\mathbf{E}_1)$ on X_{12}. So we have a lift of the local system to the correspondence and by Section 3.6 this induces homomorphisms

$$(c_1)_* c_2^* : I^{\mathbf{p}} H_j(\overline{X}_2^{BB}, \mathbf{E}) \to I^{\mathbf{p}} H_j(\overline{X}_1^{BB}, \mathbf{E}) \tag{4.6.1}$$

and

$$(c_1)_* c_2^* : I^{\mathbf{p}} H_j(\overline{X}_2^{RBS}, \mathbf{E}) \to I^{\mathbf{p}} H_j(\overline{X}_1^{RBS}, \mathbf{E}) \tag{4.6.2}$$

on intersection homology of any perversity $\mathbf{p}$. In the case of the BB compactification it is easy to see that equations (4.6.1) and (4.5.3) are compatible with the Poincaré duality isomorphism (3.2.3) and (3.2.4).

4.7 The reductive Borel-Serre compactification

In this section we do not necessarily assume that $D = G(\mathbb{R})/K_\infty$ is Hermitian. Let P be a rational parabolic subgroup of G. Let U_P be its unipotent radical and $\nu_P : P \to L_P$ be the projection to the Levi quotient. Then L_P is rationally defined. As in Section 4.1 we have a canonical decomposition $L_P(\mathbb{R}) = M_P(\mathbb{R})A_P$ where $M_P(\mathbb{R}) = {}^0L_P(\mathbb{R})$.

There is a unique θ stable lift ([BoS] Prop. 1.8), $L_P(x_0) \subset P$ of L_P to P (where θ is the Cartan involution corresponding to the basepoint $x_0 \in D$). So we obtain the *Langlands decomposition*

$$P(\mathbb{R}) = U_P(\mathbb{R})A_P M_P(\mathbb{R}). \tag{4.7.1}$$

The intersection $K_P = K^1_\infty \cap P(\mathbb{R})$ is completely contained in $M_P(\mathbb{R})$. It follows from the Iwasawa decomposition that $P(\mathbb{R})$ acts transitively on D with isotropy group $K_{P,\infty} = K_P A$. Define the *geodesic action* [BoS] of $A_P A$ on $D = P(\mathbb{R})/K_{P,\infty}$ by $(gK_{P,\infty}) \cdot a = gaK_{P,\infty}$ for $g \in P(\mathbb{R})$ and $a \in A_P A$. This action (from the right) is well defined since $A_P A$ commutes with $K_P \subset M_P(\mathbb{R})$. The group $A_G \subset A_P$ acts trivially, and the quotient

$$A'_P = A_P A / A$$

acts freely on D with orbits that are totally geodesic submanifolds of D (with respect to any invariant Riemannian metric). The geodesic action is independent of the choice of basepoint $x_0 \in D$. The (Borel-Serre) boundary component is the quotient

$$e_P = D/A_P = D/A'_P.$$

The Borel-Serre partial compactification $\overline{D}^{BS}$ is the disjoint union of D together with the boundary components e_P as P ranges over all rational parabolic subgroups, with the *Satake topology* which essentially adds a point at infinity to each A_P-geodesic orbit. This topology is arranged so that Γ acts properly on $\overline{D}^{BS}$ and the quotient $\overline{X}^{BS} = \Gamma \backslash \overline{D}^{BS}$ is the Borel-Serre compactification of X. The quotient under Γ makes identifications within each boundary component and it also identifies boundary components e_P and e_Q whenever P and Q are Γ-conjugate. The compactification $\overline{X}^{BS}$ is a finite-dimensional orbifold with boundary, whose boundary has been (homologically) stratified into "corners" which are themselves orbifolds (or manifolds, when Γ is *neat*); see [BoS, Gre2].

Using the Levi decomposition $P = U_P L_P$ we may write

$$D = U_P(\mathbb{R})L_P(\mathbb{R})A/K_P A.$$

The group K_P and the geodesic action of the group $A'_P = A_P A/A$ act (from the right) only on the factor L_P. So we obtain a diffeomorphism

$$e_P \cong U_P(\mathbb{R}) \times (L_P(\mathbb{R})A/K_P A_P A) = U_P \times D_P \tag{4.7.2}$$

where D_P is the *reductive Borel-Serre boundary component* $L_P(\mathbb{R})/K_P A_P$.

Let $\Gamma_P = \Gamma \cap P(\mathbb{R})$ and $\Gamma_L = \nu_P(\Gamma_P) \subset L_P(\mathbb{Q})$. Define the *reductive Borel-Serre stratum*

$$X_P = \Gamma_L \backslash D_P = \Gamma_L \backslash L_P(\mathbb{R})/K_P A_P \tag{4.7.3}$$

where $\Gamma_L = \nu_P(\Gamma_P) \subset L_P(\mathbb{R})) = M_P(\mathbb{R})A_P$. Then the Borel-Serre stratum Y_P is a fiber bundle over the reductive Borel-Serre stratum X_P,

$$Y_P = \Gamma_P \backslash e_P = \Gamma_P \backslash P(\mathbb{R})A/K_P A = \Gamma_P \backslash P(\mathbb{R})/K_P A_P \to X_P = \Gamma_L \backslash D_P$$

whose fiber is the compact nilmanifold $N_P = \Gamma_U \backslash U_P(\mathbb{R})$.

Define the reductive Borel-Serre partial compactification $\overline{D}^{RBS}$ (resp. the reductive Borel-Serre compactification $\overline{X}^{RBS}$) to be the quotient of $\overline{D}^{BS}$ (resp. of $\overline{X}^{BS}$) which is obtained by collapsing each e_P to D_P (resp. each Y_P to X_P).

Theorem 4.4 ([Zu1, Section 4.2, p. 190], [Gre3, Section 8.10]). *The group Γ acts on $\overline{D}^{RBS}$ with compact quotient $\Gamma \backslash \overline{D}^{RBS} = \overline{X}^{RBS}$. The boundary strata form a regular stratification of $\overline{X}^{RBS}$ and the stratum*

$$X_P = \Gamma_L \backslash M_P(\mathbb{R})/K_P = \Gamma_L \backslash L_P(\mathbb{R})/K_P A_P$$

is a locally symmetric space corresponding to the reductive group L_P. Its closure $\overline{X}_P$ in $\overline{X}^{RBS}$ is the reductive Borel-Serre compactification of X_P.

The reductive Borel-Serre compactification $\overline{X}^{RBS}$ may have strata of odd codimension so there are two "middle" perversities: the lower middle, m, with $\mathsf{m}(c) = \lfloor c/2 \rfloor - 1$ and the upper middle n, with $\mathsf{n}(c) = \lceil c/2 \rceil - 1$.

4.8 Saper's theorem

Suppose $D = G(\mathbb{R})/K_\infty$ is Hermitian, let $\Gamma \subset G(\mathbb{Q})$ be an arithmetic subgroup and $X = \Gamma \backslash D$. In [Zu2] and [Gre3] it is proven that the identity mapping $X \to X$ has a unique continuous extension

$$\mu : \overline{X}^{RBS} \to \overline{X}^{BB}. \tag{4.8.1}$$

The mapping μ takes strata to strata. It can be explicitly described, see [Gre3] Section 22 or [Gre2] Section 5, but we will not need to make use of the finer properties of μ. We will make use of the following result of L. Saper [Sa1].

Theorem 4.5. *Let $\mathbf{E}$ be the local system on X corresponding to a finite-dimensional (complex) represntation of G. Let $\mathbf{I}^\mathsf{m}\mathbf{S}^\bullet(\overline{X}^{RBS}, \mathbf{E})$, $\mathbf{I}^\mathsf{n}\mathbf{S}^\bullet(\overline{X}^{RBS}, \mathbf{E})$ and $\mathbf{I}^\mathsf{m}\mathbf{S}^\bullet(\overline{X}^{BB}, \mathbf{E})$ denote the complex of sheaves of intersection cochains on $\overline{X}^{RBS}$*

(*with upper and lower middle perversity*) *and on* $\overline{X}^{BB}$. *Then the mapping* μ *induces quasi-isomorphisms*

$$R\mu_*\mathbf{I^nS^\bullet}(\overline{X}^{RBS};\mathbf{E}) \cong \mathbf{I^mS^\bullet}(\overline{X}^{RBS};\mathbf{E}) \cong R\mu_*\mathbf{I^nS^\bullet}(\overline{X}^{RBS};\mathbf{E}) \qquad (4.8.2)$$

that is the identity $\mathbf{E} \to \mathbf{E}$ *on* X.

Theorem 4.5 is described as Conjecture 4.1 in [Rap]. It was proven in the $\mathbb{Q}$-rank one case in the appendix [SaS2] to [Rap], and in general in [Sa1].

4.9 Modular cycles

Let L/F be a quadratic extension of number fields. Let G_0 be a reductive algebraic group defined over $\mathbb{Q}$ of semisimple rank ≥ 1 and suppose that G_0 is split at all the real places of F. Define

$$H = \mathrm{Res}_{F/\mathbb{Q}} G_0 \hookrightarrow G = \mathrm{Res}_{L/\mathbb{Q}} G_0.$$

Let $D = G(\mathbb{R})/K_\infty$ where $K_\infty = K_\infty^1 A$ as in Section 4.1 and let $D_H = H(\mathbb{R})/K_{H,\infty}$ where $K_{H,\infty} = K_\infty \cap H(\mathbb{R})$. Then the embedding of symmetric spaces $D_H \to D$ induces an embedding locally symmetric spaces

$$X_H = \Gamma_H \backslash D_H \hookrightarrow X = \Gamma \backslash D \qquad (4.9.1)$$

where $\Gamma \subset G(\mathbb{Q})$ is arithmetic and $\Gamma_H = \Gamma \cap H(\mathbb{Q})$.

Theorem 4.6. *Suppose the symmetric space D is Hermitian, of real dimension r. Let $\mathbf{E}$ be the local system on X corresponding to a finite-dimensional representation of $G(\mathbb{R})$ on some complex vector space V. Let S be a flat section of $\mathbf{E}|(X_H - W)$ where W is a (possibly empty) proper subvariety of X_H. Then the compactification $\overline{X}_H^{BB} \subset \overline{X}^{BB}$ defines a class*

$$[\overline{X}_H^{BB}] \in I^m H_r(\overline{X}^{BB}, \mathbf{E})$$

in the (middle) intersection homology of the Baily-Borel compactification $\overline{X}^{BB}$.

Proof. The embedding (4.9.1) has a unique continuous extensions

$$\overline{X}_H^{RBS} \to \overline{X}^{RBS} \quad \text{and} \quad \overline{X}_H^{BB} \to \overline{X}^{BB}.$$

The flat section S determines a "fundamental" homology class $[X_H, S] \in H_r^{BM}(X:\mathbf{E})$ by Proposition 2.5. In fact this lifts canonically to a homology class

$$[\overline{X}_H^{RBS}] \in H_r(\overline{X}^{RBS};\mathbf{E})$$

where $r = \dim_{\mathbb{R}}(D)$. This is because each singular stratum (corresponding to a proper parabolic subgroup $P_H = \mathcal{U}_{P_H} A_{P_H} M_{P_H}$) of $\overline{X}_H^{\text{RBS}}$ has (real) dimension

$$r - \dim_{\mathbb{R}}(A'_{P_H}) - \dim_{\mathbb{R}}(\mathcal{U}_{P_H}(\mathbb{R})) \le r - 2.$$

We claim that this cycle satisfies the allowability conditions for upper middle intersection homology $I^n H_r(\overline{X}^{\text{RBS}}, \mathbf{E})$, that is, $\mathsf{n}(c) = \lceil c/2 \rceil - 1$.

The parabolic subgroup $P_H \subset H$ arises from some parabolic subgroup $P_0 \subset G_0$ by restriction of scalars. Consequently the corresponding boundary stratum (let us call it Z_H) of $\overline{X}_H^{\text{RBS}}$ lies in the boundary stratum (call it Z) of $\overline{X}^{\text{RBS}}$ that is associated to the parabolic subgroup $P = \text{Res}_{L/F} P_H \subset G$. The allowability condition (3.1.1) requires that

$$\dim(Z_H) \le \dim(X_H) - \text{codim}_X(Z) + \lceil \text{codim}_X(Z)/2 \rceil - 1$$

(using real dimensions and codimensions throughout), or $c + 1 \le c_H + \lceil c/2 \rceil$ where

$$c = \text{codim}_X(Z) = \dim(A'_P) + \dim(\mathcal{U}_P(\mathbb{R}))$$
$$c_H = \text{codim}_{X_H}(Z_H) = \dim(A'_{P_H}) + \dim(\mathcal{U}_{P_H}(\mathbb{R}))$$
$$= \dim(A'_P) + \tfrac{1}{2} \dim(\mathcal{U}_P(\mathbb{R})).$$

Thus, the allowability condition comes down to $1 \le \lfloor a'/2 \rfloor$, which always holds. The conclusion of Theorem 4.6 now follows from Saper's canonical isomorphism (4.8.2), viz.,

$$I^n H_r(\overline{X}^{\text{RBS}}, \mathbf{E}) \cong I^m H_r(\overline{X}^{\text{BB}}, \mathbf{E}). \qquad \square$$

We remark that flat sections arise naturally from invariant one-dimensional subspaces of V as described in Section 6.2.3 and 6.4.2. The following additional fact is an immediate consequence of (4.6.1).

Scholium 4.7. *Consider a Hecke correspondence* $(\bar{c}_1, \bar{c}_2) : \overline{X}_{12}^{\text{BB}} \to \overline{X}_1^{\text{BB}} \times \overline{X}_2^{\text{BB}}$ *that arises from commensurable arithmetic subgroups* $\Gamma_1, \Gamma_2 \subset G(\mathbb{Q})$ *and an element* $g \in G(\mathbb{Q})$. *In other words,* $X_i = \Gamma_i \backslash D$ $(i = 1, 2)$ *and* $X_{12} = \Gamma_{12} \backslash D$ *where* $\Gamma_{12} = \Gamma_1 \cap g^{-1} \Gamma_2 g$. *Then*

$$\overline{X'}_H^{\text{BB}} = \bar{c}_2 \bar{c}_1^{-1} \left(\overline{X}_H^{\text{BB}} \right)$$

is also a modular cycle and its intersection homology class satisfies

$$[\overline{X'}_H^{\text{BB}}] = (c_2)_* c_1^* \left([\overline{X}_H^{\text{BB}}] \right) \in I^m H_r(\overline{X}_2^{\text{BB}}, \mathbf{E}).$$

4.10 Integration

In Section 8.6 and Section 9.3 we will integrate a differential form over a modular cycle. Because of the Zucker conjecture (Theorem 4.2) one might expect that, given a closed L^2 differential form ω and given an algebraic cycle $Z \subset \overline{X}^{BB}$ that represents a middle intersection homology class, the integral $\int_Z \omega$ should be well defined, finite, and should equal the Kronecker product $\langle \mathcal{Z}[\omega], [Z] \rangle_K$ of the corresponding intersection (co)homology classes. See [Ch1, Ch2] for a setting in which these desiderata are realized. Unfortunately this is usually false: the integral may fail to converge and even if it does converge it may fail to equal this Kronecker product. Nevertheless in our situation the differential form ω arises from a rapidly decreasing modular form, and the cycle Z has finite volume. These ingredients turn out to be sufficient to guarantee that the integral is finite and that it coincides with the Kronecker product, as we now explain. We continue with the notation of the previous section, but in the particular case that $G_0 = \mathrm{GL}_2$, with

$$H = \mathrm{Res}_{F/\mathbb{Q}} \mathrm{GL}_2 \subset G = \mathrm{Res}_{L/\mathbb{Q}} \mathrm{GL}_2$$

and with the resulting inclusion of modular varieties $X_H \subset X$.

A choice of a $G(\mathbb{R})$-invariant Riemannian metric on $D = G(\mathbb{R})/K_\infty$ induces an $H(\mathbb{R})$-invariant metric on D_H so we obtain compatible metrics on X and X_H with finite volume, as well as pointwise inner products on differential forms on X and X_H.

Let $\mathbf{E_1}, \mathbf{E_2}$ be local systems on X that arise from representations of $G(\mathbb{R})$ on finite-dimensional complex vector spaces E_1, E_2 respectively. Choose K_∞^1-invariant Hermitian inner products on E_1, E_2 and choose a K_∞^1-invariant Hermitian form $\phi : E_1 \times E_2 \to \mathbb{C}$. These choices give rise to $G(\mathbb{R})$-invariant norms on $\mathbf{E_1}, \mathbf{E_2}$ and a $G(\mathbb{R})$-invariant product $\phi : \mathbf{E_1} \times \mathbf{E_2} \to \mathbb{C}$. In order to agree with the notation of Section 8.6 and Section 9.3 let us write $Z = X_H$ for the modular cycle. A differential form $\omega \in \Omega^r(Z, \mathbf{E_1})$ is *bounded* if there exists $M > 0$ such that $|\omega(x)| \leq M$ for all $x \in Z$. Similarly if s is a section over Z of $\mathbf{E_2}$ then we say that $\phi(\omega, s)$ is bounded if there exists $M > 0$ such that $|\phi_x(\omega(x), s(x))| < M$ for all $x \in Z$.

Let s be a flat section of $\mathbf{E_2}|(Z - W)$ where $W \subset Z$ is a (possibly empty) proper subvariety. By Proposition 2.5 the pair (Z, s) determines a Borel-Moore homology class in X which, by Theorem 4.6 lifts to an intersection homology class $[Z] = [Z, s] \in IH_r(\overline{X}^{BB}, \mathbf{E_2})$.

Proposition 4.8. *Assume there exists a character χ of $H(\mathbb{R})$ such that $|s(h.x)| \leq |\chi(h)||s(x)|$ for all $h \in H(\mathbb{R})$[1]. Let $\omega \in \Omega^r(X, \mathbf{E_1})$ be a smooth, closed differential form that is bounded. Then ω is also square summable, and by Theorem 4.2, it represents an intersection cohomology class $\mathcal{Z}([\omega]) \in IH^r(\overline{X}^{BB}, \mathbf{E_1})$. Assume that*

---
[1] See, for example Sections 6.2.3 and 6.4.2.

the product $\phi(\omega, s) \in \Omega^r(X; \mathbb{C})$ is bounded[2]. Then the integral of ω over Z is finite and is equal to the Kronecker product (see (3.4.8)) of the intersection (co)homology classes:
$$\int_Z \omega := \int_Z \phi(\omega, s) = \langle \mathcal{Z}([\omega]), [Z] \rangle_K < \infty.$$

Proof. The volume vol(Z) is finite by [BoHC], hence if ω is bounded then it is also square summable. Moreover, the integral is finite since it is bounded by $M \text{vol}(Z)$ where M is a bound for $|\phi(\omega, s)|$.

The Baily-Borel compactification of X (resp. of X_H) consists of adding finitely many cusps. A neighborhood of such a cusp is associated to a (maximal proper) rational parabolic subgroup P with $P(\mathbb{R}) = U_P(\mathbb{R}) A_P M_P(\mathbb{R})$ as described in equation (4.7.1), and such a neighborhood is diffeomorphic to the double coset space $\Gamma_P \backslash P(\mathbb{R}) / K_P A_G$ where $\Gamma_P = \Gamma \cap P(\mathbb{R})$ and $K_P = K_\infty \cap P(\mathbb{R})$. Recall (Section 4.7) that right multiplication by the one-dimensional real split torus A'_P defines the geodesic flow which moves points towards and away from the cusp. Fix a nontrivial character $\psi_P : A'_P \to \mathbb{R}_{>0}$ and let $A'_P(\geq t), A'_P(t), A'_P([s, t])$ etc. denote $\psi^{-1}(t, \infty), \psi^{-1}(t), \psi^{-1}([s, t])$ respectively. The character ψ_P can be chosen so that $A'_P(t)$ moves points towards the cusp as $t \to +\infty$. Then a (punctured) neighborhood of the cusp in X (resp. in X_H) is diffeomorphic to $L \times A'_P(\geq 1)$ (resp. $L_H \times A'_P(\geq 1)$) where L (resp. L_H) is the *link* of the cusp point in X (resp. in X_H). The character ψ_P passes to a (real) analytic mapping $\Gamma_P \backslash D \to \mathbb{R}_{>0}$ which, for $t \geq 1$ is a fiber bundle whose fiber is $\psi_P^{-1}(t) = L_t = A'_P(t) L$. In other words, the fiber is obtained from the link $L = \psi_P^{-1}(1)$ by flowing under the geodesic action for time t.

The middle-dimensional intersection homology $I^m H_r(X, \mathbf{E_2})$ is the image of (compactly supported) homology $H_r(X, \mathbf{E_2})$, so the class $[Z]$ lifts to $H_r(X, \mathbf{E_2})$. This is equivalent to saying that the homology class $[L_H] \in H_{r-1}(L, \mathbf{E_2})$ vanishes. Hence there is a subanalytic chain (even a real semi-analytic chain) R in $C_r(L, \mathbf{E_2})$ such that $\partial R = L_H$. (So the chain R consists of a subanalytic subset $|R|$ of L together with a section of $\mathbf{E_2}$ over $|R|$.) For each $n = 1, 2, \ldots$ we may construct a new cycle Z_n from Z by "truncating" at each of the cusps, and "capping" the resulting boundaries using the chain R. If there is a single cusp then
$$Z_n = \psi_P^{-1}(0, n] + R_n$$
where $R_n = A'_P(n).R$.

For each n, the Borel-Moore homology class represented by Z_n coincides with that of Z. Since Z_n has compact support, by (3.4.8) we have:
$$\int_{Z_n} \omega = \langle [\omega], [Z_n] \rangle = \langle \mathcal{Z}([\omega]), [Z] \rangle_K$$
where the right side denotes the Kronecker product between $H^r(X, \mathbf{E_1})$ and $H_r(X, \mathbf{E_2})$, and between $IH^r(X, \mathbf{E_1})$ and $IH_r(X, \mathbf{E_2})$ that is induced from ϕ :

---
[2]This condition holds if ω is rapidly decreasing.

4.10. Integration

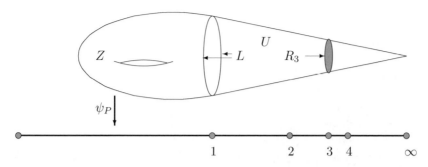

Figure 4.1: The mapping ψ_P

$\mathbf{E_1} \times \mathbf{E_2} \to \mathbb{C}$. By [Fed2] Theorem 3.4.8, and Section 4.2.28; see also [Fed3, Fed1, Hardt2, Hardt1] the chain R has finite volume so the above integral converges. Moreover,

$$Z_{n+1} = Z_n - R_n + A'_P([n, n+1]).L + R_{n+1} = Z_n + \partial(A'_P([n, n+1]).R).$$

It follows from Stokes' theorem that

$$\int_{Z_n} \omega = \int_{Z_{n+1}} \omega$$

is independent of n. This in itself does not guarantee that $\int_{Z_n} \omega = \int_Z \omega$ because these differ by $\int_{R_n} \omega$ (which is also independent of n). However, it is known that in a neighborhood of a cusp the invariant metric is a warped product ([Zu1], [Mu] Section 6) which implies that $\text{vol}(R_n) \to 0$. Since the differential form ω is bounded this implies that $\int_{Z_n} \omega \longrightarrow \int_Z \omega$. □

We remark that the Baily-Borel compactification $\overline{Z}^{BB}$ is a complex projective algebraic variety and the metric induced from an embedding in projective space (with its Fubini-Study metric) is not complete but it has finite volume. In fact, like any complex analytic variety, the variety $\overline{Z}^{BB}$ is locally and globally area minimizing (for a precise statement see [Stz, Theorem A p. 9]). If α denotes the Kähler form on projective space then, when restricted to Z, the differential form α^r becomes a volume form. These and similar results, due to Federer [Fed2, Fed3, Fed1], Bishop [Bis] and Stoll [Sto], are beautifully described in [Stz]. However, the natural metric on the modular variety $Z = X_H$ is complete, and in this case, the finite volume statement is a consequence of the structure of a fundamental domain for Γ_H.

Chapter 5
Generalities on Hilbert Modular Forms and Varieties

In this chapter we collect some known facts on Hilbert modular forms and varieties, mostly for the purpose of fixing our notation. These concepts will be used in Chapter 7, where we will recall the description of the intersection cohomology of Hilbert modular varieties in terms of Hilbert modular forms.

The motivation, details, and explanations for much of this chapter and the next are described in Chapter 6 (on automorphic vector bundles) and Appendix D (on Fourier expansions). We first deal with Hilbert modular varieties and how one passes from their adèlic realization as a finite level Shimura variety to their classical complex analytic realization and vice-versa. In Section 5.3, we discuss conventions concerning the weights which, on the one hand, parametrize spaces of automorphic forms (Section 5.4) and on the other hand, parametrize local systems (Chapter 7, Section 6.6).

We define Hilbert modular forms in Section 5.4. Our approach to Hilbert modular forms is adèlic. A review of the adèles is provided in Appendix C. It is most natural to introduce Hecke operators from this viewpoint (see Section 5.6). We then review various constructions related to these automorphic forms. In Section 5.7 we fix a normalization of the Petersson inner product which will be used in Chapter 7. Specifically, Theorem 7.11 gives the relation between the Petersson inner product of two cusp forms and the intersection pairing of the intersection cohomology classes they define. Fourier expansions of Hilbert modular forms are recalled in Section 5.9 with details in Appendix D. We will not require Fourier series until Chapters 8 and 11, where they will be required to state and prove our main theorems. Twisting of modular forms is discussed in Section 5.11 but it comes up again in Section 9.4.

Finally, in Section 5.12, we discuss some L-functions attached to Hilbert modular forms that appear later in Chapter 9. The content of this chapter is largely drawn from the work of H. Hida in [Hid3], [Hid5], and [Hid7]. In Section 5.13, we indicate how to relate Hida's notation in these papers to ours.

Throughout this chapter, L will denote a totally real number field with $[L:\mathbb{Q}] = n$ and ring of integers $\mathcal{O}_L$. Denote by $\Sigma(L)$ the set of embeddings $\sigma : L \hookrightarrow \mathbb{R}$ which we identify with the set of real places of L. Let G be the reductive algebraic $\mathbb{Q}$-group
$$G := \operatorname{Res}_{L/\mathbb{Q}}(\operatorname{GL}_2).$$
Then $G(\mathbb{R}) \cong \operatorname{GL}_2(\mathbb{R})^{\Sigma(L)}$ and $G(\mathbb{A}) \cong \operatorname{GL}_2(\mathbb{A}_L)$. Elements $b \in \mathbb{A}_L^\times$ may be identified with diagonal matrices $\left(\begin{smallmatrix} b & 0 \\ 0 & b \end{smallmatrix}\right) \in G(\mathbb{A})$.

5.1 Hilbert modular Shimura varieties

In this section we set notation for the finite-level Shimura varieties associated to G. We start by fixing the homomorphism

$$\begin{aligned} \mathbb{S}(\mathbb{R}) := \operatorname{Res}_{\mathbb{C}/\mathbb{R}}(\mathbb{G}_m)(\mathbb{R}) &\longrightarrow G(\mathbb{R}) \cong \operatorname{GL}_2(\mathbb{R})^{\Sigma(L)} \\ x + iy &\longmapsto \left(\left(\begin{smallmatrix} x & y \\ -y & x \end{smallmatrix}\right), \ldots, \left(\begin{smallmatrix} x & y \\ -y & x \end{smallmatrix}\right)\right). \end{aligned} \quad (5.1.1)$$

There is an identification of the $G(\mathbb{R})$-conjugacy class of this homomorphism with the symmetric space $D := G(\mathbb{R})/K_\infty$, where $K_\infty = K_{L,\infty}$ is the stabilizer of the image of (5.1.1). Explicitly, K_∞ is the subgroup

$$\left\{\left(\left(\begin{smallmatrix} x_1 & y_1 \\ -y_1 & x_1 \end{smallmatrix}\right), \ldots, \left(\begin{smallmatrix} x_n & y_n \\ -y_n & x_n \end{smallmatrix}\right)\right) : x_j^2 + y_j^2 > 0\right\} \cong (\mathbb{C}^\times)^{\Sigma(L)} \quad (5.1.2)$$

of $G(\mathbb{R})^0$ (the identity component of $G(\mathbb{R})$). It contains the maximal compact connected subgroup

$$K_\infty^1 := \{g \in K_\infty : \det(g_\sigma) = 1 \text{ for all } \sigma \in \Sigma(L)\} \cong \operatorname{SO}_2(\mathbb{R})^{\Sigma(L)}. \quad (5.1.3)$$

If the field L varies, we may add the subscript L, e.g., $K_{L,\infty}$. With the exception of Appendix A and Appendix E we will reserve the undecorated symbol K for a compact open subgroup of $G(\mathbb{A}_f)$.

The theory of Shimura varieties [De] provides D with the structure of a complex manifold in a canonical manner. In our case, the complex structure is given by

$$\begin{aligned} D = G(\mathbb{R})/K_\infty &\longleftrightarrow (\mathbb{C} - \mathbb{R})^{\Sigma(L)} \\ (h_1, \ldots, h_n) &\longmapsto (h_1 i, \ldots, h_n i), \end{aligned} \quad (5.1.4)$$

where, if h_j is the image of $\left(\begin{smallmatrix} a & b \\ c & d \end{smallmatrix}\right) \in \operatorname{GL}_2(\mathbb{R})$ in the quotient, we set $h_j z = \frac{az+b}{cz+d}$.

Viewing D as a $G(\mathbb{R})$-conjugacy class of homomorphisms $\mathbb{S}(\mathbb{R}) \to G(\mathbb{R})$, it can be shown that (G, D) is a Shimura datum. We now review and set notation for the finite-level Shimura varieties attached to (G, D). For each compact open subgroup $K \subseteq G(\mathbb{A}_f)$, let

$$\begin{aligned} Y_K := \operatorname{Sh}_K(G, D) &:= G(\mathbb{Q}) \backslash G(\mathbb{A}) / K_\infty K \\ &= G(\mathbb{Q}) \backslash (D \times G(\mathbb{A}_f) / K) \end{aligned} \quad (5.1.5)$$

5.1. Hilbert modular Shimura varieties

be the Hilbert modular Shimura variety associated to K and let X_K be its Bailey-Borel (Satake) compactification. We loosely refer to Y_K or X_K as a Hilbert modular variety. In either case it consists of finitely many connected components, each of which is a locally symmetric space associated to G as in Chapter 4. To make this explicit, let $T := \text{Res}_{L/\mathbb{Q}}(\mathbb{G}_m)$. Then the determinant $\det : G \to T$ induces a continuous map

$$\det : G(\mathbb{Q}) \backslash G(\mathbb{A}) / K_\infty K \longrightarrow T(\mathbb{Q}) \backslash T(\mathbb{A}) / \det(K_\infty K). \tag{5.1.6}$$

The set on the right has finitely many elements, say $h = h(K)$.

For $1 \leq j \leq h(K)$ choose $t_j = t_j(K) \in G(\mathbb{A}_f)$ such that $\det(t_1), \ldots, \det(t_h)$ form a complete set of coset representatives for the right-hand side of (5.1.6). Without loss of generality we may assume that

- We have $t_j = \begin{pmatrix} \det(t_j) & 0 \\ 0 & 1 \end{pmatrix}$.
- The fractional ideal $[\det(t_j)]$ associated to the finite part of the idèle $\det(t_j)$ is integral.

Let $G(\mathbb{Q})^+$ be the group of rational points in the identity component of $G(\mathbb{R})$. The *strong approximation* theorem says that

$$G(\mathbb{A}) = \bigcup_{j=1}^{h(K)} G(\mathbb{Q})^+ t_j G(\mathbb{R})^0 K$$

and it implies that there is an identification (see Section 4.1),

$$\bigcup_{j=1}^h \Gamma_j(K) \backslash \mathfrak{h}^{\Sigma(L)} \xrightarrow{=} Y_K = G(\mathbb{Q}) \backslash D \times G(\mathbb{A}_f) / K, \tag{5.1.7}$$

where $\mathfrak{h}$ is the upper half-plane and

$$\Gamma_j(K) := G(\mathbb{Q})^+ \cap t_j K t_j^{-1} G(\mathbb{R}) \ \text{ inside } \ G(\mathbb{A}) \tag{5.1.8}$$
$$= G(\mathbb{Q})^+ \cap t_j K t_j^{-1} \ \text{ inside } \ G(\mathbb{A}_f).$$

The map (5.1.7) is given explicitly on the jth component by

$$\Gamma_j(K) z \longmapsto G(\mathbb{Q}).(z, t_j).K$$

where we regard $\mathfrak{h}^{\Sigma(L)} = G(\mathbb{R})^0 / K_\infty = G(\mathbb{R})^0 / A_G K_\infty^1$ as a subspace of $D = G(\mathbb{R})/K_\infty$. Here, as in Chapter 4, A_G is the identity component of the real points of the maximal $\mathbb{Q}$-split torus in the center of G. The mapping (5.1.7) is a homeomorphism if $\mathfrak{h}$ and $G(\mathbb{A})$ are given their standard topologies (see [Mil, Lemma 5.13]).

5.2 Hecke congruence groups

In this book we will be interested in a particular family of compact open subgroups of $G(\mathbb{A}_f)$, chosen so as to make the Fourier series manageable. Fix an ideal $\mathfrak{c} \subset \mathcal{O}_L$. In analogy with the classical elliptic modular case, denote the standard congruence subgroup of integral matrices of "Hecke type" by

$$K_0(\mathfrak{c}) := \left\{ u \in G(\widehat{\mathbb{Z}}) : u = \begin{pmatrix} a & b \\ c & d \end{pmatrix} \text{ with } c \in \mathfrak{c}\widehat{\mathcal{O}}_L \right\}. \tag{5.2.1}$$

Here $\widehat{\mathbb{Z}} = \prod_{p<\infty} \mathbb{Z}_p$, so $G(\widehat{\mathbb{Z}}) = \mathrm{GL}_2(\widehat{\mathcal{O}}_L)$. If $\gamma \in G(\widehat{\mathbb{Z}})$ then so is γ^{-1}, and $\det(\gamma) \in \widehat{\mathcal{O}}_L^\times$. Set

$$Y_0(\mathfrak{c}) := Y_{K_0(\mathfrak{c})} \quad \text{and} \quad X_0(\mathfrak{c}) := X_{K_0(\mathfrak{c})}.$$

The number of connected components $h = h(K_0(\mathfrak{c}))$ of $Y_0(\mathfrak{c})$ turns out to be the *narrow class number* of L, and it is independent of $\mathfrak{c}$. In fact,

$$\det(K_\infty K_0(\mathfrak{c})) = \mathbb{A}_{L,\infty}^+ \widehat{\mathcal{O}}_L^\times$$

where $\mathbb{A}_{L,\infty}^+ = \prod_{v \in \Sigma(L)} L_v^{>0} \cong (\mathbb{R}^{>0})^{\Sigma(L)}$ so the right-hand side of (5.1.6) is the narrow class group, see (C.3.1).

Fix a collection of elements $t_j = t_j(\mathfrak{c}) \in G(\mathbb{A}_f)$ as in (5.1.6)–(5.1.7). We may assume that

$$\det(t_j)_v = 1 \text{ if } v|\infty \text{ or } v(\mathfrak{c}) \geq 1$$

(because $G(\mathbb{Q})$ permutes the connected components of D and elements $t \in G(\mathbb{Q})$ may be found so as to have any chosen pattern of signs $\mathrm{sgn}(\det(t))_v$ at the infinite places v). Then $Y_0(\mathfrak{c})$ is the disjoint union of the locally symmetric spaces

$$Y_0^j(\mathfrak{c}) := \Gamma_0^j(\mathfrak{c}) \backslash \mathfrak{h}^n \quad \text{where} \quad \Gamma_0^j(\mathfrak{c}) := \Gamma_j(K_0(\mathfrak{c})). \tag{5.2.2}$$

We note that $\Gamma_0^j(\mathfrak{c})$ is the set of matrices $\begin{pmatrix} a & b \\ c & d \end{pmatrix} \in \mathrm{GL}_2(L)$ such that

$$a, d \in \mathcal{O}_L, \ b \in [\det(t_j)], \ c \in \mathfrak{c}[\det(t_j)]^{-1}, \ ad - bc \in \mathcal{O}_{L,+}^\times$$

where $\mathcal{O}_{L,+}^\times$ is the group of totally positive units of $\mathcal{O}_L$.

If $L = \mathbb{Q}$ and if N is a positive integer then $h = 1$ and $Y_0(N) := Y_0^1(N\mathbb{Z}) = \Gamma_0^1(N)\backslash\mathfrak{h}$ is the classical open modular curve of level N, and $\Gamma_0^1(N)$ is the classical congruence subgroup.

The group $K_0(\mathfrak{c})$ may have torsion but it contains torsion-free normal subgroups of finite index. In particular, there exists $\mathfrak{q} \subset \mathcal{O}_L$ such that if n is sufficiently large, then

$$K^{\mathrm{fr}} := K_0(\mathfrak{c}) \cap \left\{ \gamma \in G(\widehat{\mathbb{Z}}) : \gamma \equiv I \pmod{\mathfrak{q}^n} \right\} \tag{5.2.3}$$

5.3. Weights

is such a subgroup, cf. [Gh, Section 3.1]. If $K^{fr} \trianglelefteq K_0(\mathfrak{c})$ is a torsion-free, finite index normal subgroup then

$$Y_0^{fr}(\mathfrak{c}) := G(\mathbb{Q})\backslash G(\mathbb{A})/K^{fr}K_\infty \to Y_0(\mathfrak{c})$$

is a resolution of singularities, so it is an *orbifold chart*, see Appendix B. In this way, the Hilbert modular variety $Y_0(\mathfrak{c})$ receives the structure of an orbifold, and its Baily-Borel compactification $X_0(\mathfrak{c})$ (which is obtained by attaching finitely many cusps) admits an orbifold stratification whose largest stratum is $Y_0(\mathfrak{c})$ and whose singular strata are the cusps.

5.3 Weights

As explained in Section 6.3, a rationally defined irreducible representation V of the algebraic group G on a complex vector space gives rise to a local coefficient system $\mathcal{V}$ on the Hilbert modular variety $Y_0(\mathfrak{c})$. Differential forms with coefficients in $\mathcal{V}$ correspond to Hilbert modular forms of a certain "weight," which is determined by the highest "weight" of V. Efforts to formalize these comments are complicated by the existence of torsion in the corresponding arithmetic groups $\Gamma_0^j(\mathfrak{c})$, and by the difference in the two meanings of "weight". In this section we review the solution to this problem, as described by Hida (see [Hid7]). The further goal of relating the Fourier coefficients of a newform to its Hecke eigenvalues adds another level of complexity to the combinatorics involving weights, which we address in Section 5.4.

Let V denote $\mathbb{C}^2$ with the standard representation of $\mathrm{GL}_2(\mathbb{C})$ and let $V^\vee$ be the dual vector space on which $\mathrm{GL}_2(\mathbb{C})$ acts by the *contragredient* representation, that is, $(g.\lambda)(x) = \lambda(g^{-1}.x)$ for all $\lambda \in V^\vee$ and $x \in V$. Then $\wedge^2 V$ is the one-dimensional representation on which $\mathrm{GL}_2(\mathbb{C})$ acts by the determinant, so we denote it by det. Each irreducible algebraic representation of $\mathrm{GL}_2(\mathbb{C})$ is isomorphic to $\mathrm{Sym}^k V \otimes \det^{\otimes m}$ for some $k \in \mathbb{Z}_{\geq 0}$ and some $m \in \frac{1}{2}\mathbb{Z}$, see Section 6.5. Let us refer to this as the representation with weight (k, m). Let $\kappa = (k, m) \in \mathbb{Z}_{\geq 0}^{\Sigma(L)} \times (\frac{1}{2}\mathbb{Z})^{\Sigma(L)}$. By forming the tensor product of the representations with weights (k_σ, m_σ) (for $\sigma \in \Sigma(L)$) we obtain a representation (described in more detail in Section 7.1 and Section 6.6),

$$L(\kappa, \mathbb{C}) = \bigotimes_{\sigma \in \Sigma(L)} \left(\mathrm{Sym}^{k_\sigma} V_\sigma^\vee \otimes \det_\sigma^{-m_\sigma}\right)$$

of the group $G(\mathbb{R}) \cong (\mathrm{GL}_2(\mathbb{R}))^{\Sigma(L)}$. This representation may not be algebraic, but a twist of it by a half-integral power of the determinant is algebraic. By a slight abuse of terminology, we refer to this as the representation with highest weight κ. Define $\mathcal{X}(L) \subset \mathbb{Z}_{\geq 0}^{\Sigma(L)} \times (\frac{1}{2}\mathbb{Z})^{\Sigma(L)}$ to be the subset of weights (k, m) such that

$$k + 2m \in \mathbb{Z}\mathbf{1} \tag{5.3.1}$$

where $\mathbf{1} := (1, \ldots, 1)$. In Lemma 5.1 we record the well-known fact that if $\kappa \in (\mathbb{Z}^{\geq 0})^{\Sigma(L)} \times (\frac{1}{2}\mathbb{Z})^{\Sigma(L)}$ then spaces of modular forms of weight κ are nonzero only if $\kappa \in \mathcal{X}(L)$. The same argument can be used to show that if $\kappa \in (\mathbb{Z}^{\geq 0})^{\Sigma(L)} \times (\frac{1}{2}\mathbb{Z})^{\Sigma(L)}$ then (a suitable twist of) $L(\kappa, \mathbb{C})$ gives rise to a local system only if $\kappa \in \mathcal{X}(L)$ (compare Proposition 6.3). After Lemma 5.1 below, we will always assume that κ satisfies this condition.

In Section 5.4 we discuss two weight conventions for spaces of Hilbert modular cusp forms (with nebentypus χ), which we denote by

$$S_\kappa(K_0(\mathfrak{c}), \chi) \quad \text{and} \quad S_\kappa^{\text{coh}}(K_0(\mathfrak{c}), \chi).$$

By a standard construction which we recall in Section 7.2, an element

$$f \in S_\kappa^{\text{coh}}(K_0(\mathfrak{c}), \chi| \cdot |_{\mathbb{A}_E}^{-k-2m})$$

gives a differential form with values in $\mathcal{L}(\kappa, \mathbb{C})$, the local system over $Y_0(\mathfrak{c})$ corresponding to the representation $L(\kappa, \mathbb{C})$. This makes our normalization of the weight of a Hilbert modular form seem natural when working with cohomology with respect to a local system.

5.4 Hilbert modular forms

In this section we define certain spaces of automorphic forms on G. Assume the notation of the previous section. Denote the standard generators of $\text{sl}_2(\mathbb{R}) \otimes \mathbb{C}$ as follows:

$$X := \frac{1}{2}\begin{pmatrix} 1 & i \\ i & -1 \end{pmatrix} \quad Y := \frac{1}{2}\begin{pmatrix} 1 & -i \\ -i & -1 \end{pmatrix} \quad H := \begin{pmatrix} 0 & -i \\ i & 0 \end{pmatrix}.$$

The Casimir operator for $\text{sl}_2(\mathbb{R})$ is

$$C := XY + YX + \frac{H^2}{2}. \tag{5.4.1}$$

Denote by C_σ the Casimir operator corresponding to the complexification of the σth factor of

$$G(\mathbb{R}) \cong \text{GL}_2(\mathbb{R})^{\Sigma(L)}$$

for $\sigma \in \Sigma(L)$. Let $G(\mathbb{R})^0$ denote the connected component of the identity in $G(\mathbb{R})$. For $b \in \mathbb{A}_L$ or $b \in \mathbb{A}_L^\times$, denote by $b_\infty = (b_{\sigma_1}, \ldots, b_{\sigma_n})$ (resp. b_0) the projection to the archimedean subgroup $\mathbb{A}_{L\infty}$ or $\mathbb{A}_{L\infty}^\times$ (resp. non-archimedean subgroup $\mathbb{A}_{Lf}$ or $\mathbb{A}_{Lf}^\times$). Analogously, for $\alpha \in G(\mathbb{A})$, let α_∞ (resp. α_0) be the projection to the archimedean subgroup $G(\mathbb{R})$ (resp. non-archimedean subgroup $G(\mathbb{A}_f)$). For $k \in \mathbb{Z}^{\Sigma(L)}$ define

$$b_\infty^k := \prod_{\sigma \in \Sigma(L)} b_\sigma^{k_\sigma} \in \mathbb{R}.$$

5.4. Hilbert modular forms

Fix $\kappa = (k, m) \in (\mathbb{Z}^{\geq 0})^{\Sigma(L)} \times (\frac{1}{2}\mathbb{Z})^{\Sigma(L)}$. Moreover, fix an ideal $\mathfrak{c} \subset \mathcal{O}_L$ and a (continuous) quasicharacter $\chi = \chi_0 \chi_\infty : L^\times \backslash \mathbb{A}_L^\times \to \mathbb{C}^\times$ of conductor dividing $\mathfrak{c}$, such that
$$\chi_\infty(b_\infty) = b_\infty^{-k-2m} \quad \text{for all } b \in \mathbb{A}_L^\times. \tag{5.4.2}$$
In Lemma 5.2 it is explained that we obtain characters $\chi_0 = \prod_{v < \infty} \chi_v$ and $\chi_0^\vee = \prod_{v < \infty} \chi_v^\vee$ on $K_0(\mathfrak{c})$ by setting, for each place $v < \infty$,
$$\chi_v \left(\begin{pmatrix} a_v & b_v \\ c_v & d_v \end{pmatrix} \right) = \begin{cases} \chi_v(d_v) & \text{if } \mathfrak{p}_v | \mathfrak{c} \\ 1 & \text{otherwise} \end{cases}$$
and
$$\chi_v^\vee \left(\begin{pmatrix} a_v & b_v \\ c_v & d_v \end{pmatrix} \right) = \begin{cases} \chi_v(a_v) & \text{if } \mathfrak{p}_v | \mathfrak{c} \\ 1 & \text{otherwise} \end{cases}$$
where $\mathfrak{p}_v$ is the prime ideal corresponding to the place v.

The space of *weight κ cusp forms* of nebentypus (or central character) χ on G, of level $K_0(\mathfrak{c})$, is the space
$$S_\kappa(K_0(\mathfrak{c}), \chi) \tag{5.4.3}$$
of functions $f : G(\mathbb{A}) \to \mathbb{C}$ with $f(\alpha_\infty \alpha_0)$ smooth as a function of $\alpha_\infty \in G(\mathbb{R})$ for all $\alpha_0 \in G(\mathbb{A}_f)$ satisfying the following conditions:

(1) If $\sigma \in \Sigma(L)$ then $C_\sigma f(\alpha) = \left(\frac{k_\sigma^2}{2} + k_\sigma \right) f(\alpha)$ for all $\alpha = \alpha_\infty \alpha_0 \in G(\mathbb{A})$.
(2) If $b \in \mathbb{A}_L^\times$ and $\gamma \in G(\mathbb{Q})$, we have $f(\gamma \alpha b) = \chi(b) f(\alpha)$.
(3) If $u = u_\infty^1 u_0 \in K_\infty^1 K_0(\mathfrak{c})$, then
$$f(\alpha u) = \chi_0(u_0) f(\alpha) \prod_{\sigma \in \Sigma(L)} e^{i(k_\sigma + 2)\theta_\sigma}$$

where
$$u_\infty^1 = \left(\begin{pmatrix} \cos \theta_\sigma & \sin \theta_\sigma \\ -\sin \theta_\sigma & \cos \theta_\sigma \end{pmatrix} \right)_{\sigma \in \Sigma(L)}$$

(4) The function f is cuspidal; i.e.,
$$\int_{U(\mathbb{Q}) \backslash U(\mathbb{A})} f(u\alpha) du = 0$$
for all $\alpha \in G(\mathbb{A})$, where $U(A) := \{ \begin{pmatrix} 1 & x \\ 0 & 1 \end{pmatrix} : x \in L \otimes_\mathbb{Q} A \}$ for a $\mathbb{Q}$-algebra A.

Remarks. If $\mathfrak{c}' \subset \mathfrak{c}$ then $S_\kappa(K_0(\mathfrak{c}), \chi) \subset S_\kappa(K_0(\mathfrak{c}'), \chi)$. Condition (2) implies that
$$f\left(\begin{pmatrix} b_\infty & 0 \\ 0 & b_\infty \end{pmatrix} \alpha \right) = b_\infty^{-k-2m} f(\alpha)$$
for any $b_\infty \in \mathbb{A}_{L\infty}$. Condition (1) implies that f is $\mathcal{Z}(\mathfrak{g}_\mathbb{C})$-finite, where $\mathcal{Z}(\mathfrak{g}_\mathbb{C})$ is the center of the universal enveloping algebra of the complexification of the Lie

algebra $\mathfrak{g}$ of $G(\mathbb{R})$. In condition (3) the exponent $(k_\sigma + 2)\theta_\sigma$ appears, rather than the more natural $k_\sigma \theta_\sigma$ because the modular forms defined here will eventually be multiplied by dz to give differential forms. As explained in Section 6.10, this shifts the weight by **2**.

As described in [Hid7] (3.5), the function f corresponds to a "classical" Hilbert modular form on a product of upper halfplanes. Let $z_0 := \mathbf{i} = (\sqrt{-1}, \ldots, \sqrt{-1}) \in \mathfrak{h}^{\Sigma(L)}$. For any $z \in \mathfrak{h}^{\Sigma(L)}$, pick $\alpha_z \in G(\mathbb{R})^0$ such that $\alpha_z z_0 = z$, for example $\begin{pmatrix} y & x \\ 0 & 1 \end{pmatrix} \cdot \mathbf{i} = z = x + \mathbf{i} y$ (the implied action of α_z is by Möbius transformation). With $t_j \in G(\mathbb{A}_f)$ (with $1 \leq j \leq h = h(K_0(\mathfrak{c}))$ so that j indexes the connected components of $Y_0(\mathfrak{c})$) as in equation (5.1.8), conditions (2) and (3) then imply that the expression

$$F_j(z) := \det(\alpha_z)^{m-1} j(\alpha_z, z_0)^{k+2\mathbf{1}} f(t_j \alpha_z) \qquad (5.4.4)$$

does not depend on the choice of α_z, and hence defines a function on $\mathfrak{h}^{\Sigma(L)}$ which satisfies

$$F_j(\gamma z) = \det(\gamma)^{m-1} j(\gamma, z)^{k+2\mathbf{1}} \chi_0(t_j^{-1} \gamma t_j) F_j(z)$$

for all $\gamma \in \Gamma_0^j(\mathfrak{c})$, see also Proposition 6.4. (To see this, use

$$f(\gamma_\infty \alpha_z t_j) = f(\gamma_f^{-1} \alpha_z t_j) = f(\alpha_z \gamma_f^{-1} t_j)$$

where $\gamma = (\gamma_f, \gamma_\infty)$ is diagonally embedded in $G(\mathbb{A})$.) Here, as in (6.7.2) and (6.6.1) the automorphy factor

$$j(\alpha, z)^k := \prod_{\sigma \in \Sigma(L)} j(\sigma(\alpha), z_\sigma)^{k_\sigma}$$

is defined by $j\left(\begin{pmatrix} a & b \\ c & d \end{pmatrix}, z\right) := (cz + d)$, see Chapter 6. Then, assuming conditions (2–4) for f, the holomorphicity of $F_i(z)$ for all $1 \leq i \leq h(K_0(\mathfrak{c}))$ is equivalent to condition (1) for f (see [Hid7, Section 2.4 p. 460]).

In order to incorporate Eisenstein series we need to consider a larger class of automorphic forms. Recall the definition of $t_j = t_j(\mathfrak{c})$ in (5.2.2). The space of *weight κ Hilbert modular forms of nebentypus χ on G of level $K_0(\mathfrak{c})$* is the space

$$M_\kappa(K_0(\mathfrak{c}), \chi)$$

of functions $f : G(\mathbb{A}) \to \mathbb{C}$ with $f(\alpha_\infty \alpha_0)$ smooth as a function of $\alpha_\infty \in G(\mathbb{R})$ for all $\alpha_0 \in G(\mathbb{A}_f)$ satisfying (1–3) above and the following weakening of (4):

(4') If $y_\infty^{\mathbf{1}} > C$, then $\left| y_\infty^{k/2} f\left(t_j \gamma y_\infty^{-\frac{1}{2}(\mathbf{1})} \begin{pmatrix} y_\infty & x_\infty \\ 0 & 1 \end{pmatrix}\right) \right| \leq B y_\infty^{A\mathbf{1}}$
  for all $1 \leq j \leq h$, $\gamma \in \operatorname{Res}_{L/\mathbb{Q}}(\operatorname{SL}_2)(\mathbb{Q}) \subseteq G(\mathbb{Q})$, $y \in \mathbb{A}_L^\times$, and $x \in \mathbb{A}_L$,

where $B, C \in \mathbb{R}_{>0}$, $t, A \in \mathbb{R}_{>0}^{\Sigma(L)}$, and A, B, C are independent of y and x. A function $f : G(\mathbb{A}) \to \mathbb{C}$ satisfying (5.4) is said to be *weakly increasing*. Recall the following result [Hid9, Section 2.3.2] of Hida:

Lemma 5.1. *Let $\kappa \in (\mathbb{Z}^{\geq 0})^{\Sigma(L)} \times (\frac{1}{2}\mathbb{Z})^{\Sigma(L)}$. The space $M_\kappa(K_0(\mathfrak{c}), \chi)$ is zero if $\kappa \notin \mathcal{X}(L)$.*

Proof. Assume there exists a non-zero element $f \in M_\kappa(K_0(\mathfrak{c}), \chi)$. By condition (2) in the definition of $M_\kappa(K_0(\mathfrak{c}), \chi)$, we have

$$\chi(b_0) b_\infty^{-k-2m} = 1$$

for all $b \in \mathcal{O}_L^\times$ (which we consider as an element of $\mathbb{A}_L^\times$ via the diagonal embedding). The restriction $\chi|_{\widehat{\mathcal{O}}_L^\times}$ is of finite order, say j. By raising this equation to the power j we conclude that

$$\prod_{\sigma \in \Sigma(L)} \sigma(b)^{-j(k_\sigma + 2m_\sigma)} = 1 \tag{5.4.5}$$

for all $b \in \mathcal{O}_L^\times$. Dirichlet's unit theorem says that the vectors

$$\{(\log(\sigma_1(b)), \ldots, \log(\sigma_r(b)))\}_{b \in \mathcal{O}_L^\times}$$

form an $r-1$-dimensional lattice in the trace-zero hyperplane

$$\left\{(x_1, \ldots, x_r) \in \mathbb{R}^r : \sum x_i = 0\right\}$$

(where $r = |\Sigma(L)|$). Combining this with (5.4.5) implies that $j(k+2m) \in \mathbb{Z}\mathbf{1}$ but k and $2m$ are integral by assumption, so $k + 2m \in \mathbb{Z}\mathbf{1}$. $\square$

Because of Lemma 5.1 we will henceforth assume that $\kappa \in \mathcal{X}(L)$.

5.5 Cohomological normalization

We now recall another normalization from [Hid7] for the weight of a Hilbert modular form which will be useful in Chapter 6. Let $S_\kappa^{\mathrm{coh}}(K_0(\mathfrak{c}), \chi)$ (resp. $M_\kappa^{\mathrm{coh}}(K_0(\mathfrak{c}), \chi)$) be the space of functions $f : G(\mathbb{A}) \to \mathbb{C}$ with $f(\alpha_\infty \alpha_0)$ smooth as a function of $\alpha_\infty \in G(\mathbb{R})$ for all $\alpha_0 \in G(\mathbb{A}_f)$ satisfying conditions (1) and (4) (resp. (4')) and the following modification of conditions (2) and (3):

(2^{coh}) If $b \in \mathbb{A}_L^\times$ and $\gamma \in G(\mathbb{Q})$, we have $f(\gamma \alpha b) = \chi(b)^{-1} f(\alpha)$.
(3^{coh}) If $u = u_\infty^1 u_0 \in K_\infty^1 K_0(\mathfrak{c})$, then

$$f(\alpha u) = \chi_0(u_0^{-\iota}) f(\alpha) \prod_{\sigma \in \Sigma(L)} e^{i(k_\sigma + 2)\theta_\sigma}$$

where

$$u_\infty^1 = \left(\begin{pmatrix} \cos(\theta_\sigma) & \sin(\theta_\sigma) \\ -\sin(\theta_\sigma) & \cos(\theta_\sigma) \end{pmatrix}\right)_{\sigma \in \Sigma(L)}.$$

There are natural isomorphisms

$$\begin{aligned} M_\kappa(K_0(\mathfrak{c}),\chi) &\xrightarrow{\sim} M_\kappa^{\mathrm{coh}}(K_0(\mathfrak{c}),\chi) \\ S_\kappa(K_0(\mathfrak{c}),\chi) &\xrightarrow{\sim} S_\kappa^{\mathrm{coh}}(K_0(\mathfrak{c}),\chi) \\ f &\longmapsto f^{-\iota} \end{aligned} \quad (5.5.1)$$

where $f^{-\iota}(x) := f(x^{-\iota})$. Here ι is the main involution of GL_2, so $x^\iota := \det(x)x^{-1}$ and $x^{-\iota} := \det(x)^{-1}x$.

Remark. The reason for introducing two normalizations for the weight of a modular form is that it is more natural to consider the Fourier expansions of an element of $S_\kappa(K_0(\mathfrak{c}),\chi)$ (see [Hid3] and Section 5.9) and it is more natural to associate differential forms and cohomology classes to elements of $S_\kappa^{\mathrm{coh}}(K_0(\mathfrak{c}),\chi)$ (see Section 7.2). This explains the superscript "coh." We will comment more on this in Section 5.13 below.

5.6 Hecke operators

Let $\mathfrak{c} \subset \mathcal{O}_L$ be an ideal and let $\chi = \chi_0\chi_\infty : L^\times \backslash \mathbb{A}_L^\times \to \mathbb{C}^\times$ be a Hecke character whose conductor divides $\mathfrak{c}$, see Appendix C.3. In this section we recall (from [Shim1, p. 51]) the definition of the Hecke operators associated to the spaces of automorphic forms $M_\kappa(K_0(\mathfrak{c}),\chi)$ and $M_\kappa^{\mathrm{coh}}(K_0(\mathfrak{c}),\chi)$. Let $R(\mathfrak{c})$ be the following semigroup of matrices:

$$G(\mathbb{A}_f) \cap \left\{ x \in M_2(\widehat{\mathcal{O}}_L) : x = \begin{pmatrix} a & b \\ c & d \end{pmatrix} \text{ with } c \in \mathfrak{c}\widehat{\mathcal{O}}_L \text{ and } d_v \in \mathcal{O}_v^\times \text{ whenever } \mathfrak{p}_v | \mathfrak{c} \right\}.$$

Here, $\mathfrak{p}_v$ is the prime ideal associated to the place $v < \infty$. For each place $v < \infty$ with corresponding prime ideal $\mathfrak{p}_v$ set

$$\chi_v\left(\begin{pmatrix} a_v & b_v \\ c_v & d_v \end{pmatrix}\right) = \begin{cases} \chi_v(d_v) & \text{if } \mathfrak{p}_v|\mathfrak{c} \\ 1 & \text{otherwise.} \end{cases} \quad (5.6.1)$$

$$\chi_v^\vee\left(\begin{pmatrix} a_v & b_v \\ c_v & d_v \end{pmatrix}\right) = \begin{cases} \chi_v(a_v) & \text{if } \mathfrak{p}_v|\mathfrak{c} \\ 1 & \text{otherwise.} \end{cases} \quad (5.6.2)$$

In this way the character χ_0 determines functions $\chi_0, \chi_0^\vee : R(\mathfrak{c}) \to \mathbb{C}^\times$ by setting $\chi_0 = \prod_{v<\infty} \chi_v$ and $\chi_v^\vee = \prod_{v<\infty} \chi_v^\vee$. These become characters of $K_0(\mathfrak{c})$:

Lemma 5.2. *The group $K_0(\mathfrak{c})$ is contained in $R(\mathfrak{c})$. If $x, x' \in R(\mathfrak{c})$ and $u_0 \in K_0(\mathfrak{c})$ then*

$$\chi_0(xx') = \chi_0(x)\chi_0(x') \quad \text{and} \quad \chi_0^\vee(u_0)^{-1} = \chi_0(u_0^{-\iota}). \quad (5.6.3)$$

where $u_0^{-\iota}$ denotes the main involution, cf. Section 5.5.

5.6. Hecke operators

Proof. Let $\mathfrak{p}_v | \mathfrak{c}$. If $x \in K_0(\mathfrak{c})$ then $a_v, d_v \in \mathcal{O}_v^\times$ are invertible hence $K_0(\mathfrak{c}) \subset R(\mathfrak{c})$. If $x, x' \in R(\mathfrak{c})$ then $d_v, d_v' \in \mathcal{O}_v^\times$ are invertible and

$$\chi_v \left(\begin{pmatrix} a_v & b_v \\ c_v & d_v \end{pmatrix} \begin{pmatrix} a_v' & b_v' \\ c_v' & d_v' \end{pmatrix} \right) = \chi_v \left(d_v d_v' \left(1 + c_v \frac{b_v'}{d_v d_v'} \right) \right) = \chi_v(d_v) \chi_v(d_v'). \tag{5.6.4}$$

which shows that $\chi_0(xx') = \chi_0(x)\chi_0(x')$. Equation (5.6.3) follows. $\square$

The *Hecke algebra* $\mathbb{T}_\mathfrak{c}$ is the algebra of formal $\mathbb{Z}$-linear sums of double cosets

$$[K_0(\mathfrak{c}) x K_0(\mathfrak{c})]$$

with $x \in R(\mathfrak{c})$. The multiplication in $\mathbb{T}_\mathfrak{c}$ is reviewed in Section 7.6 along with the geometric meaning of the Hecke algebra.

Any such double coset can be decomposed as a disjoint union of left cosets

$$K_0(\mathfrak{c}) x K_0(\mathfrak{c}) = \coprod_i x_i K_0(\mathfrak{c})$$

for suitable $x_i = k_i x \in R(\mathfrak{c}) \subset G(\mathbb{A}_f)$, where $k_i \in K_0(\mathfrak{c})$. We remark that $k_i x \mapsto k_i$ determines a one to one correspondence

$$K_0(\mathfrak{c}) x K_0(\mathfrak{c}) / K_0(\mathfrak{c}) \leftrightarrow K_0(\mathfrak{c}) / \left(K_0(\mathfrak{c}) \cap x K_0(\mathfrak{c}) x^{-1} \right),$$

see Section 7.6. The action of $\mathbb{T}_\mathfrak{c}$ on $M_\kappa(K_0(\mathfrak{c}), \chi)$ is then defined by

$$(f | K_0(\mathfrak{c}) x K_0(\mathfrak{c}))(g) := \sum_i \chi_0(x_i)^{-1} f(g x_i).$$

and the action of $\mathbb{T}_\mathfrak{c}$ on $M_\kappa^{\mathrm{coh}}(K_0(\mathfrak{c}), \chi)$ is defined by

$$(f | K_0(\mathfrak{c}) x K_0(\mathfrak{c}))(g) := \sum_i \chi_0(x_i)^{-1} f(g x_i^{-\iota}), \tag{5.6.5}$$

Remark. These normalizations of the Hecke action are chosen so that the isomorphism (5.5.1) induced by the main involution

$$M_\kappa(K_0(\mathfrak{c}), \chi) \cong M_\kappa^{\mathrm{coh}}(K_0(\mathfrak{c}), \chi)$$

is Hecke equivariant. As a space of functions, $M_{(k,m)}(K_0(\mathfrak{c}), \chi)$ is equal to

$$M_{(k,-m-k)}^{\mathrm{coh}}(K_0(\mathfrak{c}), \chi^{-1}),$$

but this identification of function spaces is not an isomorphism of Hecke modules.

For every ideal $\mathfrak{n} \subset \mathcal{O}_L$ set

$$T_\mathfrak{c}(\mathfrak{n}) := \sum_{[\det(x)_0] = \mathfrak{n}} [K_0(\mathfrak{c}) x K_0(\mathfrak{c})]. \tag{5.6.6}$$

Here the sum is over a set of representatives for the set of double cosets $K_0(\mathfrak{c})xK_0(\mathfrak{c})$ such that $x \in R(\mathfrak{c})$ and $\mathfrak{n} = [\det(x)_0]$ is the ideal associated to the finite idèle $\det(x)_0$. Further, for a prime $\mathfrak{p} \subset \mathcal{O}_L$ with $\mathfrak{p} \nmid \mathfrak{c}$, let $\varpi_\mathfrak{p} \in \mathcal{O}_{L,\mathfrak{p}}$ be a uniformizer. Considering $\varpi_\mathfrak{p}$ as an element of $\mathbb{A}_{Lf}^\times$ via the canonical inclusion $L_\mathfrak{p}^\times \hookrightarrow \mathbb{A}_{Lf}^\times$, set

$$T_\mathfrak{c}(\mathfrak{p}^k, \mathfrak{p}^k) := K_0(\mathfrak{c}) \begin{pmatrix} \varpi_\mathfrak{p}^k & 0 \\ 0 & \varpi_\mathfrak{p}^k \end{pmatrix} K_0(\mathfrak{c}) \tag{5.6.7}$$

for $k \geq 0$. If $\mathfrak{p}|\mathfrak{c}$, we simply set $T_\mathfrak{c}(\mathfrak{p}^k, \mathfrak{p}^k) = 0$ for $k \geq 0$. Define $T_\mathfrak{c}(\mathfrak{n}, \mathfrak{n})$ in general by multiplicativity. The Hecke operators satisfy the following identity (see [Shim4]):

$$T_\mathfrak{c}(\mathfrak{m})T_\mathfrak{c}(\mathfrak{n}) = \sum_{\mathfrak{m}+\mathfrak{n} \subseteq \mathfrak{a}} N_{L/\mathbb{Q}}(\mathfrak{a})T_\mathfrak{c}(\mathfrak{a}, \mathfrak{a})T_\mathfrak{c}(\mathfrak{a}^{-2}\mathfrak{m}\mathfrak{n}). \tag{5.6.8}$$

If $\mathfrak{c}' \subset \mathfrak{c}$ then we obtain a canonical homomorphism $\mathbb{T}_{\mathfrak{c}'} \to \mathbb{T}_\mathfrak{c}$ which takes $T_{\mathfrak{c}'}(\mathfrak{n})$ to $T_\mathfrak{c}(\mathfrak{n})$.

For our later convenience, if $f \in S_\kappa(K_0(\mathfrak{c}), \chi)$ is an eigenform for $T_\mathfrak{c}(\mathfrak{n})$ we let $\lambda_f(\mathfrak{n})$ be its $\mathfrak{n}$th Hecke-eigenvalue:

$$f|T_\mathfrak{c}(\mathfrak{n}) = \lambda_f(\mathfrak{n})f. \tag{5.6.9}$$

5.7 The Petersson inner product

In this section, we fix a normalization of the Petersson inner product. Let $G = \mathrm{Res}_{L/\mathbb{Q}}(\mathrm{GL}_2)$ as in the previous sections, and take

$$Y_K = G(\mathbb{Q}) \backslash G(\mathbb{A}) / K_\infty K$$

with $K_\infty = K_\infty^1 \cdot \mathbb{R}_+^{\Sigma(L)}$ as in equation (5.1.3) and $K \subseteq G(\mathbb{A}_f)$ a compact open subgroup. We refer to Appendix C.1 where various Haar measures and normalizations are described. Let $d\mu_\infty$ be the Haar measure on K_∞^1 (which is a product of circles) giving it total volume 1, and let

$$\frac{dx}{x} := \prod_{\sigma \in \Sigma(L)} \frac{dx_\sigma}{x_\sigma}$$

be the Haar measure on the multiplicative group $\mathbb{R}_{>0}^{\Sigma(L)}$. Let $d\mu_0$ be the Haar measure on the compact open group K giving it total volume 1. This gives a measure $d\mu_0 d\mu_\infty dx/x$ on KK_∞. By Appendix C.1 we obtain a $G(\mathbb{A})$-invariant measure $d\mu_{G(\mathbb{A}_F)/K_\infty K}$ on $G(\mathbb{A})/K_\infty K$. The discrete countable group $G(\mathbb{Q})$ acts almost freely on $G(\mathbb{A})/K_\infty K$ so we obtain a measure $d\mu_K$ on Y_K, which we refer to as the *canonical measure*. In other words, $d\mu_K$ is the restriction of $d\mu_{G(\mathbb{A})/K_\infty K}$ to a fundamental domain for $G(\mathbb{Q})$. In the special case $K = K_0(\mathfrak{c})$ we let

$$d\mu_\mathfrak{c} := d\mu_{K_0(\mathfrak{c})}. \tag{5.7.1}$$

It can be explicitly described as follows.

5.7. The Petersson inner product

The above normalizations determine a measure on the symmetric space $D = G(\mathbb{R})/K_\infty \cong (\mathbb{C}-\mathbb{R})^{\Sigma(L)}$ which coincides with the $G(\mathbb{R})$-invariant differential form

$$\prod_{\sigma \in \Sigma(L)} \frac{dx_\sigma \wedge dy_\sigma}{|y_\sigma|^2} = (-2i)^{-n} \prod_{\sigma \in \Sigma(L)} \frac{dz_\sigma \wedge d\bar{z}_\sigma}{|y_\sigma|^2} \qquad (5.7.2)$$

(where $z_\sigma = x_\sigma + iy_\sigma$). Since it is $G(\mathbb{R})$-invariant, it is the pullback of a differential form, which we denote by

$$\omega = \frac{dx \wedge dy}{|y|^2}$$

on Y_K. The normalization of $d\mu_0$ (at the finite places) amounts to assigning the counting measure on the individual components of Y_K. Consequently if $f : Y_K \to \mathbb{C}$ is a measurable function then

$$\int_{Y_K} f d\mu_K = \sum_j \int_{Y_K^j} f|_{Y_K^j} \frac{dx \wedge dy}{|y|^2}. \qquad (5.7.3)$$

where

$$Y_K^j := \Gamma_j(K) \backslash \mathfrak{h}^{\Sigma(L)}.$$

The canonical measure on Y_K can also be described in a way that facilities application of the Rankin-Selberg method (to be used in Section 10.2) as follows. Let $B' \subset G$ be the algebraic group whose points in a commutative $\mathbb{Q}$-algebra A are given by

$$B'(A) := \left\{ \begin{pmatrix} a & b \\ 0 & 1 \end{pmatrix} : a \in (L \otimes_\mathbb{Q} A)^\times \text{ and } b \in L \otimes_\mathbb{Q} A \right\}.$$

(The group B' is not unimodular.) Then $B'(\mathbb{R})$ acts transitively on D with

$$\begin{pmatrix} y & x \\ 0 & 1 \end{pmatrix} \cdot \mathbf{i} = x + \mathbf{i}y = \prod_\sigma (x_\sigma + iy_\sigma).$$

Left Haar measure on $B'(\mathbb{A})$ is given by

$$d\mu_{B'}\left(\begin{pmatrix} y & x \\ 0 & 1 \end{pmatrix}\right) = |y|_{\mathbb{A}_L}^{-1} dx d^\times y$$

where $dx := dx_\infty dx_0$ and $d^\times y := d^\times y_\infty d^\times y_0$ are the measures defined in Appendix C.1 on $\mathbb{A}_{L\infty}$, $\mathbb{A}_{Lf}$, $\mathbb{A}_{L\infty}^\times$ and $\mathbb{A}_{Lf}^\times$ respectively. In other words,

- The additive measure $dx = dx_\infty$ is Lebesgue measure on $\mathbb{R}^{\Sigma(L)}$.
- The additive measure dx_0 is the measure on $\mathbb{A}_{Lf}$ such that $\int_{\widehat{\mathcal{O}}_L} dx_0 = 1$.
- The multiplicative measure $d^\times y_\infty$ is given by $|y_\infty^{-1}| dy_\infty$.
- The multiplicative measure $d^\times y_0$ is the measure on $\mathbb{A}_{Lf}^\times$ such that $\int_{\widehat{\mathcal{O}}_L^\times} d^\times y_0 = 1$.

Here $\widehat{\mathcal{O}}_L := \mathcal{O}_L \otimes \widehat{\mathbb{Z}}$. In view of the fact that

$$B'(\mathbb{R})K_\infty^1 = G(\mathbb{R})/\mathbb{R}_+^{\Sigma(L)}$$

and[1]

$$G(\mathbb{A})/(\mathbb{R}_+)^{\Sigma(L)} = G(\mathbb{Q})B'(\mathbb{A})K_\infty^1 K,$$

the product $d\mu_{B'} d\mu_\infty d\mu_0$ induces a measure on Y_K via a choice of fundamental domain for the action of $G(\mathbb{Q})$. One checks the following lemma:

Lemma 5.3. *The measure defined in this way coincides with the canonical measure $d\mu_K$ on Y_K.* □

Definition 5.4. For $\kappa \in \mathcal{X}(L)$, a quasicharacter χ satisfying $\chi_\infty(b_\infty) = b_\infty^{-k-2m}$ and $f, g \in M_\kappa(K_0(\mathfrak{c}), \chi)$, we set

$$(f, g)_P := \int_{Y_0(\mathfrak{c})} \overline{g(\alpha)} f(\alpha) |\det(\alpha)|_{\mathbb{A}_L}^{k_\sigma + 2m_\sigma} d\mu_{\mathfrak{c}}(\alpha), \qquad (5.7.4)$$

whenever this integral is well defined. Here σ is any element of $\Sigma(L)$ (note $k_\sigma + 2m_\sigma$ is independent of this choice because $\kappa \in \mathcal{X}(L)$).

Remark. The cuspidality condition (4) in the definition of $S_\kappa(K_0(\mathfrak{c}), \chi)$ implies that a cusp form is rapidly decreasing [Harish, Section 4]. This implies in turn that the integral defining $(f, g)_P$ is absolutely convergent whenever at least one of f or g is a cusp form.

Using the main involution this gives an inner product $(f, g)_P^{\text{coh}} := (f^{-\iota}, g^{-\iota})_P$ for any $f, g \in M_\kappa^{\text{coh}}(K_0(\mathfrak{c}), \chi)$.

Lemma 5.5. *For any $f, g \in M_\kappa^{\text{coh}}(K_0(\mathfrak{c}), \chi)$,*

$$(f, g)_P^{\text{coh}} = \int_{Y_0(\mathfrak{c})} \overline{g(\alpha)} f(\alpha) |\det(\alpha)|_{\mathbb{A}_L}^{-k_\sigma - 2m_\sigma} d\mu_{\mathfrak{c}}(\alpha).$$

Proof. Since $f^{-\iota}, g^{-\iota} \in M_\kappa(K_0(\mathfrak{c}), \chi)$ we have:

$$(f^{-\iota}, g^{-\iota})_P = \int_{Y_0(\mathfrak{c})} \overline{g\left(\frac{\alpha}{\det(\alpha)}\right)} f\left(\frac{\alpha}{\det(\alpha)}\right) |\det(\alpha)|_{\mathbb{A}_L}^{k_\sigma + 2m_\sigma} d\mu_{\mathfrak{c}}(\alpha)$$

$$= \int_{Y_0(\mathfrak{c})} \overline{\chi(\det(\alpha))} \chi(\det(\alpha)) \overline{g(\alpha)} f(\alpha) |\det(\alpha)|_{\mathbb{A}_L}^{k_\sigma + 2m_\sigma} d\mu_{\mathfrak{c}}(\alpha)$$

$$= \int_{Y_0(\mathfrak{c})} \overline{g(\alpha)} f(\alpha) |\det(\alpha)|_{\mathbb{A}_L}^{-k_\sigma - 2m_\sigma} d\mu_{\mathfrak{c}}(\alpha)$$

because $|\chi(\det(\alpha))|^2 = |\det(\alpha)|_{\mathbb{A}_L}^{-2k_\sigma - 4m_\sigma}$, see equation C.3.2. □

---
[1] See [Hid5, Section 4].

5.8. Newforms

We recall that Hecke operators indexed by ideals coprime to the level are self-adjoint with respect to the Petersson inner product, at least up to a scalar:

Lemma 5.6. *If $f, g \in M_\kappa(K_0(\mathfrak{c}), \chi)$ and $\mathfrak{n} + \mathfrak{c} = \mathcal{O}_E$ then one has*

$$(f|T_\mathfrak{c}(\mathfrak{n}), g)_P = \overline{\chi}| \cdot |_{\mathbb{A}_L}^{k_\sigma + 2m_\sigma}(\mathfrak{n})(f, g|T_\mathfrak{c}(\mathfrak{n}))_P.$$

Here $\chi(\mathfrak{n}) := \chi(n_0)$ where $n \in \mathbb{A}_L^\times$ is any idèle that is 1 at the places dividing ∞ and $\mathfrak{c}$ with $[n_0] = \mathfrak{n}$.

Proof. Let $x \in R_0(\mathfrak{c})$. Then using the fact that $\mu_\mathfrak{c}$ was constructed from a Haar measure together with the invariance properties of g we obtain

$$\int_{Y_0(\mathfrak{c})} \overline{g(\alpha)} f(\alpha x) |\det(\alpha)|_{\mathbb{A}_L}^{k_\sigma + 2m_\sigma} d\mu_\mathfrak{c}(\alpha) \tag{5.7.5}$$

$$= \int_{Y_0(\mathfrak{c})} \overline{g(\alpha x^{-1})} f(\alpha) |\det(\alpha x^{-1})|_{\mathbb{A}_L}^{k_\sigma + 2m_\sigma} d\mu_\mathfrak{c}(\alpha)$$

$$= \overline{\chi}| \cdot |_{\mathbb{A}_L}^{k_\sigma + 2m_\sigma}(\det(x)^{-1}) \int_{Y_0(\mathfrak{c})} \overline{g(\alpha x^\iota)} f(\alpha) |\det(\alpha)|_{\mathbb{A}_L}^{k_\sigma + 2m_\sigma} d\mu_\mathfrak{c}(\alpha)$$

$$= \chi| \cdot |_{\mathbb{A}_L}^{k_\sigma + 2m_\sigma}(\det(x)) \int_{Y_0(\mathfrak{c})} \overline{g(\alpha x^\iota)} f(\alpha) |\det(\alpha)|_{\mathbb{A}_L}^{k_\sigma + 2m_\sigma} d\mu_\mathfrak{c}(\alpha).$$

Here in the last line we have used the fact that $\overline{\chi}| \cdot |_{\mathbb{A}_L}^{k_\sigma + 2m_\sigma}(\det(x)^{-1}) = \chi| \cdot |_{\mathbb{A}_L}^{k_\sigma + 2m_\sigma}(\det(x))$ by our assumption on the shape of χ_∞.

Now assume that $\mathfrak{n} + \mathfrak{c} = \mathcal{O}_E$ and choose $x_i \in R_0(\mathfrak{c})$ such that $(x_i)_v = \begin{pmatrix} 1 & 0 \\ 0 & 1 \end{pmatrix}$ for $v | \mathfrak{c}$ and

$$(f|T_\mathfrak{c}(\mathfrak{n}))(\alpha) = \sum_i f(\alpha x_i).$$

(see Section 5.6). From the definition of $T_\mathfrak{c}(\mathfrak{n})$ and our choice of x_i one can check that

$$(f|T_\mathfrak{c}(\mathfrak{n}))(\alpha) = \sum_i f(\alpha x_i^\iota)$$

as well. Combining this with (5.7.5) completes the proof. □

5.8 Newforms

The space of modular forms for the full modular group $SL_2(\mathbb{Z})$ admits a basis of simultaneous eigenforms for all Hecke operators T_p with the property that distinct elements in the basis have distinct sets of eigenvalues. The standard L function for each such eigenform admits an Euler product. A.O. Atkin and J. Lehner discovered [Atk] (see also [Li]) that these results can be extended to the Hecke congruence groups $\Gamma_0(N)$ provided one restricts consideration to the subspace spanned by *newforms.* The analogous results continue to hold in the Hilbert modular case, as we now describe.

If $\mathfrak{c}'|\mathfrak{c}$ are ideals in $\mathcal{O}_L$ then the natural map $\pi : Y_0(\mathfrak{c}) \to Y_0(\mathfrak{c}')$ is finite and pulling back modular forms gives

$$\pi^* : S_\kappa(K_0(\mathfrak{c}'), \chi) \to S_\kappa(K_0(\mathfrak{c}), \chi)$$

where $\pi^*(f)(x) = f(x)$. More generally, let $d \in \widehat{\mathcal{O}}_L$ and suppose that $d\mathfrak{c}'|\mathfrak{c}$. If $f \in S_\kappa(K_0(\mathfrak{c}'), \chi)$ then the function $g(x) := f\left(x\left(\begin{smallmatrix} d^{-1} & \\ & 1\end{smallmatrix}\right)\right)$ is in $S_\kappa(K_0(\mathfrak{c}), \chi)$. The vector space spanned by the images of all such mappings (for $d\mathfrak{c}'|\mathfrak{c}$) is called the *old space*. Its orthogonal complement (with respect to the Petersson inner product) is the *new space* $S_\kappa^{\mathrm{new}}(K_0(\mathfrak{c}), \chi)$. If $\mathfrak{c}'|\mathfrak{c}$ then automorphic forms can also be pushed forward using the "trace" mapping

$$\begin{aligned} \pi_* : S_\kappa(K_0(\mathfrak{c}), \chi) &\longrightarrow S_\kappa(K_0(\mathfrak{c}'), \chi) \\ f &\longmapsto \sum_{xK_0(\mathfrak{c}) \in K_0(\mathfrak{c}')/K_0(\mathfrak{c})} \chi_0(x)^{-1} f(\cdot x). \end{aligned}$$

Then $S_\kappa^{\mathrm{new}}(K_0(\mathfrak{c}), \chi)$ is equal to the intersections of the kernels of all such trace maps as $\mathfrak{c}'$ varies over the divisors of $\mathfrak{c}$ (see [La] Thm. 2.2 for the elliptic modular case). A *newform* is an element $f \in S^{\mathrm{new}}(K_0(\mathfrak{c}), \chi)$ that is a simultaneous eigenfunction of $T_\mathfrak{c}(\mathfrak{p})$ for almost all primes $\mathfrak{p}$.

Theorem 5.7 (see [Miy, Cas, Di]). *The space $S_\kappa^{\mathrm{new}}(K_0(\mathfrak{c}), \chi)$ of newforms has a basis consisting of automorphic forms that are simultaneous eigenfunctions of all Hecke operators. Moreover the eigenvalues have multiplicity one in the following sense: if $f, g \in S_\kappa^{\mathrm{new}}(K_0(\mathfrak{c}), \chi)$ and $\lambda_f(\mathfrak{p}) = \lambda_g(\mathfrak{p})$ for almost all prime ideals $\mathfrak{p}$ then $f = cg$ for some constant $c \in \mathbb{C}^\times$. Conversely, if $f, g \in S_\kappa^{\mathrm{new}}(K_0(\mathfrak{c}), \chi)$ are simultaneous eigenfunctions for $T_\mathfrak{c}(\mathfrak{p})$ for almost all prime ideals $\mathfrak{p}$ and if $f \neq cg$ for any $c \in \mathbb{C}^\times$ then f, g are orthogonal with respect to the Petersson inner product: $(f, g)_P = 0$.*

The last statement in the theorem follows from the first and the spectral theorem because each Hecke operator $T_\mathfrak{c}(\mathfrak{n})$ with $\mathfrak{c} + \mathfrak{n} = \mathcal{O}_E$ is *normal*, that is, it commutes with its adjoint with respect to the Petersson product (see Lemma 5.6). Theorem 5.7 implies that the standard L-function of any newform has an Euler product, cf. Lemma 5.12.

5.9 Fourier series

For details on this section, see Appendix D. Let $\kappa = (k, m) \in (\mathbb{Q} \times \mathbb{Q})^{\Sigma(L)}$. For each $\sigma \in \Sigma(L)$, let W_{m_σ} be the local archimedean Whittaker function

$$\begin{aligned} W_{m_\sigma} : \mathbb{R}^\times &\longrightarrow \mathbb{C} \\ y_\sigma &\longmapsto |y_\sigma|^{-m_\sigma} e^{-2\pi |y_\sigma|}. \end{aligned}$$

5.9. Fourier series

For $x \in \mathbb{A}_L$ and $y \in \mathbb{A}_L^\times$, define

$$q_\kappa(x,y) = q_\kappa(x, y_\infty) := e_L(x) \prod_{\sigma \in \Sigma(L)} W_{m_\sigma}(y_\sigma), \qquad (5.9.1)$$

where $e_L(\cdot)$ is the additive character of $\mathbb{A}_L/L$ normalized as in Appendix C; it satisfies $e_L(x_\infty) = e^{2\pi i \operatorname{Tr}_{L/\mathbb{Q}}(x_\infty)}$.

Let $\mathcal{I}_L$ denote the set of fractional ideals of $\mathcal{O}_L$, and let $y = y_0 y_\infty \in \mathbb{A}_{L,+}^\times$, the set of idèles $b \in \mathbb{A}_L^\times$ with $b_\sigma > 0$ for all $\sigma \in \Sigma(L)$. Let $\mathcal{D}_{L/\mathbb{Q}}$ be the different (see Section C.2). Assume that $\kappa \in \mathcal{X}(L)$.

Theorem 5.8. *Let $h \in M_\kappa(K_0(\mathfrak{c}), \chi)$ be a Hilbert modular form. Then h admits a Fourier series,*

$$h\left(\begin{pmatrix} y & x \\ 0 & 1 \end{pmatrix}\right) = |y|_{\mathbb{A}_L} \left(c(y) + \sum_{\substack{\xi \in L^\times \\ \xi \gg 0}} b(\xi y_0) q_\kappa(\xi x, \xi y_\infty) \right), \qquad (5.9.2)$$

valid for all $x \in \mathbb{A}_L$ and all $y = y_0 y_\infty \in \mathbb{A}_L^\times$ such that $y_\sigma > 0$ for all $\sigma \in \Sigma(L)$. Moreover, each coefficient $b(\xi y_0) \in \mathbb{C}$ depends only on the fractional ideal $[\xi y_0] \mathcal{D}_{L/\mathbb{Q}} \in \mathcal{I}_L$ and it vanishes unless this ideal is integral.

Addendum. *The constant term $c(y)$ vanishes if h is a cusp form or if $k \notin \mathbb{Z}\mathbf{1}$ or if the ideal $[y_0] \mathcal{D}_{L/\mathbb{Q}}$ is not integral. Otherwise it is a sum,*

$$c(y) = c_0(y_0)|y_\infty^{-m}| + c_1(y_0)|y_\infty^{-k-1-m}| \qquad (5.9.3)$$

of two terms. Here, $c_0(y_0)$ and $c_1(y_0)$ only depend on the (fractional) ideal $[y_0]\mathcal{D}_{L/\mathbb{Q}}$. If the functions $F_i(z)$ of (5.4.4) on $\mathfrak{h}^{\Sigma(L)}$ (corresponding to h) are holomorphic, then $c_1(\cdot) = 0$.

In what follows we will express these coefficients b, c_0, c_1 slightly differently, defining $a(\cdot, \cdot)$, $a_0(\cdot, \cdot)$ and $a_1(\cdot, \cdot)$ (respectively) to be the corresponding functions

$$\mathcal{I}_L \times M_\kappa(K_0(\mathfrak{c}), \chi) \to \mathbb{C}$$

defined by

$$a(\xi y \mathcal{D}_{L/\mathbb{Q}}, h) := b(\xi y_0)$$
$$a_0(y \mathcal{D}_{L/\mathbb{Q}}, h) := b_0(y_0)$$
$$a_1(y \mathcal{D}_{L/\mathbb{Q}}, h) := b_1(y_0).$$

Thus the fourier coefficients $a(\mathfrak{m}, h)$, $a_0(\mathfrak{m}, h)$, and $a_1(\mathfrak{m}, h)$ vanish on non-integral ideals $\mathfrak{m}$. In summary, if $h \in S_\kappa(K_0(\mathfrak{c}), \chi)$ is a cusp form, we have:

$$h\left(\begin{pmatrix} y & x \\ 0 & 1 \end{pmatrix}\right) = |y|_{\mathbb{A}_L} \sum_{\substack{\xi \in L^\times \\ \xi \gg 0}} a(\xi y \mathcal{D}_{L/\mathbb{Q}}, h) q_\kappa(\xi x, \xi y)$$

If $\mathfrak{m} = [\xi y_0]\mathcal{D}_{L/\mathbb{Q}}$ is integral, we refer to $a(\mathfrak{m}, h)$ as the $\mathfrak{m}$*th Fourier coefficient* of h. The *leading coefficient* is $a(\mathcal{O}_L, h)$ (which occurs in the sum when $\xi = 1$), and the cusp form h is said to be *normalized* if $a(\mathcal{O}_L, h) = 1$,

The Fourier expansion is an essential tool for what follows; indeed, it is built into the statement of Theorems 1.1 and 1.2. For a proof of the existence of the Fourier expansion that relies on the "classical viewpoint" of Hilbert modular forms as differential forms on $\mathfrak{h}^{\Sigma(L)}$, see [Hid5, Theorem 1.1], and for an adelic proof, see [Weil], [Hid7, Section 6]. For the convenience of the reader, we will give a proof relying on the basic theory of Whittaker models in Appendix D.

The relationship between the Fourier coefficients of a cusp form $f \in S_\kappa(K_0(\mathfrak{c}), \chi)$ and that of its Hecke-translates can be written down explicitly, (see [Hid7, Corollary 6.2]):

$$a(\mathfrak{m}, f|T_\mathfrak{c}(\mathfrak{n})) = \sum_{\substack{\mathfrak{b} \supseteq \mathfrak{m}+\mathfrak{n} \\ \mathfrak{b}+\mathfrak{c}=\mathcal{O}_L}} N_{L/\mathbb{Q}}(\mathfrak{b})\chi(\mathfrak{b})a(\mathfrak{m}\mathfrak{n}/\mathfrak{b}^2, f) \qquad (5.9.4)$$

where $\chi(\mathfrak{b}) := \chi(b)$ where $b \in \mathbb{A}_L^\times$ is an element such that:

- The component $b_v = 1$ if $v \in \Sigma(L)$ or if $\mathfrak{p}_v | \mathfrak{c}$
- The ideal $[b_0] = \mathfrak{b}$.

(Recall that $\mathfrak{p}_v$ is the prime ideal corresponding to the place v and that $[b_0]$ is the fractional ideal corresponding to the idèle b_0.) From (5.6.8) and (5.9.4) one can deduce that if f is a simultaneous eigenform for all Hecke operators normalized so that $a(\mathcal{O}_L, f) = 1$, then *the Fourier coefficients of f coincide with its Hecke eigenvalues*,

$$\boxed{a(\mathfrak{m}, f) = \lambda_f(\mathfrak{m})} \qquad (5.9.5)$$

(see [Hid7, p. 477]). With this in mind define

$$L(f, s) := \sum_{\mathfrak{m} \subset \mathcal{O}_L} a(\mathfrak{m}, f) N_{L/\mathbb{Q}}(\mathfrak{m})^{-1/2-s} \qquad (5.9.6)$$

for any $f \in S_\kappa(K_0(\mathfrak{c}), \chi)$. The function $L(f, s)$ admits an Euler product (see equation (5.11.1)) if and only if f is a simultaneous eigenform for all Hecke operators.

5.10 Killing Fourier coefficients

The lemmas in this section will be used in Section 8.3 and Section 11.1. The first lemma says that we may kill the Fourier coefficients corresponding to an ideal $\mathfrak{b}$ and still obtain a modular form, albeit with level a smaller ideal than the original form.

Lemma 5.9. *Suppose that $f \in M_\kappa(K_0(\mathfrak{c}), \chi)$ and let $\mathfrak{b} \subset \mathcal{O}_L$ be an ideal. Define*

$$a(\mathfrak{m}, f^\mathfrak{b}) = \begin{cases} a(\mathfrak{m}, f) & \text{if } \mathfrak{b} + \mathfrak{m} = \mathcal{O}_L \\ 0 & \text{otherwise.} \end{cases}$$

5.10. Killing Fourier coefficients

Then the Fourier series $f^{\mathfrak{b}}\left(\left(\begin{smallmatrix} y & x \\ 0 & 1 \end{smallmatrix}\right)\right)$ defined by

$$|y|_{\mathbb{A}_L}\left(a_0(y\mathcal{D}_{L/\mathbb{Q}}, f)|y_\infty^{-m}| + a_1(y\mathcal{D}_{L/\mathbb{Q}}, f)|y_\infty^{-k-1-m}| \right.$$
$$\left. + \sum_{\substack{\xi \in L^\times \\ \xi \gg 0}} a(\xi y \mathcal{D}_{L/\mathbb{Q}}, f^{\mathfrak{b}}) q_\kappa(\xi x, \xi y) \right)$$

is an element of $M_\kappa(K_0(\mathfrak{c}\prod_{\mathfrak{p}|\mathfrak{b}}\mathfrak{p}), \chi)$.

Proof. It suffices to show, for any prime $\mathfrak{p} \subset \mathcal{O}_L$, that $f^{\mathfrak{p}^r} \in M_\kappa(K_0(\mathfrak{c}\mathfrak{p}), \chi)$. An inductive argument then finishes the proof.

By considering Fourier series using (5.9.4), we see that

$$f^{\mathfrak{p}^r} = f^{\mathfrak{p}} = f(\alpha) - |N_{L/\mathbb{Q}}(\mathfrak{p})|^{-1} f | T_{\mathfrak{c}\mathfrak{p}}(\mathfrak{p}) \left(\alpha \begin{pmatrix} \varpi_{\mathfrak{p}}^{-1} & 0 \\ 0 & 1 \end{pmatrix} \right).$$

Here we denote by $\varpi_{\mathfrak{p}}$ an idèle that is a uniformizer for the maximal ideal of $\mathcal{O}_{L,\mathfrak{p}}$ at the place associated to $\mathfrak{p}$ and is 1 at every other place. It is then easy to see that $f^{\mathfrak{p}^r} \in M_\kappa(K_0(\mathfrak{c}\mathfrak{p}), \chi)$. $\square$

The connected components of the Hilbert modular variety $Y_0(\mathfrak{c})$ are mapped bijectively via the determinant to the elements of $T(\mathbb{Q})\backslash T(\mathbb{A})/\det(K_\infty K_0(\mathfrak{c}))$ (see Section 5.1). Such a connected component is therefore determined by an element $\mathfrak{a} \in T(\mathbb{A})$. If f is a modular form and if Y_1 is a connected component of $Y_0(\mathfrak{c})$ we may define a new modular form, $\pi_{\mathfrak{a}}(f)$ which coincides with f on Y_1 and which vanishes on all the other connected components of $Y_0(\mathfrak{c})$. The following lemma describes this "restriction" operation in terms of Fourier coefficients.

Lemma 5.10. *Suppose that* $f \in M_\kappa(K_0(\mathfrak{c}), \chi)$ *and let* $\mathfrak{a} \in \mathbb{A}_L^\times = T(\mathbb{A})$. *Define*

$$a(\xi y \mathcal{D}_{L/\mathbb{Q}}, \pi_{\mathfrak{a}} f)$$
$$:= \begin{cases} a(\xi y \mathcal{D}_{L/\mathbb{Q}}, f) & \text{if } \mathfrak{a} \sim \xi y \mathcal{D}_{L/\mathbb{Q}} \text{ in } T(\mathbb{Q})\backslash T(\mathbb{A})/\det(G(\mathbb{R})^0 K_0(\mathfrak{c})) \\ 0 & \text{otherwise,} \end{cases}$$

$$a_i(y \mathcal{D}_{L/\mathbb{Q}}, \pi_{\mathfrak{a}} f)$$
$$:= \begin{cases} a_i(y \mathcal{D}_{L/\mathbb{Q}}, f) & \text{if } \mathfrak{a} \sim y \mathcal{D}_{L/\mathbb{Q}} \text{ in } T(\mathbb{Q})\backslash T(\mathbb{A})/\det(G(\mathbb{R})^0 K_0(\mathfrak{c})) \\ 0 & \text{otherwise,} \end{cases}$$

for $i \in \{0, 1\}$ *(see (5.1.6)). Then the Fourier series* $\pi_{\mathfrak{a}} f\left(\left(\begin{smallmatrix} y & x \\ 0 & 1 \end{smallmatrix}\right)\right)$ *defined by*

$$|y|_{\mathbb{A}_L}\left(a_0(y\mathcal{D}_{L/\mathbb{Q}}, \pi_{\mathfrak{a}} f)|y_\infty^{-m}| + a_1(y\mathcal{D}_{L/\mathbb{Q}}, \pi_{\mathfrak{a}} f)|y_\infty^{-k-1-m}| \right.$$
$$\left. + \sum_{\substack{\xi \in L^\times \\ \xi \gg 0}} a(\xi y \mathcal{D}_{L/\mathbb{Q}}, \pi_{\mathfrak{a}} f) q_\kappa(\xi x, \xi y) \right)$$

is a Hilbert modular form.

Proof. The function $\pi_{\mathfrak{a}} f : G(\mathbb{A}) \to \mathbb{C}$ is defined so that

$$\pi_{\mathfrak{a}} f(g) = \begin{cases} f(g) & \text{if } \det(g) \in \left(T(\mathbb{Q})\mathfrak{a}\mathcal{D}_{L/E}^{-1}\det(G(\mathbb{R})^0 K_0(\mathfrak{c}))\right) \\ 0 & \text{otherwise.} \end{cases}$$

Lemma 5.10 now follows directly from the definition of $M_\kappa(K_0(\mathfrak{c}), \chi)$. □

5.11 Twisting

At the expense of raising the level, it is possible to "twist" a modular form f by a Hecke character η, and to obtain[2] a new modular form $f \otimes \eta$ such that the Fourier coefficient $a(\mathfrak{m}, f \otimes \eta)$ of $f \otimes \eta$ is $\eta(\mathfrak{m})a(\mathfrak{m}, f)$. This relation is usually expressed in terms of the L-function of $f \otimes \eta$, see equation (5.11.1). See also Section 9.4 where the twisting operation is interpreted geometrically.

Let $\eta : L^\times \backslash \mathbb{A}_L^\times \to \mathbb{C}^\times$ a finite-order Hecke character. For any ideal $\mathfrak{n}$ define $\eta(\mathfrak{n}) := \eta(\widetilde{n})$ where $\widetilde{n} \in \mathbb{A}_L^\times$ is an element such that $\widetilde{n}_v = 1$ if $v|\infty$ or if $\mathfrak{p}_v|\mathfrak{c}\mathfrak{b}$ and $[\widetilde{n}_0] = \mathfrak{n}$ (that is, we define the value of η on ideals as we defined the value of χ on ideals after equation (5.9.4) above). The following Proposition of [Hid5, §7.F] gives the precise properties of twisting.

Proposition 5.11. *Fix a finite-order Hecke character η as above. Let $\mathfrak{b}$ be an ideal divisible by the conductor $\mathfrak{f}(\eta)$ of the Dirichlet character associated to η, and let $w \in \frac{1}{2}\mathbb{Z}$. Let $f \in M_{(k,m)}(K_0(\mathfrak{c}), \chi)$ be a modular form with weight $\kappa = (k, m)$ and with Fourier coefficients $a(\mathfrak{m}, f)$. Then there exists a modular form*

$$f \otimes \eta| \cdot |_{\mathbb{A}_L}^w \in M_{(k, m-w\mathbf{1})}(K_0(\mathfrak{c}\mathfrak{b}^2), \chi\eta^2| \cdot |_{\mathbb{A}_L}^{2w})$$

whose Fourier coefficients are given by

$$a(\mathfrak{m}, f \otimes \eta| \cdot |_{\mathbb{A}_L}^w) := \begin{cases} \eta(\mathfrak{m})|\mathrm{N}_{L/\mathbb{Q}}(\mathfrak{m})|^w a(\mathfrak{m}, f) & \text{if } \mathfrak{m} + \mathfrak{b} = \mathcal{O}_L \\ 0 & \text{otherwise.} \end{cases}$$

and for $i = 0, 1$,

$$a_i(\mathfrak{m}, f \otimes \eta| \cdot |_{\mathbb{A}_L}^w) := \begin{cases} \eta(\mathfrak{m})|\mathrm{N}_{L/\mathbb{Q}}(\mathfrak{m})|^w a_i(\mathfrak{m}, f) & \text{if } \mathfrak{f}(\eta) = \mathcal{O}_L, \\ 0 & \text{otherwise} \end{cases}.$$

At the level of automorphic representations this corresponds to the fact that twisting a representation of GL_2 by a character η multiplies its central character by the square of η. Assuming Proposition 5.11 for the moment, we set notation for the Euler products of the twists $f \otimes \eta$.

---

[2]See [Kobl, Prop. 17 §III.3] for the classical case: Let χ, η be Dirichlet characters modulo M, N respectively. If f is a modular form of weight k, level N and Dirichlet character χ, with Fourier expansion $f(z) = \sum_{n \geq 0} a_n \exp(2\pi i n z)$ then $f \otimes \eta(z) := \sum_{n \geq 0} \eta(n) a_n \exp(2\pi i n z)$ is a modular form of weight k, level MN^2 and character $\chi\eta^2$.

5.11. Twisting

Lemma 5.12. *Fix a finite-order Hecke character η as above. (We allow the possibility that η is the trivial character.) Let $f \in M_{(k,m)}(K_0(\mathfrak{c}), \chi)$ be a newform. Then $f \otimes \eta$ is a simultaneous eigenform[3] for all Hecke operators and its L-function admits an Euler product as a product over prime ideals $\mathfrak{p}$,*

$$L(f \otimes \eta, s) := \prod_{\mathfrak{p} \subset \mathcal{O}_L} L_{\mathfrak{p}}(f \otimes \eta, s) \tag{5.11.1}$$

where $L_{\mathfrak{p}}(f \otimes \eta, s)$ is equal to

$$\begin{cases} \left(1 - \lambda_f(\mathfrak{p})\eta(\mathfrak{p}) N_{L/\mathbb{Q}}(\mathfrak{p})^{-1/2-s} + \chi(\mathfrak{p})\eta(\mathfrak{p})^2 N_{L/\mathbb{Q}}(\mathfrak{p})^{-2s}\right)^{-1} & \text{if } \mathfrak{p} \nmid \mathfrak{c} \text{ and } \mathfrak{p} \nmid \mathfrak{b} \\ \left(1 - \lambda_f(\mathfrak{p})\eta(\mathfrak{p}) N_{L/\mathbb{Q}}(\mathfrak{p})^{-1/2-s}\right)^{-1} & \text{if } \mathfrak{p} \mid \mathfrak{c} \text{ and } \mathfrak{p} \nmid \mathfrak{b} \\ 1 & \text{otherwise.} \end{cases}$$

Notice that the definition of the local Euler factor $L_{\mathfrak{p}}(f \otimes \eta, s)$ depends on $\mathfrak{b}$, the modulus of the Dirichlet character we associated to η. The following proof of Proposition 5.11 is taken from [Hid5, Section 7.F]).

Proof of Proposition 5.11. Assume first that η is trivial and $\mathfrak{b} = \mathcal{O}_L$. Then it is easy to check that we may set

$$f \otimes |\cdot|^w(g) := |N_{L/\mathbb{Q}}(\mathcal{D}_{L/\mathbb{Q}})|^w |\det(g)|_{\mathbb{A}_L}^w f(g). \tag{5.11.2}$$

Now assume that $w = 0$. Then, in view of Lemma 5.9 we may assume that $\mathfrak{b}$ is the conductor of the Dirichlet character associated to η. If $\mathfrak{b} = \mathcal{O}_L$, then it is not hard to check that we may set

$$f \otimes \eta(g) := \eta(\delta)\eta(\det(g))f(g),$$

where $\delta \in \mathbb{A}_{Lf}^\times$ is a finite idèle such that $[\delta] = \mathcal{D}_{L/\mathbb{Q}}$.

Suppose that $\mathfrak{b} \neq \mathcal{O}_L$. Let b be a finite idèle with $[b] = \mathfrak{b}$. As in [Ram3, (5.5)] and [MurR, §2.3], define the fractional ideal $\Upsilon = b^{-1}\widehat{\mathcal{O}}_L$ of $\prod_{\mathfrak{p}_v \mid \mathfrak{b}} L_v \times \prod_{\mathfrak{p}_v \nmid \mathfrak{b}} \mathcal{O}_v$ by

$$\Upsilon = \left\{ t = (t_v) \in \prod_{\mathfrak{p}_v \mid \mathfrak{b}} L_v \times \prod_{\mathfrak{p}_v \nmid \mathfrak{b}} \mathcal{O}_v : \operatorname{ord}_v(t_v) \geq -\operatorname{ord}_v(b) \text{ whenever } \mathfrak{p}_v \mid \mathfrak{b} \right\}.$$

Let $\widetilde{\Upsilon} \leftrightarrow b^{-1}\widehat{\mathcal{O}}_L/\widehat{\mathcal{O}}_L$ be a set of representatives for Υ modulo $\widehat{\mathcal{O}}_L := \prod_v \mathcal{O}_v$. Elements of $\widetilde{\Upsilon}$ are quotients x/b with $x \in \widehat{\mathcal{O}}_L$ and we consider the subset $\widetilde{\Upsilon}^\times$ consisting of quotients x/b where $x \in \widehat{\mathcal{O}}_L^\times$. Denote by $\eta_\mathfrak{b} : \widetilde{\Upsilon} \to \mathbb{C}$ the map defined by setting

$$\eta_\mathfrak{b}(t) = \begin{cases} \eta(t) & \text{if } t \in \widetilde{\Upsilon}^\times \\ 0 & \text{otherwise.} \end{cases}$$

---

[3] but it is not necessarily a newform

78 Chapter 5. Generalities on Hilbert Modular Forms and Varieties

The mapping $\eta_\mathfrak{b}$ is well defined, for if $x/b \in b^{-1}\widehat{\mathcal{O}}_L^\times$ and if $y \in \widehat{\mathcal{O}}_L$, then

$$\eta_\mathfrak{b}\left(\frac{x}{b} + b\frac{y}{b}\right) = \eta_\mathfrak{b}\left(\frac{x}{b}\left(1 + b\frac{y}{x}\right)\right) = \eta_\mathfrak{b}\left(\frac{x}{b}\right)$$

by the assumption on the conductor of η and the fact that x is a unit. For each $t \in \widetilde{\Upsilon}$, define $u(t) \in G(\mathbb{A})$ by

$$u(t)_v = \begin{cases} \begin{pmatrix} 1 & t_v \\ 0 & 1 \end{pmatrix} & \text{if } v \nmid \infty \text{ and } \operatorname{ord}_v(\beta) \geq 1 \\ \begin{pmatrix} 1 & 0 \\ 0 & 1 \end{pmatrix} & \text{otherwise.} \end{cases}$$

Finally, for $g \in G(\mathbb{A})$ define

$$h(g) := \eta(\det(g)) \sum_{t \in \widetilde{\Upsilon}} \eta_\mathfrak{b}(t) f(gu(t)). \tag{5.11.3}$$

We claim that:
(a) $h \in M_\kappa(K_0(\mathfrak{cb}^2), \chi\eta^2)$ and
(b) the function

$$f \otimes \eta := G(\eta^{-1})^{-1} h \tag{5.11.4}$$

has the desired Fourier expansion, where $G(\eta^{-1})$ is the Gauss sum

$$G(\eta^{-1}) := \sum_{t \in \widetilde{\Upsilon}} \eta(\delta^{-1}) \eta_\mathfrak{b}(t) e_L(\delta^{-1}t) \tag{5.11.5}$$

with $\delta \in \mathbb{A}_{L,f}^\times$ a finite idèle such that $[\delta] = \mathcal{D}_{L/\mathbb{Q}}$.

For part (a), let $w = \begin{pmatrix} a & b \\ c & d \end{pmatrix} \in K_0(\mathfrak{cb}^2)$. We need to check that

$$h(gw) = \chi_0(w)\eta_0^2(w)h(g) = \chi(d)\eta^2(d)h(g).$$

Define $W \in G(\mathbb{A})$ so that $W_v = w_v$ if $\mathfrak{p}_v \nmid \mathfrak{b}$ and

$$w_v.u(t_v).W_v^{-1} = \begin{pmatrix} 1 & a_v d_v^{-1} t_v \\ 0 & 1 \end{pmatrix} \quad \text{if } \mathfrak{p}_v | \mathfrak{b}$$

which gives

$$W = \begin{pmatrix} a - \frac{act}{d} & b - \frac{act^2}{d} \\ c & d + ct \end{pmatrix} = w + \begin{pmatrix} -\frac{act}{d} & -\frac{act^2}{d} \\ 0 & ct \end{pmatrix}.$$

The assumptions on $\operatorname{ord}_v(t)$ plus the assumption that $w \in K_0(\mathfrak{cb}^2)$ guarantee that ct and ct^2 are integral, from which it follows that $W \in K_0(\mathfrak{c})$. Consequently

$$h(gw) = \eta(\det(w))\eta(\det(g)) \sum_{t \in \widetilde{\Upsilon}^\times} \eta(t) f(gwu(t)W^{-1})\chi_0(W)$$

$$= \eta(a_\mathfrak{b} d_\mathfrak{b})\eta(\det(g)) \sum_{t \in \widetilde{\Upsilon}^\times} \eta(t) f(gu(ad^{-1}t))\chi(d)$$

$$= \eta(a_\mathfrak{b} d_\mathfrak{b})\eta(\det(g)) \sum_{s \in \widetilde{\Upsilon}^\times} \eta(a_\mathfrak{b}^{-1} d_\mathfrak{b} s) f(gu(s))\chi(d_\mathfrak{b}) = \eta(d_\mathfrak{b})^2 \chi(d_\mathfrak{b}) h(g)$$

which proves the claim. Here $x_\mathfrak{b}$ is the projection of $x \in \mathbb{A}_{L,f}$ to $\prod_{\mathfrak{p}_v | \mathfrak{b}} F_v$.

5.11. Twisting

For part (b) we must determine the Fourier coefficients of $f \otimes \eta := G(\eta^{-1})^{-1}h$. Let $y_{\mathfrak{b}} = (y_{\mathfrak{b},v})$ be an idèle with $y_{\mathfrak{b},v} = y_v$ if the finite place v divides $\mathfrak{b}$ and $y_{\mathfrak{b},v} = 1$ otherwise. It suffices to show that for any $y \in \mathbb{A}_{Lf}^{\times}$ the Fourier coefficients of h are given by

$$a(y,h) = \begin{cases} G(\eta^{-1})\eta(yy_{\mathfrak{b}}^{-1})a(y,f) & \text{if } y_{\mathfrak{b}} \in \otimes_{\mathfrak{p}_v|\mathfrak{b}}\mathcal{O}_{L,v}^{\times} \\ 0 & \text{otherwise} \end{cases} \quad (5.11.6)$$

$$a_i(y,h) = 0$$

for $i \in \{0,1\}$. To obtain the Fourier expansion of h we start with the Fourier expansion of f and substitute it into equation (5.11.3). If $g = \begin{pmatrix} y & x \\ 0 & 1 \end{pmatrix}$ then

$$f(gu(t)) = |y|_{\mathbb{A}_L}\left(a_0(y\mathcal{D}_{L/\mathbb{Q}},f)|y_\infty^{-m}| + a_1(y\mathcal{D}_{L/\mathbb{Q}},f)|y_\infty^{-m}| \right.$$
$$\left. + \sum_{\xi \gg 0} a(\xi y \mathcal{D}_{L/\mathbb{Q}}, f) q_\kappa(\xi x + \xi yt, \xi y) \right)$$

so that $a_i(y,h) = 0$ for all y and $i \in \{0,1\}$ and

$$a(\xi y \mathcal{D}_{L/\mathbb{Q}}, h) = \eta(y) \sum_{t \in \widetilde{\Upsilon}} \eta_\mathfrak{b}(t) a(\xi y \mathcal{D}_{L/\mathbb{Q}}) e_L(\xi yt).$$

Changing variables and using the $L^\times$-invariance of η we obtain

$$a(y,h) = \eta(\xi^{-1}y\delta^{-1}) \sum_{t \in \widetilde{\Upsilon}} \eta_\mathfrak{b}(t) a(y,f) e_L(yt\delta^{-1})$$

$$= \left(\eta(y\delta^{-1}) \sum_{t \in \widetilde{\Upsilon}} \eta_\mathfrak{b}(t) e_L(yt\delta^{-1}) \right) a(y,f)$$

Suppose that $y_\mathfrak{b} \in \otimes_{\mathfrak{p}_v|\mathfrak{b}}\mathcal{O}_{L_v}^\times$. Then $\{y_\mathfrak{b} t : t \in \widetilde{\Upsilon}\}$ is a minimal set of representatives for Υ, and hence

$$\eta(y\delta^{-1}) \sum_{t \in \widetilde{\Upsilon}} \eta_\mathfrak{b}(t) e_L(yt\delta^{-1}) = G(\eta^{-1})\eta(yy_\mathfrak{b}^{-1})$$

if $y \in \widehat{\mathcal{O}}_L - 0$ and $y_\mathfrak{b} \in \otimes_{\mathfrak{p}_v|\mathfrak{b}}\mathcal{O}_{L_v}^\times$. Thus the claim (5.11.6) and hence the proposition follows in this case. Suppose on the other hand that $y_\mathfrak{b} \notin \otimes_{\mathfrak{p}_v|\mathfrak{b}}\mathcal{O}_{L_v}^\times$. In this case we show that

$$\sum_{t \in \widetilde{\Upsilon}} \eta_\mathfrak{b}(t) e_L(yt\delta^{-1}) \quad (5.11.7)$$

is equal to zero which will complete the proof of (5.11.6) and hence complete the proof of the proposition.

80 Chapter 5. Generalities on Hilbert Modular Forms and Varieties

The expression (5.11.7) admits a factorization into a product indexed by the primes dividing $\mathfrak{b}$. Using this factorization (which we will not make precise) one uses the Chinese remainder theorem to reduce our claim that (5.11.7) is zero to the special case where $\mathfrak{b} = \mathfrak{p}^i$, $i > 0$ is divisible by a single prime $\mathfrak{p}$ of $\mathcal{O}_L$ and hence $y_\mathfrak{b} = \mathfrak{p}^j$ for some $j > 0$. We henceforth assume that we are in this special case.

Let $b \in \widehat{\mathcal{O}}_L$ be an idèle whose associated ideal is $\mathfrak{b}$, thus $[b] = \mathfrak{b} = \mathfrak{p}^i$. Let
$$\widetilde{\Upsilon}_1, \widetilde{\Upsilon}_0 \subseteq \widetilde{\Upsilon}$$
be sets of representatives for $y_\mathfrak{b}^{-1}\widehat{\mathcal{O}}_L/\widehat{\mathcal{O}}_L$ and $b^{-1}\widehat{\mathcal{O}}_L/y_\mathfrak{b}^{-1}\widehat{\mathcal{O}}_L$, respectively.

If $y_\mathfrak{b}\widehat{\mathcal{O}}_L \subseteq b\widehat{\mathcal{O}}_L$ then we take $\widetilde{\Upsilon}$ to be $\{0\}$. We then have that

$$\sum_{t \in \widetilde{\Upsilon}} \eta_\mathfrak{b}(t) e_L(y_\mathfrak{b} t \delta^{-1}) = \sum_{t_0 \in \widetilde{\Upsilon}_0} \sum_{t_1 \in \widetilde{\Upsilon}_1} \eta_\mathfrak{b}(t_0 + t_1) e_L(y_\mathfrak{b}(t_0 + t_1)\delta^{-1}) \qquad (5.11.8)$$

$$= \sum_{t_0 \in \widetilde{\Upsilon}_0} \sum_{t_1 \in \widetilde{\Upsilon}_1} \eta_\mathfrak{b}(t_0 + t_1) e_L(y_\mathfrak{b} t_0 \delta^{-1})$$

$$= \sum_{t_0 \in \widetilde{\Upsilon}_0} e_L(y_\mathfrak{b} t_0 \delta^{-1}) \sum_{t_1 \in \widetilde{\Upsilon}_1} \eta_\mathfrak{b}(t_0 + t_1).$$

At this point we note that if $y_\mathfrak{b}\widehat{\mathcal{O}}_L \subseteq b\widehat{\mathcal{O}}_L$ (i.e., $j \geq i$) then we may take $\widetilde{\Upsilon}_0 = \{0\}$. Thus in this case the sum (5.11.8) vanishes by orthogonality of characters and hence (5.11.7) vanishes as well.

We henceforth assume that
$$y_\mathfrak{b}\widehat{\mathcal{O}}_L \supsetneq b\widehat{\mathcal{O}}_L; \qquad (5.11.9)$$

i.e., $j < i$. By definition of $\eta_\mathfrak{b}$, in order for the summand indexed by t_0 in the sum (5.11.8) to be nonzero, we must have $t_0 \in b^{-1}\widehat{\mathcal{O}}_L^\times$. Moreover, if $t_0 \in b^{-1}\widehat{\mathcal{O}}_L^\times$, then $t_0 + t_1 \in b^{-1}\widehat{\mathcal{O}}_L^\times$ for all $t_1 \in y_\mathfrak{b}^{-1}\mathcal{O}_{L_v}/\mathcal{O}_{L_v}$. Therefore to show that the sum above is nonzero it suffices to show that

$$\sum_{t_1 \in \widetilde{\Upsilon}_1} \eta_\mathfrak{b}(t_0 + t_1) \qquad (5.11.10)$$

is zero for all $t_0 \in b^{-1}\widehat{\mathcal{O}}_L^\times$. Write $t_0 = \frac{a_0}{b}$. Then the sum in (5.11.10) above is equal to $\eta(b^{-1})$ times

$$\sum_{m \in \widehat{\mathcal{O}}_L/y_\mathfrak{b}\widehat{\mathcal{O}}_L} \eta(a_0 + mby_\mathfrak{b}^{-1}) = |y_\mathfrak{b}\widehat{\mathcal{O}}_L/b\widehat{\mathcal{O}}_L|^{-1} \sum_{m \in \widehat{\mathcal{O}}_L/b\widehat{\mathcal{O}}_L} \eta(a_0 + m(by_\mathfrak{b}^{-1}))$$
$$(5.11.11)$$
$$= |\mathfrak{p}^j\widehat{\mathcal{O}}_L/\mathfrak{p}^i\widehat{\mathcal{O}}_L|^{-1} \sum_{m \in \widehat{\mathcal{O}}_L/\mathfrak{p}^i\widehat{\mathcal{O}}_L} \eta(a_0 + m(\mathfrak{p}^{i-j})) =: S.$$

Here the equality is due to the fact that η is trivial on $1 + b\widehat{\mathcal{O}}_L$. We now follow the discussion after equation (3.9) of Chapter 3 in [Iwan]. Let b_1 be a generator for the ideal

$$b(y_{\mathfrak{b}}^{-1}b\widehat{\mathcal{O}}_L + b\widehat{\mathcal{O}}_L)((y_{\mathfrak{b}}^{-1}b)^2\widehat{\mathcal{O}}_L + b\widehat{\mathcal{O}}_L)^{-1} = \mathfrak{p}^{i+i-j-\min(2(i-j),i)}$$

The expression S above has the property that $\eta(1 + b_1 x)S = S$ for any $x \in \widehat{\mathcal{O}}_L$. It follows that either $S = 0$ or $b_1\widehat{\mathcal{O}}_L = \mathfrak{b}\widehat{\mathcal{O}}_L$, the conductor of η. If the latter holds, we deduce that $\mathfrak{b}\widehat{\mathcal{O}}_L \supseteq b_1\widehat{\mathcal{O}}_L$, contrary to our assumption (5.11.9). Thus $S = 0$, which implies (5.11.10) is zero. Since t_0 was arbitrary, this implies that (5.11.8) and hence (5.11.7) is zero, which completes the proof of the proposition. □

Cohomological normalization. The same procedure may be used to construct twists of modular forms $f \in M_\kappa^{\mathrm{coh}}$ rather than M_κ:

Definition 5.13. If $f \in M_\kappa^{\mathrm{coh}}(K_0(\mathfrak{c}), \chi)$ define $f \otimes \eta \in M_\kappa^{\mathrm{coh}}(K_0(\mathfrak{c}\mathfrak{b}^2), \chi\eta^2)$ by

$$(f \otimes \eta)^{-\iota} := f^{-\iota} \otimes \eta$$

Lemma 5.14. *If $f \in M_\kappa^{\mathrm{coh}}(K_0(\mathfrak{c}), \chi)$ and if η is a Hecke character as above, then*

$$f \otimes \eta = G(\eta^{-1})^{-1} h$$

where h is given by the following modification of equation (5.11.3):

$$h(g) = \eta(\det g)^{-1} \sum_{t \in \widetilde{\Upsilon}} \eta_{\mathfrak{b}}(t) f(gu(t)). \tag{5.11.12}$$

□

5.12 L-functions

In this section we define the various L-functions attached to automorphic forms on GL_2. The L-functions that arise from the periods we will later consider are imprimitive (i.e., they do not coincide with the associated "canonical" Langlands L-functions at all places). We will normalize everything so that the local L-factors are the same as the canonical Langlands L-factors at all unramified places. We warn the reader that the automorphic representation $\pi(f)$ attached to a simultaneous eigenform $f \in S_\kappa(K_0(\mathfrak{c}), \chi)$ is in general not unitary. Rather, the automorphic representation

$$\pi(f) | \cdot |_{\mathbb{A}_L}^w$$

is unitary, where $w = (k_\sigma + 2m_\sigma)/2$ (which is independent of σ). Throughout this chapter, for any L-function

$$L(s) = \sum_{\mathfrak{n} \subset \mathcal{O}_L} a(\mathfrak{n}) \mathrm{N}_{L/\mathbb{Q}}(\mathfrak{n})^{-s}$$

and any ideal $\mathfrak{c} \subset \mathcal{O}_L$, write

$$L_\mathfrak{c}(s) := a(\mathcal{O}_L) + \sum_{\mathfrak{n} = \prod_{\mathfrak{p}|\mathfrak{c}} \mathfrak{p}^{i(\mathfrak{p})}} a(\mathfrak{n}) \mathrm{N}_{L/\mathbb{Q}}(\mathfrak{n})^{-s}$$

$$L^\mathfrak{c}(s) := \sum_{\mathfrak{n} + \mathfrak{c} = \mathcal{O}_L} a(\mathfrak{n}) \mathrm{N}_{L/\mathbb{Q}}(\mathfrak{n})^{-s}.$$

Moreover, $L^{*,\mathfrak{c}}(s) =$ "Gamma factors" $\times L^\mathfrak{c}(s)$ will denote the product of the partial L-function with its Gamma factors.

To ease notation we write $\kappa = (k, m) \in \mathcal{X}(L)$ and write $[k + 2m]$ for the integer $k_\sigma + 2m_\sigma$, which is independent of $\sigma \in \Sigma(L)$. If χ (resp. π) is an character (resp. automorphic representation) we write $\mathfrak{f}(\chi)$ (resp. $\mathfrak{f}(\pi)$) for its conductor, considered as an ideal of $\mathcal{O}_L$.

5.12.1 The standard L-function

For any cusp form $f \in S_\kappa(K_0(\mathfrak{c}), \chi)$ the standard L-function is defined in equation (5.9.6) which we reproduce here:

$$L(f, s) := \sum_{\mathfrak{m} \subset \mathcal{O}_L} a(\mathfrak{m}, f) \mathrm{N}_{L/\mathbb{Q}}(\mathfrak{m})^{-1/2-s}.$$

If f is an eigenfunction of all Hecke operators then for any Hecke character η, the twist $f \otimes \eta$ is also an eigenfunction of all Hecke operators and its L-function admits an Euler product over prime ideals:

$$L(f \otimes \eta, s) := \prod_\mathfrak{p} L_\mathfrak{p}(f \otimes \eta, s)$$

where (see equation (5.11.1)) $L_\mathfrak{p}(f \otimes \eta, s)$ equals

$$\begin{cases} \left(1 - \lambda_f(\mathfrak{p})\eta(\mathfrak{p})\mathrm{N}_{L/\mathbb{Q}}(\mathfrak{p})^{-1/2-s} + \chi(\mathfrak{p})\eta(\mathfrak{p})^2 \mathrm{N}_{L/\mathbb{Q}}(\mathfrak{p})^{-2s}\right)^{-1} & \text{if } \mathfrak{p} \nmid \mathfrak{c} \\ \left(1 - \lambda_f(\mathfrak{p})\eta(\mathfrak{p})\mathrm{N}_{L/\mathbb{Q}}(\mathfrak{p})^{-1/2-s}\right)^{-1} & \text{if } \mathfrak{p} \mid \mathfrak{c} \text{ and } \mathfrak{p} \nmid \mathfrak{f}(\eta) \\ 1 & \text{otherwise.} \end{cases}$$

For each prime $\mathfrak{p} \subset \mathcal{O}_L$ choose, once and for all, $\alpha_{1,\mathfrak{p}}(f), \alpha_{2,\mathfrak{p}}(f) \in \mathbb{C}$ such that

$$L_\mathfrak{p}(f, s) = (1 - \alpha_{1,\mathfrak{p}}(f) \mathrm{N}_{L/\mathbb{Q}}(\mathfrak{p})^{-s})(1 - \alpha_{2,\mathfrak{p}}(f) \mathrm{N}_{L/\mathbb{Q}}(\mathfrak{p})^{-s}).$$

Let $\pi(f)$ be the cuspidal automorphic representation generated by f. Let $\mathfrak{f}(\pi(f))$ and $\mathfrak{f}(\eta)$ denote the corresponding conductors. At a finite place $v(\mathfrak{p})$ associated to a prime $\mathfrak{p} \nmid \mathfrak{f}(\pi(f))\mathfrak{f}(\eta)$, we have that

$$\pi(f)_{v(\mathfrak{p})} \otimes \eta_{v(\mathfrak{p})}$$

5.12. L-functions

is in the (unramified) principal series, with Satake parameters $\eta(\mathfrak{p})\alpha_{1,\mathfrak{p}}(f)$ and $\eta(\mathfrak{p})\alpha_{2,\mathfrak{p}}(f)$. It follows that

$$L_\mathfrak{p}(f \otimes \eta, s) := L(\pi(f)_{v(\mathfrak{p})} \otimes \eta_{v(\mathfrak{p})}, s)$$
$$= \left(1 - \eta(\mathfrak{p})\alpha_{1,\mathfrak{p}}(f)\mathrm{N}_{L/\mathbb{Q}}(\mathfrak{p})^{-s}\right)^{-1}\left(1 - \eta(\mathfrak{p})\alpha_{2,\mathfrak{p}}(f)\mathrm{N}_{L/\mathbb{Q}}(\mathfrak{p})^{-s}\right)^{-1}$$

for $\mathfrak{p} \nmid \mathfrak{c}\mathfrak{f}(\eta)$ (see Theorem E.4).

The automorphic representation $\pi(f)|\cdot|_{\mathbb{A}_L}^{[k+2m]/2}$ is unitary. Applying Theorem 5.3 of [JaS1], this implies that the partial Euler product

$$L^{\mathfrak{f}(\pi(f))\mathfrak{f}(\theta)}(f \otimes \eta, s)$$

is absolutely convergent for $\mathrm{Re}(s) \geq 1 - [k+2m]/2$. Actually, more is true:

Lemma 5.15. *If $f \in S_\kappa^{\mathrm{new}}(K_0(\mathfrak{c}), \chi)$ is a newform, then the Dirichlet series defining $L(f \otimes \eta, s)$ is absolutely convergent for $\mathrm{Re}(s) \geq 1 - [k+2m]/2$.*

Proof. In view of the previous paragraph, it suffices to show that $L_\mathfrak{p}(f \otimes \eta, s)$ is absolutely convergent for $\mathfrak{p} \mid \mathfrak{f}(\eta)\mathfrak{f}(\pi(f))$ for $\mathrm{Re}(s) \geq 1 - [k+2m]/2$. To see this, we apply [She, Theorem 3.3] (a generalization of some results of Atkin-Lehner-Li theory to the Hilbert modular case) to conclude that $|a(f, \mathfrak{p})| \leq \mathrm{N}_{L/\mathbb{Q}}(\mathfrak{p})^{-1/2+[k+2m]/2+1}$. This implies the desired convergence. □

5.12.2 Rankin-Selberg L-functions

Let $f \in S_\kappa(K_0(\mathfrak{c}), \chi)$ and $g \in S_\kappa(K_0(\mathfrak{c}'), \chi')$ be simultaneous eigenforms for all Hecke operators. The Rankin-Selberg L-function attached to f and g is defined by

$$L(f \times g, s) := L^{\mathfrak{c}\mathfrak{c}'}(\chi\chi', 2s) \sum_{\mathfrak{n} \subset \mathcal{O}_L} a(\mathfrak{n}, f)a(\mathfrak{n}, g)\mathrm{N}_{L/\mathbb{Q}}(\mathfrak{n})^{-1-s}.$$

The partial L-functions $L^{\mathfrak{c}\mathfrak{c}'}(f \times g, s)$ admit the following Euler product expansion:

$$L^{\mathfrak{c}\mathfrak{c}'}(f \times g, s) = \prod_{\substack{\text{prime ideals } \mathfrak{p} \subset \mathcal{O}_L \\ \mathfrak{p} \nmid \mathfrak{c}\mathfrak{c}'}} L_\mathfrak{p}(f \times g, s),$$

where

$$L_\mathfrak{p}(f \times g, s) = \prod_{i=1}^{2}\prod_{j=1}^{2}(1 - \alpha_{i,\mathfrak{p}}(f)\alpha_{j,\mathfrak{p}}(g)\mathrm{N}_{L/\mathbb{Q}}(\mathfrak{p})^{-s})^{-1}.$$

(see [HidT, (7.7)]).

It follows in particular that $L_\mathfrak{p}(f \times g, s) = L(\pi_v(f) \boxtimes \pi_v(g), s)$ for finite places $v \nmid \mathfrak{c}\mathfrak{c}'$. Thus the partial Euler product $L^{\mathfrak{c}\mathfrak{c}'}(f \times g, s)$ is absolutely convergent for $\mathrm{Re}(s) > 1 - [k+2m]$ [JaS1, Theorem 5.3]. Moreover, $L^{\mathfrak{c}\mathfrak{c}'}(f \times g, s)$ has a pole at $s = 1 - [k+2m]$ if and only if the automorphic representations $\pi(f)$ and $\pi(g)$ spanned by f and g satisfy $\pi(f) \cong \pi(g)^\vee |\cdot|^{-[k+2m]}$, where $\pi(g)^\vee$ is the contragredient of $\pi(g)$. The pole, if it exists, is simple (see [JaS2, Proposition 3.6]).

5.12.3 Adjoint L-functions

We now recall the adjoint L-functions attached to an eigenform $f \in S_\kappa^{\text{new}}(K_0(\mathfrak{c}), \chi)$. For a unitary Hecke character $\phi : L^\times \backslash \mathbb{A}_L^\times \to \mathbb{C}^\times$, set

$$L(\mathrm{Ad}(f) \otimes \phi, s) := \prod_{\mathfrak{p} \nmid \mathfrak{f}(\phi)} L_\mathfrak{p}(\mathrm{Ad}(f) \otimes \phi, s),$$

where $\mathfrak{f}(\phi)$ is the conductor of ϕ and where $L_\mathfrak{p}(\mathrm{Ad}(f) \otimes \phi, s)^{-1}$ is defined to be

$$(1 - \phi(\mathfrak{p}) \mathrm{N}_{L/\mathbb{Q}}(\mathfrak{p})^{-s})(1 - \phi(\mathfrak{p}) \frac{\alpha_{1,\mathfrak{p}}(f)}{\alpha_{2,\mathfrak{p}}(f)} \mathrm{N}_{L/\mathbb{Q}}(\mathfrak{p})^{-s})(1 - \phi(\mathfrak{p}) \frac{\alpha_{2,\mathfrak{p}}(f)}{\alpha_{1,\mathfrak{p}}(f)} \mathrm{N}_{L/\mathbb{Q}}(\mathfrak{p})^{-s})$$

if $\mathfrak{p} \nmid \mathfrak{c}$ and, in the ramified cases, we set $L_\mathfrak{p}(\mathrm{Ad}(f) \otimes \phi, s)^{-1}$ equal to

$$\begin{cases} 1 - \phi(\mathfrak{p}) \mathrm{N}_{L/\mathbb{Q}}(\mathfrak{p})^{-s} & \text{if } \mathfrak{p} \nmid \mathfrak{f}(\phi) \text{ and } \pi(f)_{v(\mathfrak{p})} \text{ is principal and minimal} \\ 1 - \phi(\mathfrak{p}) \mathrm{N}_{L/\mathbb{Q}}(\mathfrak{p})^{-1-s} & \text{if } \mathfrak{p} \nmid \mathfrak{f}(\phi) \text{ and } \pi(f)_{v(\mathfrak{p})} \text{ is special and minimal} \\ 1 & \text{otherwise} \end{cases}$$

(see [HidT, Section 7] and the corrections in [Gh, Section 5.1]). Following [HidT], we say that an admissible representation π of $\mathrm{GL}_2(L_v)$ is *minimal* if $\mathfrak{f}(\pi) \supseteq \mathfrak{f}(\pi \otimes \xi)$ for all quasicharacters $\xi : L_v^\times \to \mathbb{C}^\times$. We note that if $\pi(f) = \otimes_v' \pi(f)_v$ is the cuspidal automorphic representation attached to f and $\mathfrak{p} \nmid \mathfrak{c}$, then

$$L_\mathfrak{p}(\mathrm{Ad}(f), s) = L(\mathrm{Ad}(\pi(f)_{v(\mathfrak{p})}), s)$$

where $v(\mathfrak{p})$ is the place associated to $\mathfrak{p}$.

For $\sigma \in \Sigma(L)$ define $L_\sigma(\mathrm{Ad}(f) \otimes \phi, s)$ to be the following Γ factor:

$$(2\pi)^{-(s+k_\sigma+1)} \Gamma(s + k_\sigma + 1) \pi^{-(s+1)/2} \Gamma((s+1)/2).$$

For any ideal $\mathfrak{c}' \subset \mathcal{O}_L$ let

$$L^{*,\mathfrak{c}'}(\mathrm{Ad}(f) \otimes \phi, s) := L^{\mathfrak{c}'}(\mathrm{Ad}(f) \otimes \phi, s) \prod_{\sigma \in \Sigma(L)} L_\sigma(\mathrm{Ad}(f) \otimes \phi, s)$$

be the completed L-function. In the case that $\mathfrak{c}' = \mathcal{O}_L$, write

$$L^*(\mathrm{Ad}(f) \otimes \phi, s) = L^{*, \mathcal{O}_L}(\mathrm{Ad}(f) \otimes \phi, s).$$

The following theorem, Theorem 7.1 of [HidT] (see also [Hid5, (7.2c)]), will be crucial in the proof of Theorem 10.1 below:

Theorem 5.16 (Hida-Tilouine). *If $f \in S_\kappa(K_0(\mathfrak{c}), \chi)$ is a newform, then*

$$(f, f)_P = d_{L/\mathbb{Q}}^{[k+2m]+2} \mathrm{N}_{L/\mathbb{Q}}(\mathfrak{c}) 2^{-\{k+21\}+1} L^*(\mathrm{Ad}(f), 1),$$

where $\{k + 21\} := \sum_{\sigma \in \Sigma(L)} k_\sigma + 2$.

5.12. L-functions

Here $(f,g)_P$ is the Petersson inner product, normalized as in Section 5.7. For our later convenience, if $\phi : L^\times \backslash \mathbb{A}_L^\times \to \mathbb{C}^\times$ is a character, let

$$L^{\mathfrak{cf}(\theta)}(\operatorname{Sym}^2(f) \otimes \phi, s) := \prod_{\mathfrak{p} \nmid \mathfrak{cf}(\theta)} L_\mathfrak{p}(\operatorname{Sym}^2(f) \otimes \phi, s),$$

where $L_\mathfrak{p}(\operatorname{Sym}^2(f) \otimes \phi, s)$ is the local Euler factor given by

$$(1 - \phi(\mathfrak{p})\chi(\mathfrak{p})\mathrm{N}_{L/\mathbb{Q}}(\mathfrak{p})^{-s})(1 - \phi(\mathfrak{p})\alpha_{1,\mathfrak{p}}(f)^2 \mathrm{N}_{L/\mathbb{Q}}(\mathfrak{p})^{-s})(1 - \phi(\mathfrak{p})\alpha_{2,\mathfrak{p}}(f)^2 \mathrm{N}_{L/\mathbb{Q}}(\mathfrak{p})^{-s})$$

One checks immediately that

$$L_\mathfrak{p}(\operatorname{Sym}^2(f) \otimes (\chi)^{-1}, s) = L_\mathfrak{p}(\operatorname{Ad}(f), s) \qquad (5.12.1)$$

if χ and f are unramified at $\mathfrak{p}$.

5.12.4 Asai L-functions

Let L/E be a quadratic extension of totally real number fields and let $\langle \varsigma \rangle = \operatorname{Gal}(L/E)$. Using class field theory, identify $\langle \eta \rangle = \operatorname{Gal}(L/E)^\wedge$ with a Hecke character $\eta : E^\times \backslash \mathbb{A}_E^\times \to \mathbb{C}^\times$ trivial at the infinite places, see Section 8.1. Let $\kappa = (k, m) \in \mathcal{X}(E)$ be a weight, and define $\widehat{\kappa} = (\widehat{k}, \widehat{m}) \in \mathcal{X}(L)$ by declaring that $\widehat{k}_{\widehat{\sigma}} = k_\sigma$ (resp. $\widehat{m}_{\widehat{\sigma}} = m_\sigma$) if $\widehat{\sigma} : L \hookrightarrow \mathbb{R}$ extends $\sigma : E \hookrightarrow \mathbb{R}$. Let $\chi_E : E^\times \backslash \mathbb{A}_E^\times \to \mathbb{C}^\times$ be a quasicharacter satisfying $\chi_{E\infty}(b_\infty) = b_\infty^{-k-2m}$ for $b \in \mathbb{A}_E^\times$, and set $\chi := \chi_E \circ \mathrm{N}_{L/E}$.

For any $f \in S_{\widehat{\kappa}}(K_0(\mathfrak{c}), \chi)$ and any quasicharacter $\phi : L^\times \backslash \mathbb{A}_L^\times \to \mathbb{C}^\times$, the Asai L-function $L(\operatorname{As}(f \otimes \phi), s)$ attached to f and ϕ is defined to be the following sum over ideals $\mathfrak{n} \subset \mathcal{O}_E$,

$$L^{(\mathfrak{f}(\phi)\mathfrak{c}) \cap \mathcal{O}_E}((\phi^2 \chi)|_E, 2s) \sum_{\mathfrak{n} + \mathfrak{f}(\phi) = \mathcal{O}_E} \phi(\mathfrak{n}\mathcal{O}_L) a(\mathfrak{n}\mathcal{O}_L, f) \mathrm{N}_{E/\mathbb{Q}}(\mathfrak{n})^{-1-s},$$

where $(\phi^2\chi)|_E$ denotes the restriction of $\phi^2\chi$ to $\mathbb{A}_E^\times$ (see [Ram2] for a nice discussion of this L-function). The associated partial L-function admits an Euler product

$$L^{(\mathfrak{f}(\phi)\mathfrak{c}) \cap \mathcal{O}_E}(\operatorname{As}(f \otimes \phi), s) = \prod_{\mathfrak{p} \nmid (\mathfrak{c} \cap \mathcal{O}_E)} L_\mathfrak{p}(\operatorname{As}(f \otimes \phi), s)$$

where, if $\mathfrak{p} \nmid d_{L/E}((\mathfrak{f}(\theta)\mathfrak{c}) \cap \mathcal{O}_E)$, we have that $L_\mathfrak{p}(\operatorname{As}(f \otimes \phi), s)^{-1}$ is equal to

$$\prod_{i=1}^{2} \prod_{j=1}^{2} \left(1 - \phi(\mathfrak{p}) \alpha_{i,\mathfrak{P}}(f) \alpha_{j,\overline{\mathfrak{P}}}(f) \mathrm{N}_{E/\mathbb{Q}}(\mathfrak{p})^{-s} \right)$$

if $\mathfrak{p} = \mathfrak{P}\overline{\mathfrak{P}}$ splits, and

$$\left(1 - \phi(\mathfrak{p})\alpha_{1,\mathfrak{p}}(f)\mathrm{N}_{E/\mathbb{Q}}(\mathfrak{p})^{-s}\right)\left(1 - \phi(\mathfrak{p})\alpha_{2,\mathfrak{p}}(f)\mathrm{N}_{E/\mathbb{Q}}(\mathfrak{p})^{-s}\right)$$
$$\times \left(1 - \phi(\mathfrak{p})^2 \chi(\mathfrak{p}) \mathrm{N}_{E/\mathbb{Q}}(\mathfrak{p})^{-2s}\right)$$

if $\mathfrak{p}$ is inert (see [HarL, Section 2]). Here we have abused notation and set $\phi(\mathfrak{p}) := \phi(\mathfrak{p}\mathcal{O}_L)$ and $\chi(\mathfrak{p}) := \chi(\mathfrak{p}\mathcal{O}_L)$.

Suppose that $f \in S_{\widehat{\kappa}}^{\mathrm{new}}(K_0(\mathfrak{c}), \chi)$ is a newform. Then there is another newform $f^{\varsigma} \in S_{\widehat{\kappa}}^{\mathrm{new}}(K_0(\mathfrak{c}'), \chi)$ uniquely determined by the fact that for $\mathfrak{m} \subset \mathcal{O}_L$ coprime to $\mathfrak{cc}'$ we have

$$a(f^{\varsigma}, \mathfrak{m}) = a(f, \varsigma(\mathfrak{m})),$$

where $\mathfrak{c}' \subset \mathcal{O}_L$ is some other ideal. We see directly from the local Euler factors given above that

$$L^{\mathcal{D}_{L/E}\mathfrak{c}}(f \times f^{\varsigma}, s) = L^{d_{L/E}(\mathfrak{c} \cap \mathcal{O}_E)}(\mathrm{As}(f), s) L^{d_{L/E}(\mathfrak{c} \cap \mathcal{O}_E)}(\mathrm{As}(f) \otimes \eta, s). \quad (5.12.2)$$

Remark. Ramakrishnan shows in [Ram2] that there is an isobaric automorphic representation $\mathrm{As}(\pi(f))$ of $\mathrm{Res}_{E/\mathbb{Q}}(\mathrm{GL}_4)$ whose L-function is equal to $L(\mathrm{As}(f), s)$ up to finitely many Euler factors. Similarly, in [Ram1], he shows that there is an isobaric automorphic representation $\pi(f) \boxtimes \pi(f^{\varsigma})$ of $\mathrm{Res}_{L/\mathbb{Q}}(\mathrm{GL}_4)$ whose L-function is equal to $L(f \times f^{\varsigma}, s)$ up to finitely many places. One can show that $\pi(f) \boxtimes \pi(f^{\varsigma})$ is the base change of $\mathrm{As}(\pi(f))$ to $\mathrm{Res}_{L/\mathbb{Q}}(\mathrm{GL}_4)$; this is a substantial refinement of (5.12.2).

Let $\mathfrak{c}_E \subset \mathcal{O}_E$ be an ideal and let $f \in S_\kappa(K_0(\mathfrak{c}_E), \chi_E)$. As in the Introduction, we write $\widehat{f}$ for the unique newform on L generating the automorphic representation of $\mathrm{GL}_2(\mathbb{A}_L)$ that is the base change to L of the automorphic representation of $\mathrm{GL}_2(\mathbb{A}_E)$ generated by f. Thus $\widehat{f} \in S_{\widehat{\kappa}}^{\mathrm{new}}(K_0(\mathfrak{c}), \chi)$ for some ideal $\mathfrak{c} \subset \mathcal{O}_L$. (*Note:* it is not necessarily true that $\mathfrak{c} \cap \mathcal{O}_E = \mathfrak{c}_E$.) We say that a newform $g \in S_{\widehat{\kappa}}^{\mathrm{new}}(K_0(\mathfrak{c}), \chi)$ is a *base change* from E if $g = \widehat{f}$ for some newform f on E. We will require the following proposition in the proof of Theorem 10.1 below.

Proposition 5.17. *Fix a quasi-character* $\vartheta : L^\times \backslash \mathbb{A}_L^\times \to \mathbb{C}^\times$ *(resp.* $\theta : L^\times \backslash \mathbb{A}_L^\times \to \mathbb{C}^\times$*) such that* $\vartheta| \cdot |_{\mathbb{A}_L}^{[k+2m]/2}$ *(resp.* $\theta| \cdot |_{\mathbb{A}_L}^{[k+2m]/2}$*) is unitary and its restriction to* $E^\times \backslash \mathbb{A}_E^\times$ *is* χ_E *(resp.* $\chi_E \eta$*). If* $\widehat{f} \in S_{\widehat{\kappa}}^{\mathrm{new}}(K_0(\mathfrak{c}), \chi)$ *is a newform that is a base change of a newform* $f \in S_\kappa(K_0(\mathfrak{c}_E), \chi_E)$ *for some ideal* $\mathfrak{c}_E \subset \mathcal{O}_E$*, then*

$$L^{d_{L/E}}(\mathrm{As}(\widehat{f} \otimes \vartheta^{-1}), s) = L^{d_{L/E}\mathfrak{cf}(\vartheta) \cap \mathcal{O}_E}(\eta, s) L^{d_{L/E}\mathfrak{f}(\vartheta) \cap \mathcal{O}_E}(\mathrm{Ad}(f), s)$$

and

$$L^{\mathfrak{b}'}(\mathrm{As}(\widehat{f} \otimes \theta^{-1}), s) = \zeta^{d_{L/E}(\mathfrak{cf}(\theta) \cap \mathcal{O}_E)}(s) L^{d_{L/E}\mathfrak{f}(\theta) \cap \mathcal{O}_E}(\mathrm{Ad}(f) \otimes \eta, s),$$

where $\mathfrak{b}' := \prod_{\substack{\mathfrak{p} | d_{L/E} \\ \mathfrak{p} \nmid \mathfrak{f}(\theta) \cap \mathcal{O}_E}} \mathfrak{p}$. *Thus*

$$\mathrm{Res}_{s=1} L^{\mathfrak{b}'}(\mathrm{As}(\widehat{f} \otimes \theta^{-1}), s) = L^{d_{L/E}\mathfrak{f}(\theta) \cap \mathcal{O}_E}(\mathrm{Ad}(f) \otimes \eta, 1) \mathrm{Res}_{s=1} \zeta^{d_{L/E}(\mathfrak{cf}(\theta) \cap \mathcal{O}_E)}(s).$$

This proposition is stated, with a typographical error, on p. 4 of [Ram2]. For a proof that one can always find characters ϑ and θ satisfying the requirements of the proposition, see Lemma 2.1 of [Hid8].

5.12. *L*-functions

Remark. Let π be the automorphic representation generated by a simultaneous eigenform $f \in S_\kappa(K_0(\mathfrak{c}), \chi)$ for almost all Hecke operators and let $\mathrm{Ad}(\pi)$ denote the automorphic representation with *L*-function $L(\mathrm{Ad}(f), s)$ [GelJ]. Regardless of κ, $\mathrm{Ad}(\pi)$ is unitary. The unramified character $|\cdot|_{\mathbb{A}_E}^{[k+2m]/2}$ is present in the proposition to make the automorphic representation attached to $\mathrm{As}(\widehat{f} \otimes \theta^{-1})$ (resp. $\mathrm{As}(\widehat{f} \otimes \vartheta^{-1})$) unitary as well.

Proof. We will use the theory of quadratic base change for GL_2 as developed in [Lan] freely in this proof (see Appendix E for a synopsis). First, for all primes $\mathfrak{p} \subset \mathcal{O}_E$ we have

$$\prod_{\mathfrak{P}|\mathfrak{p}} L_\mathfrak{P}(\widehat{f}, s) = L_\mathfrak{p}(f, s) L_\mathfrak{p}(\underline{f \otimes \eta}, s) \tag{5.12.3}$$

where $\underline{f \otimes \eta}$ is the newform with $\lambda_{\underline{f \otimes \eta}}(\mathfrak{p}) = \lambda_{f \otimes \eta}(\mathfrak{p})$ for almost all primes $\mathfrak{p} \subset \mathcal{O}_E$. Moreover, $a(\mathfrak{P}, \widehat{f}) = a(\mathfrak{P}^c, \widehat{f})$ for all primes $\mathfrak{P} \subset \mathcal{O}_L$. Thus, for $i \in \{1, 2\}$, without loss of generality we have:

(1) $\alpha_{i,\mathfrak{P}}(\widehat{f}) = \alpha_{i,\mathfrak{P}^c}(\widehat{f})$,
(2) $\alpha_{i,\mathfrak{P}}(\widehat{f}) = \alpha_{i,\mathfrak{p}}(f)$ if $\mathfrak{p}$ splits as $\mathfrak{p} = \mathfrak{P}\overline{\mathfrak{P}}$ in L/E, and
(3) $\alpha_{i,\mathfrak{p}}(\widehat{f}) = \alpha_{i,\mathfrak{p}}(f)^2$ if $\mathfrak{p}$ is inert in L/E.

Using these facts, it is easy to deduce the equalities

$$L_\mathfrak{p}(\mathrm{As}(\widehat{f} \otimes \vartheta^{-1}), s) = (1 - \eta(\mathfrak{p}) \mathrm{N}_{E/\mathbb{Q}}(\mathfrak{p})^{-s})^{-1} L_\mathfrak{p}(\mathrm{Ad}(f), s) \text{ and}$$
$$L_\mathfrak{p}(\mathrm{As}(\widehat{f} \otimes \theta^{-1}), s) = (1 - \mathrm{N}_{E/\mathbb{Q}}(\mathfrak{p})^{-s})^{-1} L_\mathfrak{p}(\mathrm{Ad}(f) \otimes \eta, s)$$

for $\mathfrak{p} \nmid d_{L/E}(\mathfrak{c} \cap \mathcal{O}_E)$ from the local Euler factors for $L(\mathrm{As}(f \otimes \phi), s)$ and $L(\mathrm{Ad}(f), s)$ ((5.12.1) is also helpful).

Now assume that $\mathfrak{p} \mid (\mathfrak{c} \cap \mathcal{O}_E)$ but $\mathfrak{p} \nmid d_{L/E}(\mathfrak{f}(\theta) \cap \mathcal{O}_E)$. In this case we wish to prove the local equalities

$$L_\mathfrak{p}(\mathrm{As}(\widehat{f} \otimes \vartheta^{-1}), s) = L_\mathfrak{p}(\mathrm{Ad}(f), s) \tag{5.12.4}$$
$$L_\mathfrak{p}(\mathrm{As}(\widehat{f} \otimes \theta^{-1}), s) = L_\mathfrak{p}(\mathrm{Ad}(f) \otimes \eta, s).$$

By Proposition E.9, $\mathfrak{p} \mid \mathfrak{c}_E$. Thus for any prime $\mathfrak{P}|\mathfrak{p}$ we have

$$\mathfrak{f}(\pi(f)_{v(\mathfrak{p})}) \neq \mathcal{O}_{E,v(\mathfrak{p})}, \ \mathfrak{f}(\pi(\widehat{f})_{v(\mathfrak{P})}) \neq \mathcal{O}_{L,v(\mathfrak{P})}, \text{ and } \mathfrak{f}(\chi_{E,v(\mathfrak{p})}) = \mathcal{O}_{E,v(\mathfrak{p})}. \tag{5.12.5}$$

Assume first that $\pi(f)_{v(\mathfrak{p})}$ is not special; this implies that the same is true of $\pi(\widehat{f})_{v(\mathfrak{P})}$. In this case it follows from the classification of irreducible admissible representations of GL_2 over a local field that $L(\pi(f)_{v(\mathfrak{p})}) = L(\pi(\widehat{f})_{v(\mathfrak{P})}) = 1$ which implies that $a(\mathfrak{p}^n, f) = a(\mathfrak{p}^n \mathcal{O}_L, \widehat{f}) = 0$ for all $n > 0$ (see, e.g., Section E.6

for a table of the L-functions of admissible representations of GL_2, keeping in mind that any supercuspidal representation π of GL_2 satisfies $L(\pi, s) = 1$ and remains supercuspidal when base changed along an unramified extension). Thus the left-hand side of each of the equations in (5.12.4) is 1. On the other hand, (5.12.5) also implies that $\pi_{v(\mathfrak{p})}$, if it is principal, is not minimal. This by the discussion in Section 5.12.3 implies the right-hand side of each of the equalities in (5.12.4) is 1 as well. Now suppose that $\pi(f)_{v(\mathfrak{p})}$ and hence $\pi(\widehat{f})_{v(\mathfrak{P})}$ is special. By (5.12.5) this implies that $\pi(f)_{v(\mathfrak{p})}$ and $\pi(f)_{v(\mathfrak{P})}$ are unramified twists of the Steinberg representation (and hence minimal). This implies $a(\mathfrak{p}^n, f) = a(\mathfrak{p}^n \mathcal{O}_L, \widehat{f}) = 1$ for all $n > 0$. In view of the discussion in Section 5.12.3, the asserted identity follows in this case as well.

For the rest of the primes, the first two equalities in the proposition are tautologies.

The last statement in the proposition, namely the residue formula, is a consequence of the fact that $L^{d_{L/E} \mathfrak{f}(\theta) \cap \mathcal{O}_E}(\mathrm{Ad}(f) \otimes \eta, s)$ is necessarily analytic and nonzero at $s = 1$. This fact follows from Proposition 5.18 below and the fact that $L^{d_{L/E}(\mathfrak{f}(\theta) \cap \mathcal{O}_E)}(\mathrm{Ad}(f), s)$ and $L^{\mathfrak{f}(\theta) \mathcal{D}_{L/E}}(\mathrm{Ad}(\widehat{f}), s)$ are both holomorphic at $s = 1$ [GelJ]. $\square$

Proposition 5.18. *If $f \in S_\kappa^{\mathrm{new}}(K_0(\mathfrak{c}_E), \chi_E)$ is a newform and $\widehat{f} \in S_{\widehat{\kappa}}^{\mathrm{new}}(K_0(\mathfrak{c}), \chi)$ is the unique normalized newform that is a base change of f, then*

$$L^{\mathfrak{f}(\theta) \mathcal{D}_{L/E}}(\mathrm{Ad}(\widehat{f}), s) = L^{d_{L/E}(\mathfrak{f}(\theta) \cap \mathcal{O}_E)}(\mathrm{Ad}(f), s) L^{d_{L/E}(\mathfrak{f}(\theta) \cap \mathcal{O}_E)}(\mathrm{Ad}(f) \otimes \eta, s),$$

where θ is as in Proposition 5.17.

Proof. We again use the theory of quadratic base change for GL_2 freely. For unexplained facts on minimal and non-minimal representations, see [HidT, p. 243][4]. We will prove that for $\mathfrak{p} \nmid d_{L/E}(\mathfrak{f}(\theta) \cap \mathcal{O}_E)$ we have

$$\prod_{\mathfrak{P} | \mathfrak{p}} L_{\mathfrak{P}}(\mathrm{Ad}(\widehat{f}), s) = L_{\mathfrak{p}}(\mathrm{Ad}(f), s) L_{\mathfrak{p}}(\mathrm{Ad}(f) \otimes \eta, s). \qquad (5.12.6)$$

Notice first that the set of primes dividing $\mathfrak{c}_E$ is contained in $d_{L/E}(\mathfrak{c} \cap \mathcal{O}_E)$. Thus if $\mathfrak{p} \nmid d_{L/E}(\mathfrak{c} \cap \mathcal{O}_E)$, then the equality follows immediately from the definition of the Euler factors above combined with our observations on the relationship of $\alpha_{i,\mathfrak{p}}(f)$ and $\alpha_{i,\mathfrak{P}}(\widehat{f})$ from the proof of Proposition 5.17.

Assume now that $\mathfrak{p} \nmid d_{L/E}$ but $\mathfrak{p} | \mathfrak{c} \cap \mathcal{O}_E$ (still under the assumption that $\mathfrak{p} \nmid \mathfrak{f}(\theta) \cap \mathcal{O}_E$). Then (5.12.5) is valid. Suppose that $\pi(f)_{v(\mathfrak{p})}$ is not special, and hence $\pi(\widehat{f})_{v(\mathfrak{P})}$ is not special. The conductor relations in (5.12.5) together with the description of the adjoint L-factors given in Section 5.12.3 imply that in this case both sides of the asserted equation (5.12.6) are 1. Now assume that $\pi(f)_{v(\mathfrak{p})}$ and hence $\pi(\widehat{f})_{v(\mathfrak{P})}$ is special. It is easy to see that $\pi(f)_{v(\mathfrak{p})}$ is minimal if and only

---
[4]Hida and Tilouine use notation for the local representations that differs from ours. Our notation is the same as that in [JaLan].

if the same is true of $\pi(\widehat{f})_{v(\mathfrak{P})}$, and twisting by χ_E preserves the minimality at $\mathfrak{p}$ since $\mathfrak{p} \nmid d_{L/E}(\mathfrak{f}(\theta) \cap \mathcal{O}_E)$. Thus in this case the discussion in Section 5.12.3 implies (5.12.6) □

5.13 Relationship with Hida's notation

As mentioned above, the content of this chapter is drawn from Hida's papers [Hid3], [Hid5], and [Hid7]. For the convenience of the reader, we describe how the notation for spaces of Hilbert modular forms in these papers relates to ours. For this purpose, let

$$K_{11}(\mathfrak{c}) := \left\{ \begin{pmatrix} a & b \\ c & d \end{pmatrix} \in K_0(\mathfrak{c}) : \begin{pmatrix} a & b \\ c & d \end{pmatrix} \equiv \begin{pmatrix} 1 & * \\ 0 & 1 \end{pmatrix} \pmod{\mathfrak{c}} \right\}.$$

Fix a weight $\kappa \in \mathcal{X}(L)$. Let $S_\kappa(K_{11}(\mathfrak{c}))$ be the space obtained by replacing $K_0(\mathfrak{c})$ by $K_{11}(\mathfrak{c})$ in the definition of $S_\kappa(K_0(\mathfrak{c}), \chi)$ and replacing (2) by

(2') If $\gamma \in G(\mathbb{Q})$, we have $f(\gamma \alpha) = f(\alpha)$.

in Section 5.4 above. Define $S_\kappa^{\mathrm{coh}}(K_{11}(\mathfrak{c}))$ similarly. We then summarize the notation of [Hid3], [Hid5], and [Hid7] in the following list:

- Hida uses the letter "k" to denote a weight that differs from our use of the letter "k" by a shift of 2. To distinguish these, write $\kappa = (k', m) \in \mathcal{X}(L)$ and set $k = k' + 2\mathbf{1}$ and $w = k' + m + \mathbf{1}$. Then in [Hid3], the space $S_\kappa(K_{11}(\mathfrak{c}))$ is denoted by $S_{k,w}^*(K_{11}(\mathfrak{c}))$ and the space $S_\kappa^{\mathrm{coh}}(K_{11}(\mathfrak{c}))$ is denoted by $S_{k,w}(K_{11}(\mathfrak{c}))$.
- Let $\kappa = (k', m)$ and set $k = k + 2\mathbf{1}$ as above. Set $w = \mathbf{1} - m$. Then in [Hid5], the space $S_\kappa(K_{11}(\mathfrak{c}))$ is denoted by $S_{k,w}(K_{11}(\mathfrak{c}))$.
- In [Hid7], the space $S_\kappa(K_0(\mathfrak{c}), \chi)$ is denoted by $S_{(k,(m,0))}(K_0(\mathfrak{c}), \chi)$, where the "0" in the weight reflects the fact that we only consider totally real fields in this paper, whereas Hida treats automorphic forms on $\mathrm{Res}_{F/\mathbb{Q}}(\mathrm{GL}_2)$ for arbitrary number fields F in [Hid7].

Finally, we explain the purpose of introducing two notations

$$S_\kappa(K_0(\mathfrak{c}), \chi) \quad \text{and} \quad S_\kappa^{\mathrm{coh}}(K_0(\mathfrak{c}), \chi)$$

for the same Hecke module. On the one hand, it is more convenient to relate Fourier expansions to Hecke eigenvalues if we define the Hecke action as it is defined on $S_\kappa(K_0(\mathfrak{c}), \chi)$. On the other hand, it is easier to relate the action of the Hecke algebra on $S_\kappa^{\mathrm{coh}}(K_0(\mathfrak{c}), \chi)$ to the action of the Hecke algebra on certain sheaves via Hecke correspondences. This is especially the case if one wants the Hecke correspondences to preserve local systems of A-modules where A is not necessarily a field (see Section 7.6). The two Hecke modules $S_\kappa(K_0(\mathfrak{c}), \chi)$ and $S_\kappa^{\mathrm{coh}}(K_0(\mathfrak{c}), \chi)$ are then seen to be isomorphic using the isomorphism (5.5.1) given above.

Chapter 6
Automorphic Vector Bundles and Local Systems

In this section we begin with the general theory of local systems, automorphic vector bundles, and automorphy factors. After describing the finite-dimensional representation theory of GL_2 we determine the explicit equations relating modular forms and differential forms.

In modern terminology, a Hilbert modular form f may be identified with a section of a certain vector bundle $\mathcal{L}(\kappa, \chi_0)$ on $Y_0(\mathfrak{c})$. It may also be identified with a differential form on $Y_0(\mathfrak{c})$ with coefficients in $\mathcal{L}(\kappa, \chi_0)$. Each of these identifications takes some work. In the first place, the modular form f takes values in the complex numbers, while the fibers of the vector bundle $\mathcal{L}(\kappa, \chi_0)$ are vector spaces of dimension ≥ 1, so we need a way to convert a complex number into a vector in the appropriate vector space. This is accomplished by the mapping P_z of equation (6.7.1). Secondly, the modular form f is a function on the group $G(\mathbb{A})$ whereas we are looking for a section of a vector bundle on the Hilbert modular variety $Y_0(\mathfrak{c})$. The translation between these two descriptions of f involves an automorphy factor. These ideas are combined in Proposition 6.4 which gives the precise correspondence between Hilbert modular forms and sections of $\mathcal{L}(\kappa, \chi_0)$. In order to obtain a holomorphic differential n-form (with coefficients in $\mathcal{L}(\kappa, \chi_0)$) we essentially need to tensor with the top exterior power of the tangent bundle of $Y_0(\mathfrak{c})$. This has the effect of raising the "weight" by 2. So the final result, stated in Proposition 6.5 (see [Hid3] Section 2 and Section 6) starts with a Hilbert modular form $f \in S_\kappa^{\mathrm{coh}}(K_0(\mathfrak{c}), \chi)$ of weight $\kappa = (k, m)$ and central character $\chi : L^\times \backslash \mathbb{A}_L^\times \to \mathbb{C}$ and constructs a differential form

$$\omega(f) \in \Omega^n\left(Y_0(\mathfrak{c})), \mathcal{L}(\kappa, \chi_0)\right)$$

on $Y_0(\mathfrak{c}) = G(\mathbb{Q}) \backslash G(\mathbb{A}) / K_0(\mathfrak{c}) K_\infty$, with coefficients in the local system $\mathcal{L}(\kappa, \chi_0)$ that corresponds to the representation $L(\kappa, \chi_0) = \mathrm{Sym}^k(V^\vee) \otimes \det_\infty^{-m} \otimes \mathbf{E}(\chi_0^\vee)$ of $K_0(\mathfrak{c}) K_\infty \subset G(\mathbb{A})$. (The shift by 2 in the weight was already incorporated into the definition of the weight of the modular form f so it appears in Proposition

6.4 rather than Proposition 6.5.) In Section 6.11 it is explained that the group of connected components $G(\mathbb{R})/G(\mathbb{R})^0$ acts on $Y_0(\mathfrak{c})$ by complex conjugation on certain coordinates. The induced action on $H^*(Y_0(\mathfrak{c}), \mathcal{L}(\kappa, \chi_0))$ changes the Hodge (p,q) type of the cohomology class.

6.1 Generalities on local systems

A rank n topological *complex vector bundle* on a topological space X is a surjective mapping $\pi : \mathcal{E} \to X$ together with an atlas $\mathcal{C} = \{(U, \phi_U)\}$ of local trivializations. Here $U \subset X$ is an open set (the collection of which are required to cover X) and

$$\phi_U : \pi^{-1}(U) \to U \times \mathbb{C}^n$$

is a homeomorphism that commutes with the projection to U. These local trivializations are required to have linear transition functions, that is, if $\phi_V : \pi^{-1}(V) \to V \times \mathbb{C}^n$ then on $U \cap V$ the resulting transition function

$$\phi_V \circ \phi_U^{-1} : (U \cap V) \times \mathbb{C}^n \to (U \cap V) \times \mathbb{C}^n$$

is given by $(x, v) \mapsto (x, h(x)v)$ where $h_{U,V} : U \cap V \to \mathrm{GL}_n(\mathbb{C})$. This linearity condition is equivalent to the existence of globally defined, continuous addition $\mathcal{E} \times_X \mathcal{E} \to \mathcal{E}$ and scalar multiplication $\mathbb{C} \times \mathcal{E} \to \mathcal{E}$ mappings with respect to which each of the local trivializations ϕ_U is linear on the fibers of π. If X and E are a smooth (resp. complex) manifolds and all the ϕ_U are smooth (resp. holomorphic) then E is referred to as a smooth (resp. holomorphic) vector bundle.

Let $\pi : E \to X$ be a topological vector bundle. Suppose there exists an atlas $\mathcal{C} = \{(U, \phi_U)\}$ of local trivializations and a subgroup $\Gamma \subset \mathrm{GL}_n(\mathbb{C})$ such that the image of each of the transition mappings $h : U \cap V \to \mathrm{GL}_n(\mathbb{C})$ lies in Γ. Then we say that the *structure group* of E can be reduced to Γ. For example, if $\pi : E \to X$ is a smooth vector bundle (over a smooth manifold X), then a choice of Hermitian metric on E gives a reduction of E to the unitary group $U(n) \subset \mathrm{GL}_n(\mathbb{C})$. If the structure group of E can be reduced to a discrete group $\Gamma \subset GL_n(\mathbb{C})$ then we say that E is a *local coefficient system* or, equivalently, that it has a *discrete structure group*. If $\pi : E \to X$ is a smooth vector bundle then a *flat connection* (meaning a smooth connection whose Riemannian curvature vanishes everywhere) on E determines a reduction to a discrete structure group. Conversely, if the structure group of a smooth vector bundle E is discrete, then E admits a connection whose Riemann curvature vanishes everywhere. Thus, a local system on X is the "same thing" as a smooth vector bundle with a flat connection.

Let $\pi : E \to X$ be a smooth (complex) vector bundle and let $\Omega^r(X, E)$ be the vector space of smooth differential r-forms with values in E. It is the space of smooth sections of the vector bundle $\wedge^r T^*X \otimes_\mathbb{R} E$. In order to define the exterior derivative $d\omega$ of a differential form $\omega \in \Omega^r(X, E)$ it is necessary to have a connection ∇_E on E. In this case, $dd\omega = \mathfrak{R} \wedge \omega$ where $\mathfrak{R} \in \Omega^2(X, \mathrm{Hom}(E, E))$ is the

6.1. Generalities on local systems

curvature of ∇_E. Consequently, if (E, ∇_E) is a flat vector bundle then $dd\omega = 0$ and the (de Rham) cohomology $H^*(X, E)$ is defined.

Let G be a topological group and let $K \subset G$ be a closed subgroup. A representation $\psi : K \to \mathrm{GL}(V)$ on some (complex) vector space determines a (complex, topological, homogeneous) vector bundle

$$\mathcal{V} := G \times_K V$$

over $X = G/K$, which consists of equivalence classes $[g, v]$ under the equivalence relation

$$[gk, v] \sim [g, \psi(k)v] \text{ for all } k \in K.$$

The structure group of $\mathcal{V}$ is the image $\psi(K) \subset \mathrm{GL}(V)$. (In particular, if this group is discrete then $\mathcal{V}$ is a local system.) The group G acts on $\mathcal{V}$ by $h \cdot [g, v] := [hg, v]$. A continuous mapping $f : G \to V$ corresponds to a section $s : G/K \to \mathcal{V}$ with $s(gK) = [g, f(g)]$ if and only if the mapping f satisfies the following equivariance condition:

$$f(gk) = \psi(k)^{-1} f(g) \text{ for all } k \in K. \tag{6.1.1}$$

Let $K_1 \subset K$ and let $\pi : X_1 = G/K_1 \to X = G/K$ be the natural projection. The pullback $\mathcal{V}_1 := \pi^*(\mathcal{V})$ is the vector bundle $\mathcal{V}_1 = G \times_{K_1} V$ associated to the representation $\psi_1 := \psi|K_1$ of K_1. Let $s_1(g) = [g, f_1(g)]$ be a section of $\mathcal{V}_1$ (so that $f_1(gk_1) = \psi_1^{-1}(k_1) f_1(g)$ for all $k_1 \in K_1$). If K/K_1 is finite, then we obtain a section $s = \pi_*(s_1)$ by setting $s(g) = [g, f(g)]$ with

$$f(g) := \sum_{k \in K/K_1} \psi(k) f_1(gk) \tag{6.1.2}$$

which is easily seen to satisfy (6.1.1). We will use this in Section 9.2 and again in Section 9.4. (In the latter case, ψ_1 is the trivial representation.)

Suppose $\psi_1 : K \to \mathrm{GL}(V_1)$ and $\psi_2 : K \to \mathrm{GL}(V_2)$ are representations that give rise to vector bundles $\mathcal{V}_1, \mathcal{V}_2$ respectively on G/K. To give a bilinear pairing $\mathcal{B} : \mathcal{V}_1 \times \mathcal{V}_2 \to \mathbb{C}$ is the same as to give a family of bilinear mappings $B_g : V_1 \times V_2 \to \mathbb{C}$ for $g \in G$ such that

$$B_{gk}(v_1, v_2) = B_g(\psi_1(k)v_1, \psi_2(k)v_2) \text{ for all } k \in K, v_1 \in V_1, v_2 \in V_2. \tag{6.1.3}$$

Consequently, if $f_1 : G \to V_1$ and $f_2 : G \to V_2$ define sections s_1, s_2 of $\mathcal{V}_1, \mathcal{V}_2$ respectively then $\mathcal{B}(s_1(gK), s_2(gK))$ is well defined, that is,

$$B_{gk}(f_1(gk), f_2(gk)) = B_g(f_1(g), f_2(g)) \text{ for all } g \in G, k \in K$$

Suppose $K_1 \subset K$ has finite index, and $\pi : X_1 = G/K_1 \to X = G/K$ as in the previous paragraph. Let $f_1 : G \to V_1$ define a section s_1 of $\mathcal{V}_1 = G \times_{K_1} V_1$ on X_1 and let $f_2 : G \to V_2$ define a section s_2 of $\mathcal{V}_2 = G \times_K V_2$ on X. Then we have sections $\pi_*(s_1)$ of $\mathcal{V}_1$ on X and $\pi^*(s_2)$ of $\mathcal{V}_2$ on X_1 and

$$\mathcal{B}(\pi_* s_1(gK), s_2(gK)) = [K : K_1] \mathcal{B}(s_1(gK_1), \pi^* s_2(gK_1)) \tag{6.1.4}$$

because

$$B_g(\pi_* f_1(g), f_2(g)) = B_g\left(\sum_{k \in K/K_1} \psi_1(k) f_1(gk), f_2(g)\right)$$
$$= \sum_{k \in K/K_1} B_g(\psi_1(k) f_1(gk), \psi_2(k) f_2(gk))$$
$$= \sum_{k \in K/K_1} B_{gk}(f_1(gk), f_2(gk)) = [K : K_1] B_g(f_1(g), f_2(g)).$$

6.2 Classical description of automorphic vector bundles

6.2.1 Representations of Γ

Let H be a reductive algebraic group defined over $\mathbb{Q}$, let $K_\infty \subset H(\mathbb{R}) = K_\infty^1 A$ be the product of a compact subgroup $K_\infty^1 \subset H(\mathbb{R})$ which has finite index in a maximal compact subgroup, with the identity component A of the real points of a central torus S containing the maximal $\mathbb{Q}$-split torus in the center of H. Set $D := H(\mathbb{R})/K_\infty$ with its basepoint $x_0 = 1 \cdot K_\infty$. In the semisimple case, D is the symmetric space associated to H. In the reductive case, dividing by A has the effect of removing the extraneous copies of $\mathbb{R}^\times$ that come from the center of H, and D will be connected if K_∞^1 is a maximal compact subgroup of $H(\mathbb{R})$. Let $\Gamma \subset H(\mathbb{Q})$ be an arithmetic subgroup which acts freely on D. (Equivalently, $\Gamma \cap gKg^{-1} = \{1\}$ for all $g \in H(\mathbb{R})$.) Set

$$Y := \Gamma \backslash D = \Gamma \backslash H(\mathbb{R})/K_\infty.$$

Bundles over Y may be constructed using representations of Γ or of K_∞. Since Γ is discrete, a finite-dimensional representation $\psi : \Gamma \to \mathrm{GL}(E)$ on a complex vector space E determines an $H(\mathbb{R})$-equivariant local system (that is, a homogeneous vector bundle with an $H(\mathbb{R})$-invariant flat connection),

$$\mathcal{E}_\psi := E \times_\Gamma H(\mathbb{R})$$

on $\Gamma \backslash H(\mathbb{R})$ whose structure group is the image of ψ. The local system $\mathcal{E}_\psi$ is the set of equivalence classes of pairs $[e, h]$ with $e \in E$ and $h \in H(\mathbb{R})$, where $[e, \gamma h] \sim [\psi^{-1}(\gamma) e, h]$. A section $s : \Gamma \backslash H(\mathbb{R}) \to \mathcal{E}_\psi$ is a mapping $s : H(\mathbb{R}) \to \mathbb{C}$ such that $s(\gamma h) = \psi(\gamma) s(h)$ (for all $h \in H(\mathbb{R})$ and $\gamma \in \Gamma$). The group $H(\mathbb{R})$ acts on $\mathcal{E}_\psi$ (from the right) by $[e, h].g = [e, hg]$. Consequently this local system passes to a local system, also denoted $\mathcal{E}_\psi$, on Y.

If we drop the assumption that Γ acts freely on D then the bundle $\mathcal{E}_\psi$ will not be well defined unless we also assume that for each point $x \in D$,

$$\text{the representation } \psi|\mathrm{Stab}_\Gamma(x) \text{ is trivial.} \tag{6.2.1}$$

6.2. Classical description of automorphic vector bundles

where $\mathrm{Stab}_\Gamma(x) = \{\gamma \in \Gamma : \gamma.x = x\}$. More generally, $\mathcal{E}_\psi$ will be an *orbifold local system*, or a sheaf in the sense of orbifolds (see Section B.5) if

$$\psi | \mathrm{Stab}_\Gamma(x) \text{ acts through a finite group} \tag{6.2.2}$$

for all $x \in D$.

6.2.2 Representations of K_∞

A finite-dimensional representation $\lambda : K_\infty \to \mathrm{GL}(V)$ on a complex vector space V determines a homogeneous vector bundle

$$\mathcal{V}_\lambda := H(\mathbb{R}) \times_{K_\infty} V \longrightarrow D.$$

It is the set of equivalence classes of pairs $[h, v]$ where $[hk, v] \sim [h, \lambda(k)v]$ for all $k \in K$. A section $D \to \mathcal{V}_\lambda$ is a mapping $s : H(\mathbb{R}) \to V$ such that $s(hk) = \lambda(k)^{-1}s(h)$ (for all $h \in H(\mathbb{R})$ and $k \in K_\infty$. The group $H(\mathbb{R})$ acts (from the left) on $\mathcal{V}_\lambda$ by $g.[h, v] = [gh, v]$. Assume that Γ acts freely on D. Then, after dividing by Γ, the vector bundle $\mathcal{V}_\lambda$ passes to an *automorphic vector bundle* on Y whose (continuous, resp. smooth) sections can therefore be identified with (continuous, resp. smooth) functions $s : H(\mathbb{R}) \to V$ such that

$$s(\gamma h k) = \lambda(k)^{-1} s(h) \text{ for all } \gamma \in \Gamma, \ h \in H(\mathbb{R}), \ k \in K_\infty. \tag{6.2.3}$$

6.2.3 Flat vector bundles

The vector bundle $\mathcal{V}_\lambda$ described above may fail to admit a flat connection. It carries a canonical connection (due to K. Nomizu [No]), which is induced from the Cartan decomposition, but it is not necessarily flat. Suppose however, that λ is the restriction to K_∞ of a representation $H(\mathbb{R}) \to \mathrm{GL}(V)$. Then there is a flat connection on the vector bundle $\mathcal{V}_\lambda$: the mapping $[g, v] \mapsto (gK_\infty, \lambda(g)v)$ defines an $H(\mathbb{R})$ equivariant trivialization $H(\mathbb{R}) \times_{K_\infty} V \to D \times V$, so the trivial connection on $D \times V$ passes to a flat connection after dividing by Γ. So in this case, $\mathcal{V}_\lambda$ is an automorphic vector bundle with two canonical connections, one of which is flat[1].

Flat sections of $\mathcal{V}_\lambda \to D$ arise from invariant one-dimensional subspaces as follows. Suppose there is a nonzero vector $v_0 \in V$ and a character $\chi : G(\mathbb{R}) \to \mathbb{C}^\times$ such that $\lambda(g)v_0 = \chi(g)v_0$ for all $g \in G(\mathbb{R})$. Then we obtain a section S of $\mathcal{V}_\lambda$ given by $S(gK) = [g, \chi(g)^{-1}v_0]$. This section is flat with respect to the above connection, and in fact, with respect to the above trivialization the section is constant. If $\chi|\Gamma$ is trivial then the section also passes to a flat section on $\Gamma \backslash G(\mathbb{R})/K_\infty$.

The flat connection in this case may also be described by the automorphy factor (see Section 6.3 below) $J(h, x) = \lambda(h)$ which is independent of x, for in this case the J automorphic action (6.3.1) of $\gamma \in \Gamma \subset H(\mathbb{Q}) \subset H(\mathbb{R})$ on $D \times V$ preserves the factors, so it preserves horizontal and vertical subspaces.

---
[1] See [Gre7] Section 5 for a more complete discussion.

6.2.4 Orbifold local systems

If we drop the assumption that Γ acts freely on D then the vector bundle $\mathcal{V}_\lambda \to D$ may fail to pass to a vector bundle on the quotient $\Gamma \backslash D$. For any $x = hK_\infty \in D$ the group $\mathrm{Stab}_{H(\mathbb{R})}(x) = hK_\infty h^{-1}$ acts on the fiber $(\mathcal{V}_\lambda)_x$ by the rule

$$hkh^{-1}.[h,v] = [hkh^{-1}h, v] = [h, \lambda(k)v] \qquad (6.2.4)$$

for any $k \in K_\infty$. If the element $\gamma := hkh^{-1}$ also lies in Γ and if $\lambda(k)$ is not trivial, then this equation implies that the vector $[h,v]$ will become identified with the vector $\gamma.[h,v] = [h, \lambda(k)v]$ when dividing by Γ. Therefore, a necessary and sufficient condition for the bundle to pass to a vector bundle on the quotient $\Gamma \backslash D$ is the following: for every $x \in D$, the action of $\Gamma \cap \mathrm{Stab}_{H(\mathbb{R})}(x)$ on the fiber $(\mathcal{V}_\lambda)_x$ must be trivial, or equivalently, $\lambda|(K_\infty \cap h\Gamma h^{-1})$ is trivial for all $h \in H(\mathbb{R})$. Similarly, we have the following.

Lemma 6.1. *Suppose the representation* $\lambda : K_\infty \to \mathrm{GL}(V)$ *extends to a representation of* $H(\mathbb{R})$ *and suppose that*

$$\lambda|(K \cap h^{-1}\Gamma h) \text{ acts through a finite group.} \qquad (6.2.5)$$

for all $h \in H(\mathbb{R})$, *or equivalently, that for all* $x \in D$ *the group* $\Gamma \cap \mathrm{Stab}_{H(\mathbb{R})}(x)$ *acts on the fiber* $(\mathcal{V}_\lambda)_x$ *through a finite group, cf.* (6.2.2). *Then the local system* $\mathcal{V}_\lambda$ *passes to a local system in the orbifold sense on* $\Gamma \backslash D$. $\square$

Let us put these two constructions together: let $\lambda : K_\infty \to \mathrm{GL}(V)$ be a finite-dimensional representation and let $\psi : \Gamma \to \mathrm{GL}(E)$ be a representation on a finite-dimensional vector space E. Suppose that Γ acts freely on D, or more generally, suppose that $\lambda|(K \cap h^{-1}\Gamma h)$ is trivial for all $h \in H(\mathbb{R})$ and that $\psi|\mathrm{Stab}_\Gamma(x)$ is trivial for all $x \in D$. Then the representations λ, ψ determine a vector bundle

$$\mathcal{V}_\lambda \otimes \mathcal{E}_\psi$$

on $\Gamma \backslash H(\mathbb{R})/K_\infty$, which is "flat in the $\mathcal{E}_\psi$ direction", and whose sections may be identified with mappings $s : H(\mathbb{R}) \to V \otimes E$ such that

$$s(\gamma h k) = \psi(\gamma)\lambda(k)^{-1} s(h) \text{ for all } h \in H(\mathbb{R}), \gamma \in \Gamma, k \in K_\infty. \qquad (6.2.6)$$

In the case of primary interest to this paper, Γ will not necessarily act freely, but λ will extend to a representation on all of $H(\mathbb{R})$ and $\lambda|(K \cap h^{-1}\Gamma h)$ and $\psi|\mathrm{Stab}_\Gamma(x)$ will act through finite groups. Thus still have a flat vector bundle (i.e., local system in the sense of orbifolds) $\mathcal{V}_\lambda \otimes \mathcal{E}_\psi$ over $\Gamma \backslash D$ whose sections are given by mappings $s : H(\mathbb{R}) \to V \otimes E$ satisfying (6.2.6).

6.3 Classical description of automorphy factors

We continue to study the quotient $Y = \Gamma\backslash H(\mathbb{R})/K_\infty$ of the preceding paragraph. Let $\mathcal{V}_\lambda$ be an automorphic vector bundle corresponding to a representation $\lambda : K_\infty \to \mathrm{GL}(V)$. A continuous (resp. smooth, resp. holomorphic) *automorphy factor* $J : H(\mathbb{R}) \times D \to \mathrm{GL}(V)$ for $\mathcal{V}_\lambda$ is a continuous (resp. smooth, resp. holomorphic in x) mapping such that

(1) $J(hh', x) = J(h, h'x)J(h', x)$ for all $h, h' \in H(\mathbb{R})$ and $x \in D$
(2) $J(k, x_0) = \lambda(k)$ for all $k \in K_\infty$

where $x_0 \in D$ is the basepoint. The first condition is known as the *cocycle condition*. It follows (by taking $h = 1$) that $J(1, x) = I$. The automorphy factor J is determined by its values $J(h, x_0)$ at the basepoint: any smooth mapping $j : H(\mathbb{R}) \to \mathrm{GL}(V)$ such that $j(hk) = j(h)\lambda(k)$ (for all $k \in K_\infty$ and $h \in H(\mathbb{R})$) extends in a unique way to an automorphy factor $J : H(\mathbb{R}) \times D \to \mathrm{GL}(V)$ by setting $J(g, hx_0) = j(gh)j(h)^{-1}$.

An automorphy factor J determines a continuous (resp. smooth, resp. holomorphic) trivialization

$$\Phi_J : H(\mathbb{R}) \times_{K_\infty} V \to D \times V$$

by $[h, v] \mapsto (h.x_0, J(h, x_0)v)$. With respect to this trivialization the action of $\gamma \in H(\mathbb{R})$ is given by

$$\gamma \cdot (x, v) = (\gamma x, J(\gamma, x)v). \tag{6.3.1}$$

Conversely any smooth trivialization $\Phi : \mathcal{V}_\lambda \cong (H(\mathbb{R})/K_\infty) \times V$ of $\mathcal{V}_\lambda$ determines a unique automorphy factor J such that $\Phi = \Phi_J$. If J is a holomorphic automorphy factor then it gives rise to a holomorphic trivialization of the automorphic vector bundle $\mathcal{V}_\lambda$, and conversely.

A continuous (resp. smooth, resp. holomorphic) automorphy factor

$$J : H(\mathbb{R}) \times D \to \mathrm{GL}(V)$$

for a representation $\lambda : K_\infty \to \mathrm{GL}(V)$ determines a canonical identification between continuous (resp. smooth, resp. holomorphic) sections of the vector bundle $\mathcal{V}_\lambda \otimes \mathcal{E}_\psi$ (where $\psi : \Gamma \to \mathrm{GL}(E)$ is a finite-dimensional representation) on $\Gamma\backslash D$ with continuous (resp. smooth. resp. holomorphic) functions $S : D \to V \otimes E$ that are equivariant under Γ, i.e., that satisfy the familiar equation

$$S(\gamma x) = \psi(\gamma)J(\gamma, x)S(x) \tag{6.3.2}$$

for all $\gamma \in \Gamma$ and $x \in D$. The corresponding section $s : H(\mathbb{R}) \to V \otimes E$ is

$$s(h) = J(h, x_0)^{-1} S(h.x_0) \tag{6.3.3}$$

which is easily seen to be well defined, and to satisfy equation (6.2.6). An automorphy factor $J : H(\mathbb{R}) \times D \to \mathrm{GL}(V)$ that is independent of $x \in D$ is the same as an extension of the representation λ to $H(\mathbb{R})$ in which case (see Section 6.2 above) the vector bundle $\mathcal{V}_\lambda \otimes \mathcal{E}_\psi$ is flat.

If $\gamma = \left(\begin{smallmatrix} a & b \\ c & d \end{smallmatrix}\right) \in \mathrm{GL}_2(\mathbb{R})$ and $D = \mathfrak{h}^\pm = \mathbb{C} - \mathbb{R}$ then setting

$$j(\gamma, z) = (cz + d)^k$$

we have that $\det(\gamma)^{-k/2} j(\gamma, z)^k$ is an automorphy factor for the representation

$$\lambda\left(t \left(\begin{smallmatrix} \cos\theta & \sin\theta \\ -\sin\theta & \cos\theta \end{smallmatrix}\right)\right) = e^{-ik\theta} \in \mathrm{GL}_1(\mathbb{C}) \tag{6.3.4}$$

of $K_\infty \cong \mathbb{R}^{>0}\mathrm{SO}(2)$, which we identify with $\mathbb{C}^\times$. An example of a section of the resulting vector bundle on D is the following mapping,

$$s : \mathrm{GL}_2(\mathbb{R}) \to \mathbb{C}, \quad \text{with} \quad s\left(\left(\begin{smallmatrix} a & b \\ c & d \end{smallmatrix}\right)\right) = (ad - bc)^{k/2}(ci + d)^{-k}. \tag{6.3.5}$$

We remark that the center of $G(\mathbb{R})$, which is "detected" by the determinant, acts trivially on $D = \mathfrak{h}^\pm$.

6.4 Adèlic automorphic vector bundles

6.4.1 Definitions

Let H be a connected reductive group over $\mathbb{Q}$. Let $K_\infty \subset H(\mathbb{R})$ be defined as in Section 6.2, and let $K_0 \subset H(\mathbb{A}_f)$ be a compact open subgroup. Let $K = K_\infty K_0$. We wish to describe automorphic vector bundles over the adèlic quotient

$$Y := H(\mathbb{Q}) \backslash H(\mathbb{A})/K = H(\mathbb{Q}) \backslash D \times H(\mathbb{A}_f)/K_0$$

where $D = H(\mathbb{R})/K_\infty$. Let $\mu : K \to \mathrm{GL}(W)$ be a finite-dimensional representation. It gives rise to a vector bundle $\mathcal{W}_\mu$ on $H(\mathbb{A})/K$ consisting of equivalence classes of pairs $[h, v]$ where $h \in H(\mathbb{A})$, $v \in W$ and where $[hk, v] \sim [h, \mu(k)v]$ for all $k \in K$. For each $x = hK \in H(\mathbb{A})/K$ (where $h \in H(\mathbb{A})$) the stabilizer $\mathrm{Stab}_{H(\mathbb{A})}(x) = hKh^{-1}$ acts on the fiber $(\mathcal{W}_\mu)_x$ by

$$hkh^{-1}.[h, v] = [hk, v] = [h, \mu(k)v]$$

as in (6.2.4). If $\gamma := hkh^{-1}$ is in $H(\mathbb{Q})$ then dividing by $H(\mathbb{Q})$ will equate $[h, v]$ with $\gamma.[h, v] = [h, \mu(k)v]$. Thus, a necessary and sufficient condition that the bundle $\mathcal{W}_\mu$ should pass to the quotient Y is that for all $x \in H(\mathbb{A})/K$, the group

$$\mathrm{Stab}_{H(\mathbb{A})}(x) \cap H(\mathbb{Q}) \text{ acts trivially on } (\mathcal{W}_\mu)_x \tag{6.4.1}$$

or equivalently, that $\mu|(K \cap h^{-1}H(\mathbb{Q})h)$ is trivial for all $h \in H(\mathbb{A})$. Assume this to be true. Then (continuous) sections of $\mathcal{W}_\mu \to Y$ can be identified with (continuous) mappings $s : H(\mathbb{A}) \to W$ such that

$$s(\gamma h k) = \mu(k)^{-1} s(h) \tag{6.4.2}$$

for all $\gamma \in H(\mathbb{Q})$, $h \in H(\mathbb{A})$ and $k \in K$.

6.4.2 Flat bundles

If the representation $K \to \mathrm{GL}(W)$ extends to a representation $H(\mathbb{A}) \to \mathrm{GL}(W)$ then, as in Section 6.2.3 the resulting vector bundle carries a flat connection. If there exists a vector $w_0 \in W$ and a character $\chi : H(\mathbb{A}) \to \mathbb{C}^\times$ such that $\mu(h)w_0 = \chi(h)w_0$ for all $h \in H(\mathbb{A})$ then, as in Section 6.2.3, the vector w_0 determines a flat section of the homogeneous vector bundle $H(\mathbb{A}) \times_K W$. Moreover, if the character χ is trivial on $H(\mathbb{Q})$ then this section passes to a flat section of the vector bundle $\mathcal{W}_\mu \to Y$.

Suppose the representation μ is a tensor product of representations $\lambda : K_\infty \to \mathrm{GL}(V)$ and $\psi : K_0 \to \mathrm{GL}(E)$. If there exists an automorphy factor $J : H(\mathbb{R}) \times D \to \mathrm{GL}(V)$ for λ then continuous sections of $\mathcal{W}_\mu = \mathcal{V}_\lambda \otimes \mathcal{E}_\psi$ can equivalently be described as continuous mappings $S : D \times H(\mathbb{A}_f) \to V \otimes E$ such that

$$S(\gamma z, h_0 k_0) = (J(\gamma, z) \otimes \psi(k_0)^{-1}) S(z) \tag{6.4.3}$$

for all $\gamma \in H(\mathbb{Q})$, $z \in D$, $h_0 \in H(\mathbb{A}_f)$, and $k_0 \in K_0$, by setting

$$S(h_\infty.x_0, h_0) = J(h_\infty, x_0) s(h)$$

where $x_0 = 1.K_\infty \in D$ is the basepoint and where $h = h_\infty h_0 \in H(\mathbb{A})$.

If the representation $\lambda : K_\infty \to \mathrm{GL}(V)$ extends to a representation $H(\mathbb{R}) \to \mathrm{GL}(V)$ then the vector bundle $\mathcal{V}_\lambda \otimes \mathcal{E}_\psi$ is flat: the factor $\mathcal{V}_\lambda$ is flat as described in Section 6.2 above, and the factor $\mathcal{E}_\psi$ is always flat, by the following fact:

Proposition 6.2. *Let $\psi : K_0 \to \mathrm{GL}(E)$ be a finite-dimensional representation. Then the resulting vector bundle $\mathcal{E}_\psi \to Y$ is flat.*

Proof. As described in Section 4.1, there is a homeomorphism

$$\iota : \coprod_i \Gamma_i \backslash D \longrightarrow Y \tag{6.4.4}$$

given explicitly on the ith component by $\Gamma_i x \mapsto H(\mathbb{Q})(x, t_i) K_0$, where the t_i are a minimal set of representatives for the finite set $H(\mathbb{Q})^+ \backslash H(\mathbb{A}_f) / K_0$ and $\Gamma_i := H(\mathbb{Q})^+ \cap t_i K_0 t_i^{-1}$. For each i, the representation ψ determines a representation $\psi_i : \Gamma_i \to \mathrm{GL}(E)$ by conjugation by t_i, so we obtain a flat vector bundle

$$\mathcal{E}_i = E \times_{\Gamma_i} D \longrightarrow \Gamma_i \backslash D$$

because Γ_i is a discrete subgroup of $H(\mathbb{R})$. The mapping

$$[v, x] \longrightarrow H(\mathbb{Q})[(x, t_i), v] \in H(\mathbb{Q}) \backslash (D \times H(\mathbb{A}_f) \times_{K_0} E)$$

defines an isomorphism of vector bundles $\mathcal{E}_i \cong \iota^* \mathcal{E}_\psi$, which completes the proof. $\square$

6.4.3 Orbifold bundles

In this work, we will consider vector bundles $\mathcal{W}_\mu = \mathcal{V}_\lambda \otimes \mathcal{E}_\psi$ where λ extends to a representation of $H(\mathbb{R})$ but where condition (6.4.1) does not necessarily hold. Rather, the weaker assumption

$$\mu|(K \cap h^{-1}H(\mathbb{Q})h) \text{ acts through a finite group for all } h \in H(\mathbb{A}) \qquad (6.4.5)$$

will hold (as proven in Proposition 6.3). In this case, by Lemma 6.1, the flat vector bundle $\mathcal{W}_\mu$ passes to a local system in the sense of orbifolds on Y. The sections of this vector bundle can be described as functions $s : H(\mathbb{A}) \to W$ satisfying (6.4.2) or alternately functions $S : D \times H(\mathbb{A}_f) \to W$ satisfying (6.4.3).

6.5 Representations of GL_2

Let $V_\mathbb{Q}$ be the standard representation of $\mathrm{GL}_2(\mathbb{Q})$ and set $V = V_\mathbb{R} \otimes \mathbb{C}$. Recall (for example, from [Fu] Section 8 or [FuH] Section 15.5) that a partition $\mu = (\mu_1 \geq \mu_2 \geq 0)$ of $|\mu| = \mu_1 + \mu_2$ corresponds to a Young diagram with two rows, having μ_1 and μ_2 boxes respectively, and also to a Schur module $\mathbb{S}_\mu = \mathbb{S}_\mu(V)$, which is an irreducible representation of $\mathrm{GL}_2(\mathbb{Q})$ and which has a basis whose elements correspond to *Young tableaux*, that is, fillings of the Young diagram with entries from $\{1, 2\}$ that are weakly increasing along each row and strictly increasing along each column. The partition $|\mu| = 1+1$ gives the determinant representation $\wedge^2 V = \det$. The partition $|\mu| = k + 0$ gives the irreducible representation $\mathrm{Sym}^k(V)$ consisting of the space of homogeneous polynomials of degree k in the coordinates x_1, x_2.

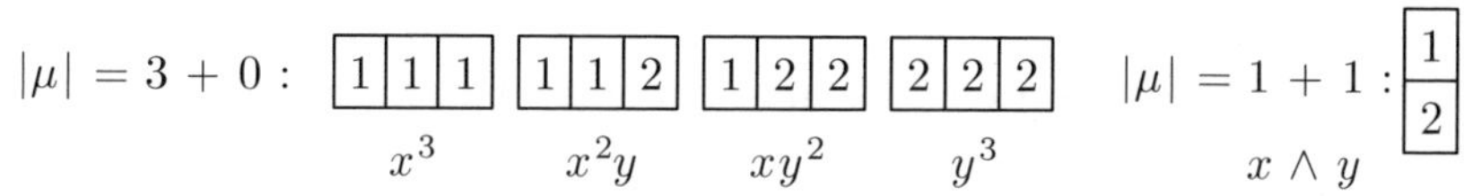

$|\mu| = 3 + 0$: [1|1|1] [1|1|2] [1|2|2] [2|2|2] $|\mu| = 1 + 1$: [1/2]
$\qquad\qquad\quad\;\; x^3 \qquad x^2y \qquad xy^2 \qquad y^3 \qquad\qquad x \wedge y$

Figure 6.2: Bases of $\mathbb{S}_\mu$, $\mu = (3, 0)$ and $\mu = (1, 1)$.

Moreover, every algebraic irreducible representation of $\mathrm{GL}_2(\mathbb{Q})$ is isomorphic to $\mathbb{S}_\mu \otimes \det^m$ for some $m \in \mathbb{Z}$. The representations $\mathbb{S}_\mu \otimes \det^m$ and $\mathbb{S}_{\mu'} \otimes \det^{m'}$ are isomorphic if and only if $\mu_i + m = \mu'_i + m'$ for $i = 1, 2$ (which allows us to define $\mathbb{S}_\mu$ when $\mu_i < 0$). Consequently we may always take $|\mu| = k + 0$ (that is, $\mu_2 = 0$). For $\kappa = (k, m) \in \mathbb{Z}^{\geq 0} \times \frac{1}{2}\mathbb{Z}$ set

$$M(\kappa) = \mathbb{S}_\mu \otimes \det^m = \mathrm{Sym}^k V \otimes \det^m. \qquad (6.5.1)$$

By a slight abuse of terminology we refer to this as the representation with highest weight κ. This is an algebraic representation of $\mathrm{GL}_2(\mathbb{Q})$ if and only if $m \in \mathbb{Z}$; in general it is a twist of an algebraic representation by a half-integral power of the determinant. We remark that the action of $K^1_\infty = \{ \begin{pmatrix} \cos\theta & \sin\theta \\ -\sin\theta & \cos\theta \end{pmatrix} \}$ decomposes

$\operatorname{Sym}^k(V)$ into a sum of one-dimensional representations $\mathbb{C}.z^a\bar{z}^{k-a}$ for $0 \le a \le k$ where $z = e^{i\theta}$.

Using the standard basis and its dual basis for the standard representation and its dual, or contragredient, the mapping $V \to V^\vee$ given by $\binom{x}{y} \mapsto \binom{y}{-x}$ determines an isomorphism of representations $V \cong V^\vee \otimes \det$. Similarly, the contragredient of the representation $\operatorname{Sym}^k V \otimes \det^m$ is isomorphic to

$$\operatorname{Sym}^k(V^\vee) \otimes \det^{-m} \cong \operatorname{Sym}^k(V) \otimes \det^{-m-k}. \tag{6.5.2}$$

6.6 Representations of $G = \operatorname{Res}_{L/\mathbb{Q}}\mathbf{GL}_2$

Now let L be a totally real number field and let $G = \operatorname{Res}_{L/\mathbb{Q}}\mathrm{GL}_2$. Then $G(\mathbb{R}) \cong \mathrm{GL}_2(\mathbb{R})^{\Sigma(L)}$ and $G(\mathbb{A}) \cong \mathrm{GL}_2(\mathbb{A}_L)$. As in equation (5.1.2) let $K_\infty \subset G(\mathbb{R})^0$ be the subgroup

$$K_\infty = \left\{ \left(\left(\begin{smallmatrix} x_\sigma & y_\sigma \\ -y_\sigma & x_\sigma \end{smallmatrix}\right)\right)_{\sigma \in \Sigma(L)} : x_\sigma^2 + y_\sigma^2 > 0 \right\} \cong (\mathbb{C}^\times)^{\Sigma(L)}$$

and let K_∞^1 be the maximal compact subgroup with $x_\sigma^2 + y_\sigma^2 = 1$ for all $\sigma \in \Sigma(L)$. Set $D = G(\mathbb{R})/K_\infty \cong (\mathfrak{h}^\pm)^{\Sigma(L)}$ where $\mathfrak{h}^\pm = \mathbb{C} - \mathbb{R}$, and the action of $G(\mathbb{R})$ on D is given by fractional linear transformations in each factor. Then K_∞ is the stabilizer of the base point $\mathbf{i} = (\sqrt{-1}, \ldots, \sqrt{-1}) \in D$. Let $k \in (\mathbb{Z}^{\geq 0})^{\Sigma(L)}$. If $\alpha = (\alpha_\sigma) \in G(\mathbb{R}) \cong (\mathrm{GL}_2(\mathbb{R}))^{\Sigma(L)}$ and if $z = (z_\sigma) \in D$ set

$$j(\alpha, z)^k = \prod_{\sigma \in \Sigma(L)} j_\sigma(\sigma(\alpha), z_\sigma)^{k_\sigma} \tag{6.6.1}$$

where $j_\sigma\left(\left(\begin{smallmatrix} a & b \\ c & d \end{smallmatrix}\right), z_\sigma\right) := (cz_\sigma + d)$. This function $j^k : G(\mathbb{R}) \times D \to \mathbb{C}^\times$ is an automorphy factor for the one-dimensional representation of K_∞^1,

$$\mu_k(u_\infty^1) := e^{-ik\theta} = \prod_{\sigma \in \Sigma(L)} e^{-ik_\sigma \theta_\sigma}$$

where

$$u_\infty^1 := \left(\left(\begin{smallmatrix} \cos(\theta_\sigma) & \sin(\theta_\sigma) \\ -\sin(\theta_\sigma) & \cos(\theta_\sigma) \end{smallmatrix}\right)\right)_{\sigma \in \Sigma(L)} \in K_\infty^1. \tag{6.6.2}$$

For each real place $\sigma \in \Sigma(L)$ let λ_σ be the standard representation of $\mathrm{GL}_2(\mathbb{R})$ on the vector space $V_\sigma = \mathbb{R}^2 \otimes \mathbb{C}$. This determines a representation $\lambda_{\sigma,k}^\vee$ of $\mathrm{GL}_2(\mathbb{R})$ on the space $\operatorname{Sym}^k(V_\sigma^\vee)$ of homogeneous polynomial functions of degree k on V_σ in which an element $\gamma \in \mathrm{GL}_2(\mathbb{R})$ acts by

$$\lambda_{\sigma,k}^\vee(\gamma)p_\sigma\left(\begin{smallmatrix} X_\sigma \\ Y_\sigma \end{smallmatrix}\right) = p_\sigma\left(\gamma^{-1}\left(\begin{smallmatrix} X_\sigma \\ Y_\sigma \end{smallmatrix}\right)\right) \tag{6.6.3}$$

for any polynomial function $p_\sigma\left(\begin{smallmatrix} X_\sigma \\ Y_\sigma \end{smallmatrix}\right)$. Similarly, for each $k \in (\mathbb{Z}^{\geq 0})^{\Sigma(L)}$ we obtain a representation $\lambda_k^\vee = \bigotimes_{\sigma \in \Sigma(L)} \lambda_{\sigma,k}^\vee$ of $G(\mathbb{R}) \cong \mathrm{GL}_2(\mathbb{R})^{\Sigma(L)}$ on the complex vector

space
$$\operatorname{Sym}^k(V^\vee) := \bigotimes_{\sigma \in \Sigma(L)} \operatorname{Sym}^{k_\sigma}(V_\sigma^\vee).$$

If $p = \bigotimes_{\sigma \in \Sigma(L)} p_\sigma \in \operatorname{Sym}^k(V^\vee)$ and $\gamma = (\gamma_\sigma)_{\sigma \in \Sigma(L)} \in G(\mathbb{R})$ then

$$\lambda_k^\vee(\gamma).p\left(\bigotimes_{\sigma \in \Sigma(L)} \left(\begin{smallmatrix} X_\sigma \\ Y_\sigma \end{smallmatrix}\right)\right) := \bigotimes_{\sigma \in \Sigma(L)} p_\sigma\left(\gamma_\sigma^{-1}\left(\begin{smallmatrix} X_\sigma \\ Y_\sigma \end{smallmatrix}\right)\right).$$

6.7 The section P_z

Let $G = \operatorname{Res}_{L/\mathbb{Q}}(\operatorname{GL}_2)$ as in the preceding section. Let $\kappa = (k, m) \in (\mathbb{Z}^{\geq 0})^{\Sigma(L)} \times (\frac{1}{2}\mathbb{Z})^{\Sigma(L)}$. We have a representation $\lambda_k \otimes \det^m$ of $G(\mathbb{R}) \cong \operatorname{GL}_2(\mathbb{R})^{\Sigma(L)}$ on the vector space

$$M(\kappa) = \operatorname{Sym}^k(V) \otimes \det^m = \bigotimes_{\sigma \in \Sigma(L)} \operatorname{Sym}^{k_\sigma}(V_\sigma) \otimes \det_\sigma^{m_\sigma}$$

and its dual, $\lambda_k^\vee \otimes \det^{-m}$ on the vector space

$$L(\kappa) = \operatorname{Sym}^k(V^\vee) \otimes \det^{-m} \cong \operatorname{Sym}^k(V) \otimes \det^{-m-k}.$$

The key technique for turning a complex-valued modular form into a vector-valued modular form involves the function

$$P = P^{(k)} : D = (\mathfrak{h}^\pm)^{\Sigma(L)} \to \operatorname{Sym}^k(V^\vee)$$

that assigns to any $z = (z_\sigma)_{\sigma \in \Sigma(L)} \in D$ the polynomial

$$P_z\left(\begin{smallmatrix} X \\ Y \end{smallmatrix}\right) := \prod_{\sigma \in \Sigma(L)} (-X_\sigma + z_\sigma Y_\sigma)^{k_\sigma}, \qquad (6.7.1)$$

which we abbreviate by writing $P_z\left(\begin{smallmatrix} X \\ Y \end{smallmatrix}\right) = (-X + zY)^k$.

For any $\gamma = \left(\begin{smallmatrix} a_\sigma & b_\sigma \\ c_\sigma & d_\sigma \end{smallmatrix}\right)_{\sigma \in \Sigma(L)} \in G(\mathbb{R})$ the following equation holds:

$$P_{\gamma \cdot z} = \det(\gamma)^k j(\gamma, z)^{-k} \lambda_k^\vee(\gamma) \cdot P_z \qquad (6.7.2)$$

where

$$j(\gamma, z)^{-k} = \prod_{\sigma \in \Sigma(L)} (c_\sigma z_\sigma + d_\sigma)^{-k_\sigma} \quad \text{and} \quad \det(\gamma)^{-k} = \prod_{\sigma \in \Sigma(L)} \det(\gamma_\sigma)^{-k_\sigma},$$

because

$$\begin{aligned}
(\lambda_k^\vee(\gamma) \cdot P_z)\left(\begin{smallmatrix} X \\ Y \end{smallmatrix}\right) &= P_z\left(\frac{1}{\det(\gamma)}\begin{pmatrix} dX - bY \\ -cX + aY \end{pmatrix}\right) \\
&= \det(\gamma)^{-k}(cz+d)^k\left(-X + \frac{az+b}{cz+d}Y\right)^k \\
&= \det(\gamma)^{-k} j(\gamma, z)^k P_{\gamma \cdot z}.
\end{aligned}$$

We would like to say that the function $P = P^{(k)}$ gives a section of the local system corresponding to $\mathrm{Sym}^k(V^\vee)$, but there are two problems with this. The representation $\mathrm{Sym}^k(V^\vee)$ is not necessarily rational, and the function P does not have the correct equivariance properties until it is multiplied by an appropriate automorphy factor as in Proposition 6.4 below.

Regarding the rationality problem, this can be addressed under certain hypotheses by tensoring the representation $\mathrm{Sym}^k(V^\vee)$ by a power of the determinant. More precisely, for each real place σ, the torus $\begin{pmatrix} a_\sigma & 0 \\ 0 & b_\sigma \end{pmatrix}$ acts on $\mathrm{Sym}^k(V_\sigma)$ with weights
$$a_\sigma^{k_\sigma}, a_\sigma^{k_\sigma-1} b_\sigma, \ldots, a_\sigma b_\sigma^{k_\sigma-1}, b_\sigma^{k_\sigma}.$$
So in the representation $\mathrm{Sym}^k(V_\sigma) \otimes \det^m$ the center of $\mathrm{GL}_2(\mathbb{R})$ acts through the character
$$\phi_\kappa \left(\begin{pmatrix} a_\sigma & 0 \\ 0 & a_\sigma \end{pmatrix} \right) = a_\sigma^{k_\sigma + 2 m_\sigma}$$
In order that $\mathrm{Sym}^k(V) \otimes \det^m$ be a twist of a rational representation of $G = \mathrm{Res}_{L/\mathbb{Q}} \mathrm{GL}_2$ by $\det^{w\mathbf{1}}$ for some $w \in \frac{1}{2}\mathbb{Z}$, it is necessary that these characters coincide at all real places, that is, $k_\sigma + 2 m_\sigma = k_\tau + 2 m_\tau$ for all $\sigma, \tau \in \Sigma(L)$, a condition that we abbreviate by writing
$$k + 2m \in \mathbb{Z}\mathbf{1}. \tag{6.7.3}$$

6.8 The local system $\mathcal{L}(\kappa, \chi_0)$

As above let $G = \mathrm{Res}_{L/\mathbb{Q}}(\mathrm{GL}_2)$. Set $\det_\infty(g) = \det(g_\infty) = \prod_{\sigma \in \Sigma(L)} \det(g_\sigma)$ for $g \in G(\mathbb{A})$. If $g \in K_\infty$ or if $g \in G(\mathbb{Q})$ we will sometimes write $\det(g)$ for $\det_\infty(g)$.

Let $\mathfrak{c} \subset \mathcal{O}_L$ be an ideal. As in Section 5.1 the Hilbert modular variety is
$$Y_0(\mathfrak{c}) := G(\mathbb{Q}) \backslash G(\mathbb{A}) / K_\infty K_0(\mathfrak{c})$$
where
$$K_0(\mathfrak{c}) = \left\{ \gamma \in G(\widehat{\mathbb{Z}}) : \gamma \equiv \begin{pmatrix} * & * \\ 0 & * \end{pmatrix} \pmod{\mathfrak{c}} \right\}.$$

Let $\chi = \chi_0 \chi_\infty : L^\times \backslash \mathbb{A}_L^\times \to \mathbb{C}^\times$ be a Hecke character whose conductor divides $\mathfrak{c}$, see Section C.3. By Lemma 5.2, the character χ_0 determines representations χ_0 and $\chi_0^\vee$ of $K_0(\mathfrak{c})$ on one-dimensional vector spaces $E(\chi_0) = E_{\chi_0}$ and $E(\chi_0^\vee) = E_{\chi_0^\vee}$, which therefore determine one-dimensional local systems[2] $\mathcal{E}(\chi_0)$ and $\mathcal{E}(\chi_0^\vee)$ on $Y_0(\mathfrak{c})$. Define
$$\mathcal{X}(L) \subset (\mathbb{Z}^{\geq 0})^{\Sigma(L)} \times (\tfrac{1}{2}\mathbb{Z})^{\Sigma(L)}$$
to be the set of weights (k, m) defined by the following condition:
$$k + 2m \in \mathbb{Z}\mathbf{1}. \tag{6.8.1}$$

---

[2] As explained in Section 6.4, $\mathcal{E}(\chi_0) = (G(\mathbb{Q}) \backslash G(\mathbb{A}) / K_\infty) \times_{K_0(\mathfrak{c})} E(\chi_0)$ consists of equivalence classes of pairs $[g, v]$ where $g \in G(\mathbb{A})$, $v \in E(\chi_0)$ and where $[\gamma g k, v] \sim [g, \chi_0(k)v]$ for all $\gamma \in G(\mathbb{Q})$ and all $k \in K_0(\mathfrak{c})$.

Proposition 6.3. *Let $\kappa = (k, m) \in \mathcal{X}(L)$ be a weight. Fix a Hecke character,*

$$\chi = \chi_0 \chi_\infty : L^\times \backslash \mathbb{A}_L^\times \to \mathbb{C}^\times$$

such that the conductor of χ divides $\mathfrak{c}$ (see Section C.3) and such that

$$\chi_\infty(b_\infty) = b_\infty^{-k-2m} \tag{6.8.2}$$

for all $b \in \mathbb{A}_L^\times$. As in Sections 6.6 and 6.7 we also have a representation

$$L(\kappa) = \lambda_k^\vee \otimes \det_\infty^{-m} = \mathrm{Sym}^k(V^\vee) \otimes \det_\infty^{-m}$$

of $G(\mathbb{R})$. Then the representations $L(\kappa, \chi_0) = L(\kappa) \otimes E(\chi_0^\vee)$ and $L^\vee(\kappa, \chi_0) = L(\kappa) \otimes E(\chi_0)$ determine local systems

$$\mathcal{L}(\kappa, \chi_0) = \mathcal{L}(\kappa) \otimes \mathcal{E}(\chi_0^\vee) = G(\mathbb{Q}) \backslash G(\mathbb{A}) \times_{K_\infty K_0(\mathfrak{c})} L(\kappa) \otimes E(\chi_0^\vee) \tag{6.8.3}$$

$$\mathcal{L}^\vee(\kappa, \chi_0) = \mathcal{L}(\kappa) \otimes \mathcal{E}(\chi_0) = G(\mathbb{Q}) \backslash G(\mathbb{A}) \times_{K_\infty K_0(\mathfrak{c})} L(\kappa) \otimes E(\chi_0). \tag{6.8.4}$$

on the orbifold $Y_0(\mathfrak{c}) = G(\mathbb{Q}) \backslash G(\mathbb{A})/K_\infty K_0(\mathfrak{c})$.

Remark. The condition that $(k, m) \in \mathcal{X}(L)$ is necessary to ensure condition (6.4.5) holds for all $h \in H(\mathbb{A})$; compare the proof below and the proof of Lemma 5.1.

Proof. Let $K^{\mathrm{fr}} \triangleleft K_0(\mathfrak{c})$ be a torsion-free normal subgroup of finite index as in Section 5.2, and consider the manifold

$$Y_0^{\mathrm{fr}}(\mathfrak{c}) := Y_{K^{\mathrm{fr}}} := G(\mathbb{Q}) \backslash G(\mathbb{A})/K^{\mathrm{fr}} K_\infty.$$

If we assume that the local system

$$\mathcal{L}(\kappa, \chi_0) = \mathcal{L}(\kappa) \otimes \mathcal{E}(\chi_0^\vee) = G(\mathbb{Q}) \backslash G(\mathbb{A}) \times_{K_\infty K^{\mathrm{fr}}} L(\kappa) \otimes E(\chi_0^\vee)$$

is well defined then the projection $Y_0^{\mathrm{fr}}(\mathfrak{c}) \to Y_0(\mathfrak{c})$ gives rise to a system of orbifold charts defining the local system $\mathcal{L}(\kappa) \otimes \mathcal{E}(\chi_0^\vee)$ over $Y_0(\mathfrak{c})$ as in the statement of the proposition. Therefore we need only check that the local system $\mathcal{L}(\kappa, \chi_0) \to Y_0^{\mathrm{fr}}(\mathfrak{c})$ is well defined.

For this purpose we recall from Lemma 6.1 that it is enough to check that the stabilizer of a given point $x \in D = G(\mathbb{R})/K_\infty$ in $G(\mathbb{Q}) \cap K^{\mathrm{fr}}$ acts trivially on $L(\kappa) \otimes E(\chi_0^\vee)$. The group $\mathrm{Stab}_{G(\mathbb{R})}(x)$ is a conjugate of K_∞ so it is a product of a compact group and a central subgroup of $G(\mathbb{A})$. Since K^{fr} is torsion free, the intersection $\mathrm{Stab}_{G(\mathbb{Q})}(x) \cap K^{\mathrm{fr}}$ is contained in the center of $G(\mathbb{Q}) \cap K^{\mathrm{fr}}$ which is a subgroup of the group of diagonal matrices in $\mathrm{GL}_2(\mathcal{O}_L)$ (compare [Mil, Proposition 3.1]). Identifying the group of diagonal matrices in $\mathrm{GL}_2(\mathcal{O}_L)$ with $\mathcal{O}_L^\times$ in the natural manner, we see that it suffices to check that $\mathcal{O}_L^\times$ acts trivially on $L(\kappa) \otimes E(\chi_0^\vee)$.

In the representation $L(\kappa) \otimes E(\chi_0^\vee)$, an element $\left(\begin{smallmatrix} \epsilon & 0 \\ 0 & \epsilon \end{smallmatrix} \right) \in \mathcal{O}_L^\times$ acts by multiplication by

$$\det\left(\left(\begin{smallmatrix} \epsilon_\infty & 0 \\ 0 & \epsilon_\infty \end{smallmatrix} \right) \right)^{-m} \epsilon_\infty^{-k} \chi_0(\epsilon_0) \in \mathbb{C}$$

where ϵ is regarded as an element of $\mathbb{A}_L^\times$ via the diagonal embedding $\mathcal{O}_L^\times \hookrightarrow \mathbb{A}_L^\times$ and where $\chi = \chi_0 \chi_\infty$. Therefore it suffices to show that

$$\chi_0(\epsilon_0) \prod_{\sigma \in \Sigma(L)} \sigma(\epsilon)^{-k_\sigma - 2m_\sigma} = 1. \tag{6.8.5}$$

But $\chi(\epsilon) = \chi_0(\epsilon_0)\chi_\infty(\epsilon_\infty) = 1$, so (6.8.5) follows from (6.8.2). □

By a similar argument we obtain a well-defined local system on $Y_0(\mathfrak{c})$:

$$\mathcal{L}^\vee(\kappa, \chi_0) = \mathcal{L}(\kappa) \otimes \mathcal{E}(\chi_0) = G(\mathbb{Q}) \backslash G(\mathbb{A}) \times_{K_\infty K_0(\mathfrak{c})} L(\kappa) \otimes E(\chi_0).$$

6.9 Adèlic geometric description of automorphic forms

An automorphic form f has two possible geometric interpretations: (a) it determines a section of a certain automorphic vector bundle or (b) it determines a differential form with coefficients in a certain automorphic vector bundle. In this section we describe interpretation (a), while interpretation (b) is described in Section 6.10. In either case, we will need to put together all the ingredients that have been developed so far.

Let $G = \mathrm{Res}_{L/\mathbb{Q}}(\mathrm{GL}_2)$ and let $\mathfrak{c} \subset \mathcal{O}_L$ be an ideal. Let $\kappa = (k,m) \in (\mathbb{Z}^{\geq 0})^{\Sigma(L)} \times (\frac{1}{2}\mathbb{Z})^{\Sigma(L)}$. Let $\chi : L^\times \backslash \mathbb{A}_L^\times \to \mathbb{C}^\times$ be a Hecke character such that $\chi_\infty(b_\infty) = b_\infty^{-k-2m}$. Let

$$L(\kappa, \chi_0) = \mathrm{Sym}^k(V^\vee) \otimes \det_\infty^{-m} \otimes E(\chi_0^\vee)$$

be the vector space on which we have the representation $\lambda_k^\vee \otimes \det_\infty^{-m} \otimes \chi_0^\vee$ of $K_\infty \times K_0(\mathfrak{c})$. Let

$$\mathcal{L}(\kappa, \chi_0) := G(\mathbb{Q}) \backslash G(\mathbb{A}) \times_{K_0(\mathfrak{c}) K_\infty} L(\kappa, \chi_0)$$

be the resulting vector bundle on $Y_0(\mathfrak{c})$. Set $P_z = P_z^{(k)}$.

Proposition 6.4. *Let $\kappa = (k,m) \in \mathcal{X}(L)$ and set $\kappa' = (k-21, m+1) \in (\mathbb{Z}^{\geq 0})^{\Sigma(L)} \times (\frac{1}{2}\mathbb{Z})^{\Sigma(L)}$. Suppose $f : G(\mathbb{A}) \to \mathbb{C}$ satisfies the equivariance conditions of a Hilbert modular form in $S_{\kappa'}^{\mathrm{coh}}(K_0(\mathfrak{c}), \chi)$ as in Section 5.5, that is,*

$$f(\gamma b g u_\infty^1 u_0) = \chi(b)^{-1} \chi_0(u_0^{-\iota}) e^{ik\theta} f(g) \tag{6.9.1}$$

for all $b \in \mathbb{A}_L^\times$, $\gamma \in G(\mathbb{Q})$, $g \in G(\mathbb{A})$, $u_0 \in K_0(\mathfrak{c})$ and $u_\infty^1 = \begin{pmatrix} \cos\theta & \sin\theta \\ -\sin\theta & \cos\theta \end{pmatrix} \in K_\infty^1$. Then the following function $s : G(\mathbb{A}) \to \mathrm{Sym}^k(V^\vee)$

$$s(g) := \det(g_\infty)^{-k} j(g_\infty, \mathbf{i})^k f(g) \lambda_k^\vee(g_\infty)^{-1} P_{g_\infty \cdot \mathbf{i}} \tag{6.9.2}$$

defines a section of the flat vector bundle $\mathcal{L}(\kappa, \chi_0)$ over $Y_0(\mathfrak{c})$.

Proof. According to equation (6.4.2) we need to check that

$$s(\gamma g u_0 u_\infty) = \chi_0^\vee(u_0)^{-1} \det(u_\infty)^m \lambda_k^\vee(u_\infty)^{-1} s(g) \qquad (6.9.3)$$

for all $g \in G(\mathbb{A})$, $\gamma \in G(\mathbb{Q})$, $u_0 \in K_0(\mathfrak{c})$ and $u_\infty \in K_\infty$. Let $u_\infty = b_\infty u_\infty^1$ where $b_\infty \in G(\mathbb{R}) \cap \mathbb{A}_L^\times$ is identified with $\begin{pmatrix} b_\infty & 0 \\ 0 & b_\infty \end{pmatrix}$ and $u_\infty^1 = \begin{pmatrix} \cos\theta & \sin\theta \\ -\sin\theta & \cos\theta \end{pmatrix} \in K_\infty^1$. Set $b = b_0 b_\infty \in \mathbb{A}_L^\times$ with $b_0 = 1$ so that $\chi(b)^{-1} = b_\infty^{k+2m}$. Let $z = g_\infty.\mathbf{i}$. Using the cocycle condition for j, (see Section 6.3), the equivariance property (6.7.2) for P, and equation (5.6.3), equation (6.9.3) then follows from these calculations:

$$\det(\gamma b_\infty g_\infty u_\infty^1)^{-k} = \det(\gamma)^{-k} b_\infty^{-2k} \det(g_\infty)^{-k}$$
$$j(\gamma b_\infty g_\infty u_\infty^1, \mathbf{i})^k = b_\infty^k j(\gamma, z)^k e^{-ik\theta} j(g_\infty, \mathbf{i})^k$$
$$f(\gamma b g u_0 u_\infty^1) = \chi(b)^{-1} \chi_0^\vee(u_0)^{-1} e^{ik\theta} f(g)$$
$$\lambda_k^\vee(\gamma g_\infty u_\infty)^{-1} = \lambda_k^\vee(u_\infty)^{-1} \lambda_k^\vee(g_\infty)^{-1} \lambda_k^\vee(\gamma)^{-1}$$
$$P_{\gamma b_\infty g_\infty u_\infty^1.\mathbf{i}} = P_{\gamma g_\infty.\mathbf{i}} = \det(\gamma)^k j(\gamma, z)^{-k} \lambda_k^\vee(\gamma) P_z$$

which hold for any $b \in \mathbb{A}_L^\times$, $\gamma \in G(\mathbb{Q})$, $g \in G(\mathbb{A})$, $u_0 \in K_0(\mathfrak{c})$, and $u_\infty^1 \in K_\infty^1$. $\square$

Using the automorphy factor $\lambda_k^\vee \otimes \det_\infty^{-m}$ for $G(\mathbb{R})$, this section may also be described as the mapping $S : D \times G(\mathbb{A}_f) \to \mathrm{Sym}^k(V^\vee)$ given by

$$S(z, x) := \det(\alpha_z)^{-m} \lambda_k^\vee(\alpha_z) s(x\alpha_z) \qquad (6.9.4)$$
$$= \det(\alpha_z)^{-m-k} j(\alpha_z, \mathbf{i})^k f(x\alpha_z) P_z \qquad (6.9.5)$$

where $\alpha_z \in G(\mathbb{R})$ is any element such that $\alpha_z.\mathbf{i} = z$. Then S is well defined, for if $u_\infty \in K_\infty$, say, $u_\infty = b_\infty u_\infty^1$ with $b_\infty \in G(\mathbb{R}) \cap \mathbb{A}_L^\times$ and $u_\infty^1 \in K_\infty^1$ as in the previous paragraph, then replacing α_z by $\alpha_z u_\infty$ in equation (6.9.5) gives

$$\det(\alpha_z u_\infty)^{-m-k} j(\alpha_z u_\infty, \mathbf{i})^k f(b_\infty x \alpha_z u_\infty^1) P_z$$
$$= \det(\alpha_z)^{-m-k} b_\infty^{-2m-2k} b_\infty^k j(\alpha_z, \mathbf{i})^k j(u_\infty^1, \mathbf{i})^k e^{ik\theta} \chi(b)^{-1} f(x\alpha_z) P_z$$
$$= S(z, x)$$

where $b = b_0 b_\infty \in \mathbb{A}_L^\times$ with $b_0 = 1$. Similarly one checks that

$$S(\gamma z, \gamma x u_0) = \chi_0^\vee(u_0)^{-1} \det(\gamma)^{-m} \lambda_k^\vee(\gamma) S(z, x)$$

for all $z \in D$, $x \in G(\mathbb{A}_f)$, $\gamma \in G(\mathbb{Q})$ and $u_0 \in K_0(\mathfrak{c})$, which verifies equation (6.4.3).

Remarks.
(1) If $f \in S_{\kappa'}(K_0(\mathfrak{c}), \chi)$ (rather than $S_{\kappa'}^{\mathrm{coh}}(\mathcal{L}_0(\mathfrak{c}), \chi)$) then equation (6.9.2) determines a section of the vector bundle $\mathcal{L}^\vee(\kappa, \chi_0^{-1}) = \mathcal{L}(\kappa) \otimes \mathcal{E}(\chi_0^{-1})$.
(2) The shift by 2 in the weight k is a consequence of the exponent $(k_\sigma + 2)\theta_\sigma$ that occurs in condition (3) of Section 5.4 and in condition (3^{coh}) of Section 5.5. These definitions are chosen so that the differential form corresponding to f will take values in the local system $\mathcal{L}(\kappa, \chi_0)$, see Proposition 6.5.

6.10 Differential forms

If $\mathcal{L}$ is a local system of complex vector spaces on a Hilbert modular variety Y then we denote by $\Omega^r(Y, \mathcal{L})$ the vector space of smooth differential forms on Y (as an orbifold) with coefficients in $\mathcal{L}$. It consists of smooth orbifold sections of the vector bundle $\wedge^r T^*Y \otimes \mathcal{L}$. If $\omega \in \Omega^r(Y, \mathcal{L})$ is also holomorphic and satisfies certain growth conditions then it corresponds to a Hilbert modular form of a certain weight. In this section we make the correspondence explicit.

Let G be a reductive algebraic group over $\mathbb{Q}$, let $K_\infty = K_\infty^1 A_G(\mathbb{R})^0$ and let $D = G(\mathbb{R})/K_\infty$ as in Section 6.2. The Cartan involution $\theta : \mathfrak{g} \to \mathfrak{g}$ determines a decomposition $\mathfrak{g}(\mathbb{R}) \cong \mathfrak{k}_\infty \oplus \mathfrak{p}$ into ± 1 eigenspaces, with $\mathfrak{k}_\infty = \mathrm{Lie}(K_\infty)$. Although the subspace $\mathfrak{p}$ is not a Lie subalgebra, it is preserved by the adjoint action of K_∞ so this representation $ad : K_\infty \to \mathrm{GL}(\mathfrak{p})$ determines an automorphic vector bundle $\mathcal{V}_{\mathrm{ad}}$ on D. In fact, the tangent space $T_{x_0} D$ at the basepoint may be identified with the corresponding quotient of Lie algebras $\mathfrak{g}(\mathbb{R})/\mathfrak{k}_\infty \cong \mathfrak{p}$ so the vector bundle $\mathcal{V}_{\mathrm{ad}}$ is the tangent bundle of D. It is not a flat bundle because the adjoint action of $G(\mathbb{R})$ does not preserve $\mathfrak{p}$. Let $\mathfrak{p}^* = \mathrm{Hom}(\mathfrak{p}, \mathbb{R})$ be the dual vector space with its corresponding representation of K_∞. Then sections of the automorphic vector bundle associated to $\wedge^r \mathfrak{p}^*$ are differential r-forms on D.

Remark. This construction of automorphic vector bundles leads naturally to the subject of $(\mathfrak{g}, K_\infty)$-cohomology, see [BoW].

In the case $G = \mathrm{GL}_2$, the Cartan involution is $\theta(g) = {}^t g^{-1}$. Let K_∞ be the identity component of the fixed point set of θ, so

$$K_\infty = \{ \begin{pmatrix} x & y \\ -y & x \end{pmatrix} : x^2 + y^2 > 0 \} \cong \mathbb{C}^\times,$$
$$\mathfrak{p} = \{ \begin{pmatrix} a & b \\ b & -a \end{pmatrix} : a, b \in \mathbb{R} \} \cong \mathbb{C},$$

and $D = G(\mathbb{R})/K_\infty \cong \mathbb{C} - \mathbb{R}$ is disconnected. Identifying $z \in K_\infty$ with $z = x+iy \in \mathbb{C}^\times$ and $u \in \mathfrak{p}$ with $u = a+ib \in \mathbb{C}$ we find that the tangent bundle of D is the homogeneous vector bundle that corresponds to the representation $\lambda : K_\infty \to \mathrm{GL}(\mathfrak{p})$ that is given by $\lambda(z)(u) = |z|^{-2} z^2 u$. Therefore the cotangent bundle corresponds to the representation $\lambda(z)(u) = |z|^2 z^{-2} u$. From equation (6.3.4) we see that a section of the cotangent bundle of $Y = \Gamma \backslash G(\mathbb{R})/K_\infty$ is therefore a mapping $S : D \to \mathbb{C}$ such that $S(\gamma z) = \det(\gamma) j(\gamma, z)^{-2} S(z)$. We remark that the "function" $S(z) = dz$ satisfies

$$S(gz) = \det(g) j(g, z)^{-2} S(z) \tag{6.10.1}$$

(for all $g \in G(\mathbb{R})$) so dz defines a section of T^*Y.

If $G = \mathrm{Res}_{L/\mathbb{Q}} \mathrm{GL}_2$ then we have differential forms dz_σ and $d\bar{z}_\sigma$ for each real place $\sigma \in \Sigma(L)$. We will be primarily interested in differential n-forms where $n = [L : \mathbb{Q}]$. Choose an ordering for the real places, $\Sigma(L) = \{\sigma_1, \ldots, \sigma_n\}$, and write dz_i in place of dz_{σ_i}. Let

$$dz := dz_1 \wedge \cdots \wedge dz_n.$$

In the following proposition, we use equations (6.7.1) (for P_z), (6.10.1) (for dz) to obtain a differential form ω_f and hence a mapping

$$\omega : S^{\mathrm{coh}}_\kappa(K_0(\mathfrak{c}), \chi) \to \Omega^n(Y_0(\mathfrak{c}), \mathcal{L}(\kappa, \chi_0)).$$

The computation required to prove the proposition is entirely analogous to that given in the proof of Proposition 6.4 and hence is omitted.

Proposition 6.5. *Fix a weight $\kappa = (k, m) \in \mathcal{X}(L)$ and a Hecke character χ as in Proposition 6.4. Let $f \in M^{\mathrm{coh}}_\kappa(K_0(\mathfrak{c}), \chi)$ or more generally, let $f : G(\mathbb{A}) \to \mathbb{C}$ be a smooth function that satisfies the equivariance conditions of a Hilbert modular form in $M^{\mathrm{coh}}_\kappa(K_0(\mathfrak{c}), \chi_0)$, namely,*

$$f(\gamma b g u_0 u^1_\infty) = \chi(b)^{-1} \chi_0(u_0^{-\iota}) f(g) \prod_{\sigma \in \Sigma(L)} e^{i(k_\sigma + 2)\theta_\sigma}$$

where $b \in \mathbb{A}_L^\times$, $\gamma \in G(\mathbb{Q})$, $g \in G(\mathbb{A})$, $u_0 \in K_0(\mathfrak{c})$ and

$$u^1_\infty = \left(\left(\begin{smallmatrix} \cos\theta_\sigma & \sin\theta_\sigma \\ -\sin\theta_\sigma & \cos\theta_\sigma \end{smallmatrix}\right)\right)_{\sigma \in \Sigma(L)} \in K^1_\infty.$$

Then the function $\omega(f) = \omega_f : D \times G(\mathbb{A}_f) \to \mathrm{Sym}^k(V^\vee) \otimes \wedge^n T^*D$ given by

$$\omega_f(z, x_0) := \det(\alpha_z)^{-m-k-1} j(\alpha_z, \mathbf{i})^{k+2} f(x_0 \alpha_z) (-X + zY)^k \, dz \qquad (6.10.2)$$

(where $\alpha_z \in G(\mathbb{R})$; $\alpha_z.\mathbf{i} = z$) defines a smooth differential form

$$\omega_f \in \Omega^n(Y_0(\mathfrak{c}), \mathcal{L}(\kappa, \chi_0)). \qquad \square$$

Proposition 6.6. *If $f \in S^{\mathrm{coh}}_\kappa(K_0(\mathfrak{c}), \chi)$ then the differential form ω_f is closed ($d\omega_f = 0$), square integrable, and holomorphic. It therefore defines a class $[\omega_f] \in H^n_{(2)}(Y_0(\mathfrak{c}), \mathcal{L}(\kappa, \chi_0))$ in L^2-cohomology.*

Proof. In equation (6.10.2) write $\omega_f = F(z) dz$ where $dz = dz_1 \wedge \cdots \wedge dz_n$. The function $F(z)$ is holomorphic by the remarks in Section 5.4. With respect to this coordinate system on $D = G(\mathbb{R})/K_\infty$ we have

$$d\omega_f = \sum_{i=1}^n \frac{\partial F(z)}{\partial \bar{z}_i} d\bar{z}_i \wedge dz + \sum_{i=1}^n \frac{\partial F(z)}{\partial z_i} dz_i \wedge dz$$

The first term vanishes since F is analytic in z and the second term vanishes since dz is a top degree (in the holomorphic variables) differential form. The differential form ω_f is square integrable because $(f, f)_P < \infty$. $\square$

6.11 Action of the component group

In this section we explain that the group $G(\mathbb{R})/G(\mathbb{R})^0$ of connected components of $G(\mathbb{R})$ acts by complex conjugation with respect to certain coordinates. Let $I, J \subset \Sigma(L)$ be disjoint sets with $I \cup J = \Sigma(L)$ and let

$$dz_I \wedge d\bar{z}_J = \bigwedge_{i=1}^{n} dw_i \quad \text{where} \quad dw_i = \begin{cases} dz_i & \text{if } \sigma_i \in I \\ d\bar{z}_i & \text{if } \sigma_i \in J. \end{cases}$$

Then $dz_I \wedge d\bar{z}_J$ is a section of the vector bundle corresponding to the representation of $K_\infty^1 \cong (S^1)^n$ on $\mathbb{C}$ given by

$$\mu_J = \bigotimes_{\sigma \in \Sigma L} \mu_{J,\sigma} \quad \text{where} \quad \mu_{J,\sigma}\left(\begin{pmatrix} \cos\theta_\sigma & \sin\theta_\sigma \\ -\sin\theta_\sigma & \cos\theta_\sigma \end{pmatrix}\right) := \begin{cases} e^{2i\theta_\sigma} & \text{if } \sigma \in I \\ e^{-2i\theta_\sigma} & \text{if } \sigma \in J. \end{cases}$$

The differential form $dz_I \wedge d\bar{z}_J$ may also be obtained as the pullback

$$dz_I \wedge d\bar{z}_J = \iota_J^*(dz_1 \wedge \cdots \wedge dz_n)$$

where $\iota_J : G(\mathbb{R})/K_\infty \to G(\mathbb{R})/K_\infty$ is defined on each factor by

$$\iota_J(z_i) := \begin{cases} z_i & \text{if } \sigma_i \in I \\ \bar{z}_i & \text{if } \sigma_i \in J. \end{cases}$$

The mapping ι_J yields an involution of $Y_0(\mathfrak{c})$ and of its Baily-Borel compactification $X_0(\mathfrak{c})$, which is a stratum preserving orbifold morphism on each nonempty stratum. This morphism is proper, finite and surjective, but it may not be orientation preserving (in fact it will usually change Hodge types). For $J \subset \Sigma(L)$ define

$$w_J = (\gamma_\sigma)_{\sigma \in \Sigma(L)} \quad \text{where} \quad \gamma_\sigma = \begin{cases} \begin{pmatrix} -1 & 0 \\ 0 & 1 \end{pmatrix} & \sigma \in J \\ \begin{pmatrix} 1 & 0 \\ 0 & 1 \end{pmatrix} & \sigma \notin J. \end{cases}$$

Then $\{w_J : J \subset \Sigma(L)\}$ is a collection of representatives for the component group $G(\mathbb{R})/G(\mathbb{R})^0$, and multiplication from the right by w_J gives the mapping $\iota_J : G(\mathbb{R})/K_\infty \to G(\mathbb{R})/K_\infty$. In fact, if $\sigma \in J$ and if

$$z_\sigma = \begin{pmatrix} a_\sigma & b_\sigma \\ c_\sigma & d_\sigma \end{pmatrix}.i = \frac{a_\sigma i + b_\sigma}{c_\sigma i + d_\sigma}$$

then

$$\bar{z}_\sigma = \begin{pmatrix} -a_\sigma & b_\sigma \\ -c_\sigma & d_\sigma \end{pmatrix}.i = \begin{pmatrix} a_\sigma & b_\sigma \\ c_\sigma & d_\sigma \end{pmatrix} \gamma_\sigma.i.$$

Lemma 6.7. *Fix a weight $\kappa = (k, m) \in \mathcal{X}(L)$ and a Hecke character χ. Let $J \subset \Sigma(L)$. Right multiplication by w_J induces isomorphisms of local systems on $Y_0(\mathfrak{c})$,*

$$\iota_J^* \left(\mathcal{L}(\kappa, \chi_0) \right) \cong \mathcal{L}(\kappa, \chi_0) \quad \text{and} \quad \iota_J^* \left(\mathcal{L}(\kappa, \chi_0^\vee) \right) \cong \mathcal{L}(\kappa, \chi_0^\vee)$$

so it also induces an isomorphism of smooth differential forms,

$$\iota_J^* : \Omega^k(Y_0(\mathfrak{c}), \mathcal{L}(\kappa, \chi_0)) \to \Omega^k(Y_0(\mathfrak{c}), \mathcal{L}(\kappa, \chi_0)).$$

Proof. Suppose first that $\lambda : K = K_0 K_\infty \to \mathrm{GL}(V)$ is a representation giving rise to a homogeneous vector bundle $M(\lambda) = G(\mathbb{Q}) \backslash G(\mathbb{A}) \times_K V$ on $Y = G(\mathbb{Q}) \backslash G(\mathbb{A}) / K$. Let $w \in G(\mathbb{A})$. Suppose that w normalizes K, and that the representation λ extends to a representation of the group generated by K and w. Then the mapping

$$[g, v] \mapsto [gw, \lambda(w)^{-1} v] \qquad (6.11.1)$$

defines a mapping of homogeneous vector bundles $\phi : M(\lambda) \to M(\lambda)$ which covers the mapping $Y \to Y$ that is given by $G(\mathbb{Q}) g K \mapsto G(\mathbb{Q}) g w K$. To see this, it suffices to check that the mapping (6.11.1) is well defined. If $k \in K$ then

$$[gk, v] \mapsto [gkw, \lambda(w)^{-1} v] = [gw(w^{-1} k w), \lambda(w)^{-1} v]$$
$$= [gw, \lambda(w^{-1} k w) \lambda(w)^{-1} v] = [gw, \lambda(w)^{-1} \lambda(k) v]$$

which is the image of $[g, \lambda(k)(v)]$. Suppose now that $M(\lambda)$ is a local system in the sense of orbifolds. In other words, $M(\lambda)$ is flat and (6.4.5) (but not necessarily (6.4.1)) holds for all $h \in H(\mathbb{A})$ holds. In this case, the same argument provides a lift of right multiplication by w to $M(\lambda)$. The lemma is a special case of this statement. $\square$

Definition 6.8. *Fix a weight $\kappa = (k, m) \in \mathcal{X}(L)$ and a Hecke character χ as in Proposition 6.4. Let $f \in S_\kappa^{\mathrm{coh}}(K_0(\mathfrak{c}), \chi_0)$ and let $\omega_f \in \Omega^n(Y_0(\mathfrak{c}), \mathcal{L}(\kappa, \chi_0))$ be the corresponding differential form as defined in equation (6.10.2). Let $J \subset \Sigma(L)$. Define*

$$\omega_J(f) := \iota_J^*(\omega_f). \qquad (6.11.2)$$

Lemma 6.7 above implies that $\omega_J(f) \in \Omega^n(Y_0(\mathfrak{c}), \mathcal{L}(\kappa, \chi_0))$.

Chapter 7

The Automorphic Description of Intersection Cohomology

In this chapter we use Proposition 6.6 to construct a map
$$\text{Hilbert modular forms} \xrightarrow{\omega} \text{intersection cohomology}$$
which takes
$$\begin{aligned}\text{weight, nebentypus} &\longrightarrow \text{local coefficient system}\\ \text{Hecke operator} &\longrightarrow \text{action of Hecke correspondence}\\ \text{Petersson product} &\longrightarrow \text{intersection product}\end{aligned}$$
The compatibility between the left and right sides of this chart involve a series of technical complications including the following

- The Hilbert modular variety $Y_0(\mathfrak{c})$ is an orbifold, rather than a manifold.
- The local coefficient systems that arise are slightly more general than those usually considered in the literature on intersection cohomology.
- The Hecke correspondences must be lifted to the local systems.
- The normalization of the Petersson product does not agree with the usual normalization of the intersection product.
- As mentioned in Section 6.9 and Section 6.10 the transition from modular forms to differential forms involves a shift in the weight.

Agreement between the weight conventions is achieved using the *main involution* which converts modular forms in $S_\kappa(Y_0(\mathfrak{c}), \chi_0)$ into modular forms in $S_\kappa^{\text{coh}}(Y_0(\mathfrak{c}), \chi_0)$, as discussed in Section 5.5 and Section 5.7. Agreement between the normalizations for the Petersson product and the intersection product is achieved using the *Atkin Lehner operator*, which converts the local system $\mathcal{E}(\chi_0)$ into the local system $\mathcal{E}(\chi_0^\vee)$, see Section 7.4, and the *complex conjugation* involution, see Section 7.3. Besides these actions, the component group of G (see Section 6.11) and the Hecke algebra (see Sections 7.6 and 7.7) act on modular forms and on cohomology.

In Section 7.1, the local coefficient system $\mathcal{L}(\kappa, \chi_0)$ is recalled from Chapter 6. Using the Atkin-Lehner operator (Section 7.4), the intersection pairing is defined and normalized in Section 7.5. Hecke correspondences are defined and lifted to the local systems in Section 7.6.

In Proposition 6.5 the process of converting a complex-valued Hilbert modular form into a differential form with coefficients in the local system $\mathcal{L}(\kappa, \chi_0)$ was explained. This is applied in Section 7.2 to the Hilbert modular forms f of Section 5.4 to obtain intersection cohomology classes

$$[\omega_f] \in I^m H^*(X_0(\mathfrak{c}), \mathcal{L}(\kappa, \chi_0)).$$

We show in this section that ω is Hecke equivariant and use it to give a description, due to G. Harder, of this intersection cohomology group as a Hecke module. The automorphic description of the intersection cohomology groups gives us, in particular, a complete picture of these cohomology groups as Hecke modules. This will be crucial in the proof of Theorem 8.4, the full version of the "first main theorem" of the introduction, in Chapter 8 below.

In Theorem 7.11 we show how the pairings on intersection cohomology described in Section 7.5 relate to the Petersson inner product. This will be a key tool in the proof of Theorem 8.5, the full version of the "second main theorem" of the introduction, in Chapter 8.

Finally in Section 7.9 we indicate how these constructions may be performed using integral coefficients. Our presentation owes much to Hida, who used automorphic forms to fix integral normalizations of the isomorphisms occurring in Harder's work (see [Hid7], [Hid3] and [Gh]).

Throughout this chapter we use the notation established in Chapter 5 (see also Section 6.6): L is a totally real number field of degree $n = [L : \mathbb{Q}]$; $G = \operatorname{Res}_{L/\mathbb{Q}}(\operatorname{GL}_2)$; $\mathfrak{c} \subset \mathcal{O}_L$ is an ideal and $K_0(\mathfrak{c}) \subset G(\widehat{\mathbb{Z}})$ is the corresponding Hecke congruence subgroup.

The Hilbert modular variety is $Y_0(\mathfrak{c}) = G(\mathbb{Q})\backslash G(\mathbb{A})/K_\infty K_0(\mathfrak{c})$.

7.1 The local system $\mathcal{L}(\kappa, \chi_0)$

The following definitions are recalled from Chapter 6. Let $\mathcal{X}(L) \subset (\mathbb{Z}^{\geq 0})^{\Sigma(L)} \times (\frac{1}{2}\mathbb{Z})^{\Sigma(L)}$ be the allowable set of weights, as described in Section 5.3 equation (5.3.1) and Section 6.8 equation (6.8.1). Fix $\kappa = (k, m) \in \mathcal{X}(L)$. As in Section 6.6 the group GL_2 acts via the standard representation on a complex vector space $V = \mathbb{C}^2$, and this determines a representation of $G(\mathbb{R}) \cong \operatorname{GL}_2(\mathbb{R})^{\Sigma(L)}$ on the space

$$\operatorname{Sym}^k(V^\vee) := \bigotimes_{\sigma \in \Sigma(L)} \operatorname{Sym}^{k_\sigma}(V_\sigma^\vee)$$

7.1. The local system $\mathcal{L}(\kappa,\chi_0)$

of homogeneous polynomial functions of degree k. Let $\det_\infty(\gamma_\infty) = \prod_{\sigma \in \Sigma(L)} \det(\gamma_\sigma)$ for $\gamma_\infty \in G(\mathbb{R})$. We obtain a representation of $G(\mathbb{R})$,

$$L(\kappa) = \mathrm{Sym}^k(V^\vee) \otimes \det_\infty^{-m}.$$

Let $\chi : \mathbb{A}_L^\times \to \mathbb{C}^\times$ be a quasicharacter whose conductor divides $\mathfrak{c}$ and satisfies $\chi_\infty(b_\infty) = b_\infty^{-k-2m}$ for all $b_\infty \in \mathbb{A}_{L\infty}^\times$. As in Lemma 5.2 the finite part χ_0 of χ determines one-dimensional representations, $\chi_0, \chi_0^\vee : K_0(\mathfrak{c}) \to \mathbb{C}^\times$ as the product $\chi_0 := \prod_{v<\infty} \chi_v$ of local characters defined by setting

$$\chi_v\left(\begin{pmatrix} a_v & b_v \\ c_v & d_v \end{pmatrix}\right) = \begin{cases} \chi_v(d_v) & \text{if } \mathfrak{p}_v | \mathfrak{c} \\ 1 & \text{otherwise.} \end{cases}$$

$$\chi_v^\vee\left(\begin{pmatrix} a_v & b_v \\ c_v & d_v \end{pmatrix}\right) = \begin{cases} \chi_v(a_v) & \text{if } \mathfrak{p}_v | \mathfrak{c} \\ 1 & \text{otherwise.} \end{cases}$$

for each place $v < \infty$ with $\mathfrak{p}_v$ the corresponding prime ideal, see Section 6.8. Denote the complex numbers $\mathbb{C}$ with this representation by E_{χ_0} or $E(\chi_0)$. So we obtain a representation

$$L(\kappa,\chi_0) := L(\kappa) \otimes E(\chi_0^\vee) = \mathrm{Sym}^k(V^\vee) \otimes \det_\infty^{-m} \otimes E(\chi_0^\vee)$$

of $K_0(\mathfrak{c}).K_\infty$ (in fact, a representation of $K_0(\mathfrak{c}).G(\mathbb{R})$). In other words, as a vector space, $L(\kappa,\chi_0)$ is isomorphic to $\mathrm{Sym}^k(V^\vee)$ but the action of $\gamma = \gamma_0\gamma_\infty \in K_0(\mathfrak{c}).K_\infty$ on a polynomial function $p\left(\begin{pmatrix} X \\ Y \end{pmatrix}\right)$ is given by

$$\gamma.p\left(\begin{pmatrix} X_\sigma \\ Y_\sigma \end{pmatrix}_{\sigma \in \Sigma(L)}\right) := \det_\infty(\gamma_\infty)^{-m} \chi_0^\vee(\gamma_0) p\left(\begin{pmatrix} \gamma_\sigma^{-1}\begin{pmatrix} X_\sigma \\ Y_\sigma \end{pmatrix} \end{pmatrix}_{\sigma \in \Sigma(L)}\right). \quad (7.1.1)$$

The representation $L(\kappa,\chi_0)$ gives rise to an orbifold local system

$$\mathcal{L}(\kappa,\chi_0) = \mathcal{L}(\kappa) \otimes \mathcal{E}(\chi_0^\vee) := G(\mathbb{Q})\backslash G(\mathbb{A}) \times_{K_0(\mathfrak{c})K_\infty} L(\kappa,\chi_0)$$

on the orbifold $Y_0(\mathfrak{c})$, see Proposition 6.3. Similar remarks apply to the construction of the orbifold local system $\mathcal{L}^\vee(\kappa,\chi_0)$ that is associated to the following representation of $K_0(\mathfrak{c})K_\infty$:

$$L^\vee(\kappa,\chi_0) = \mathrm{Sym}^k(V^\vee) \otimes \det_\infty^{-m} \otimes E(\chi_0)$$

(that is, the character $\chi_0^\vee$ is replaced by χ_0).

7.2 The automorphic description of intersection cohomology

As above, let $\kappa = (k,m) \in \mathcal{X}(L)$ and let $\chi : L^\times \backslash \mathbb{A}_L^\times \to \mathbb{C}^\times$ be a quasicharacter satisfying $\chi_\infty(b_\infty) = b_\infty^{-k-2m}$ for $b_\infty \in \mathbb{A}_{L\infty}^\times$. Assume that the conductor of χ divides the ideal $\mathfrak{c} \subset \mathcal{O}_L$. In this section we describe a basis of differential forms for the intersection cohomology groups $I^m H^*(X_0(\mathfrak{c}), \mathcal{L}(\kappa, \chi_0))$.

By Proposition 6.6 a modular form $f \in S_\kappa^{\mathrm{coh}}(K_0(\mathfrak{c}), \chi)$ gives rise to an L^2 holomorphic differential form $\omega_f \in \Omega^n(Y_0(\mathfrak{c}), \mathcal{L}(\kappa, \chi_0))$. We wish to use the isomorphism $\mathcal{Z}$ (of the Zucker conjecture, see Theorem 4.2) to associate an intersection cohomology class to such a differential form. We cannot apply this isomorphism as it stands since the representation $L(\kappa, \chi_0)$ does not necessarily extend to a representation of G. To overcome this, define

$$K_{11}(\mathfrak{c}) := \{\begin{pmatrix} a & b \\ c & d \end{pmatrix} \in K_0(\mathfrak{c}) : \begin{pmatrix} a & b \\ c & d \end{pmatrix} \equiv \begin{pmatrix} 1 & * \\ 0 & 1 \end{pmatrix} \pmod{\mathfrak{c}}\};$$

it is a finite-index subgroup of $K_0(\mathfrak{c})$. One then has a covering map

$$p : Y_{K_{11}(\mathfrak{c})} \longrightarrow Y_0(\mathfrak{c}).$$

Since χ_0 is trivial on $K_{11}(\mathfrak{c})$, one sees that there is a natural isomorphism

$$p^* \mathcal{L}(\kappa, \chi_0) = G(\mathbb{Q}) \backslash G(\mathbb{A}) \times_{K_\infty K_{11}(\mathfrak{c})} L(\kappa, \chi_{\mathrm{triv}}).$$

In particular, $p^* \mathcal{L}(\kappa, \chi_0)$ is the local system associated to the representation $L(\kappa, \chi_{\mathrm{triv}})$ of G and the Zucker isomorphism gives

$$\mathcal{Z} : H_{(2)}^*(Y_{K_{11}(\mathfrak{c})}, p^* \mathcal{L}(\kappa, \chi_0)) \xrightarrow{\sim} IH^*(X_{K_{11}(\mathfrak{c})}, p^* \mathcal{L}(\kappa, \chi_0)). \tag{7.2.1}$$

We can then define the Zucker isomorphism

$$\mathcal{Z} : H_{(2)}^*(Y_0(\mathfrak{c}), \mathcal{L}(\kappa, \chi_0)) \xrightarrow{\sim} IH^*(X_0(\mathfrak{c}), \mathcal{L}(\kappa, \chi_0)) \tag{7.2.2}$$

by $\mathcal{Z} := \frac{1}{[K_0(\mathfrak{c}):K_{11}(\mathfrak{c})]} p_* \circ \mathcal{Z} \circ p^*$. We therefore have for each $J \subseteq \Sigma(L)$ a homomorphism

$$\begin{aligned} \omega_J : S_\kappa^{\mathrm{coh}}(K_0(\mathfrak{c}), \chi) &\to I^m H^n(X_0(\mathfrak{c}), \mathcal{L}(\kappa, \chi_0)) \\ f &\mapsto \mathcal{Z}([\omega_J(f)]) = \mathcal{Z}([\iota_J^*(\omega_f)]). \end{aligned} \tag{7.2.3}$$

The space of *cuspidal cohomology classes* is defined as follows:

$$I^m H_{\mathrm{cusp}}^n(X_0(\mathfrak{c}), \mathcal{L}(\kappa, \chi_0)) := \bigoplus_\chi \bigoplus_{J \subset \Sigma(L)} \omega_J\left(S_\kappa^{\mathrm{coh}}(K_0(\mathfrak{c}), \chi)\right).$$

Here, the first sum is over all Hecke characters $\chi : L^\times \backslash \mathbb{A}_L^\times \to \mathbb{C}^\times$ such that $\chi_\infty(a_\infty) = a_\infty^{-k-2m}$ and $\chi|\mathcal{O}_L^\times = \chi_0$. If $j \neq n$ we set $I^m H_{\mathrm{cusp}}^j(X_0(\mathfrak{c}), \mathcal{L}(\kappa, \chi_0)) = 0$. In Proposition 7.13 we will prove that the mapping ω_J is Hecke-equivariant.

7.2. The automorphic description of intersection cohomology

We next isolate the subspace of invariant classes as in Section 4.4. These can only occur in the case that $\kappa = (0, m) \in \mathcal{X}(L)$ for some $m \in \mathbb{Z}^{\Sigma(L)}$. As explained in Section 4.4 for each $1 \subseteq j \subseteq h(K_0(\mathfrak{c}))$ there is a canonical injection

$$\varsigma_j : H^i(\mathfrak{g}, K_\infty; L(\kappa, \chi_{\text{triv}})) \hookrightarrow H^i_{(2)}(Y_0^j(\mathfrak{c}), \mathcal{L}(\kappa, \chi_{\text{triv0}})) \hookrightarrow H^i(Y_0(\mathfrak{c}), \mathcal{L}(\kappa, \chi_{\text{triv0}})),$$

Here $H^i(\mathfrak{g}, K_\infty; L(\kappa, \chi_{\text{triv0}}))$ is the relative Lie algebra cohomology of the Lie algebra $\mathfrak{g}$ of $G(\mathbb{R})$ and χ_{triv} is the trivial character. Translating this into the adèlic language using (5.1.6) we thus have a subspace of *invariant cohomology classes*

$$I^m H^i_{\text{inv}}(X_0(\mathfrak{c}), \mathcal{L}((0, m), \chi_0)) := \mathcal{Z}_\varsigma \left(\bigoplus_\phi H^i(\mathfrak{g}, K_\infty; \mathbb{C}\phi \otimes L((0, m), \chi_{\text{triv0}})) \right)$$
(7.2.4)

sitting inside $I^m H^i(X_0(\mathfrak{c}), \mathcal{L}(\kappa, \chi_{\text{triv0}}))$. The sum is over the set of functions $\phi : G(\mathbb{Q}) \backslash G(\mathbb{A}) \to \mathbb{C}^\times$ of the form $\phi = \chi \circ \det$ for some quasi-character $\chi : L^\times \backslash \mathbb{A}_L^\times \to \mathbb{C}^\times$ that is unramified at all finite places; this implies that ϕ is trivial on $K_0(\mathfrak{c})$. In other words, ϕ factors through the determinant map (5.1.6). For a more explicit description of these classes, see Section 11.2 below. We set

$$I^m H^*_{\text{inv}}(X_0(\mathfrak{c}), \mathcal{L}(\kappa, \chi_0)) = 0$$

if $\kappa \neq (0, m)$ for some $m \in \mathbb{Z}\mathbf{1}$ or $\chi_0 \neq \chi_{\text{triv0}}$.

The following theorem of Harder [Har] (as rephrased by Hida [Hid7, Section 3]) says that these two subspaces of the intersection cohomology actually fill the space:

Theorem 7.1 (Harder). *The middle intersection cohomology decomposes as follows:*

$$I^m H^*(X_0(\mathfrak{c}), \mathcal{L}(\kappa, \chi_0)) = I^m H^*_{\text{inv}}(X_0(\mathfrak{c}), \mathcal{L}(\kappa, \chi_0)) \oplus I^m H^*_{\text{cusp}}(X_0(\mathfrak{c}), \mathcal{L}(\kappa, \chi_0)).$$

Remark. We note that this theorem is not phrased in the language of intersection cohomology in either [Har] or [Hid7]. However, it is straightforward to deduce the given statement using the identification

$$I^m H^n(X_0(\mathfrak{c}), \mathcal{L}(\kappa, \chi_0)) = \text{Image}\left(H^n_c(Y_0(\mathfrak{c}), \mathcal{L}(\kappa, \chi_0)) \to H^n(Y_0(\mathfrak{c}), \mathcal{L}(\kappa, \chi_0))\right)$$

which holds since $X_0(\mathfrak{c})$ has isolated singularities at the cusps and orbifold singularities in the interior.

7.3 Complex conjugation

There are several different notions of "complex conjugation" in the current setting.

(1) If f, g are (holomorphic) modular forms with values in $\mathbb{C}$, the Petersson product involves the values of f and $\bar{g}$.

(2) If f is a (holomorphic) modular form with values in $\mathbb{C}$ then $\bar{f}$ will be anti-holomorphic, but $f_c(z) = \bar{f}(\bar{z})$ will be holomorphic in z.

(3) If ω_f is the holomorphic differential form that corresponds to a (holomorphic) modular form f then $\iota_J^*(\omega_f)$ will be a differential form of type $(n - |J|, |J|)$ with $n - |J|$ factors of type dz_i and $|J|$ factors of type $d\bar{z}_i$.

(4) If a modular form f has a Fourier expansion

$$f((\begin{smallmatrix} y & x \\ 0 & 1 \end{smallmatrix})) = |y|_{\mathbb{A}_L} \sum_{\xi \gg 0} a(\mathfrak{m}, f) q_\kappa(\xi x, \xi y)$$

(where $\mathfrak{m} = \xi y \mathcal{D}_{L/\mathbb{Q}}$) then one might consider the function obtained by replacing each Fourier coefficient $a(\mathfrak{m})$ by its complex conjugate.

In [Shim4, Thm. 1.5] Shimura constructs an action of a certain Galois group on the space of Hilbert modular forms by acting on the Fourier coefficients of the modular form. This action is further described in [Hid3, Thm. 4.4] and [Hid5, (2.1)]. For the Galois group element that is complex conjugation, it may be described as follows. Define $w \in \mathrm{GL}_2(\mathbb{A}_L)$ by $w_v = I$ for $v < \infty$ and $w_\sigma = (\begin{smallmatrix} -1 & 0 \\ 0 & 1 \end{smallmatrix})$ for $\sigma \in \Sigma(L)$. Let $f : G(\mathbb{A}) \to \mathbb{C}$ be a modular form. Then the *complex conjugate modular form* is

$$f_c(g) := \bar{f}(gw).$$

If we consider the function f to be a section of a vector bundle over $D \times G(\mathbb{A}_f)$ where $D = (\mathbb{C} - \mathbb{R})^{\Sigma(L)}$ then by taking $g = (\begin{smallmatrix} y & x \\ 0 & 1 \end{smallmatrix}) \in G(\mathbb{R})$ and $z = x + \mathbf{i}y$ we see that $f_c(z, g_0) = \bar{f}(\bar{z}, g_0)$ for all $(z, g_0) \in D \times G(\mathbb{A}_f)$.

Lemma 7.2. *Let $\kappa = (k, m) \in \mathcal{X}(L)$ be a weight and set $[k + 2m] := k_\sigma + 2m_\sigma$ (which is independent of $\sigma \in \Sigma(L)$). Let $\chi : L^\times \backslash \mathbb{A}_L^\times \to \mathbb{C}^\times$ be a Hecke character such that $\chi_\infty(b_\infty) = b_\infty^{-k-2m}$, whose conductor divides $\mathfrak{c}$. Then complex conjugation of modular forms defines isomorphisms*

$$S_\kappa(K_0(\mathfrak{c}), \chi) \to S_\kappa(K_0(\mathfrak{c}), \chi^{-1} |\cdot|^{-2[k+2m]})$$
$$S_\kappa^{\mathrm{coh}}(K_0(\mathfrak{c}), \chi) \to S_\kappa^{\mathrm{coh}}(K_0(\mathfrak{c}), \chi^{-1} |\cdot|^{-2[k+2m]})$$

where $|\cdot|^{-2[k+2m]}$ denotes the character $t \mapsto |t|_{\mathbb{A}_L}^{-2[k+2m]}$.

Proof. Let $f \in S_\kappa(K_0(\mathfrak{c}), \chi)$. Let $t \in \mathbb{A}_L^\times$, $\gamma \in G(\mathbb{Q})$, $g \in G(\mathbb{A})$, $u_\infty^1 = (\begin{smallmatrix} \cos\theta & \sin\theta \\ -\sin\theta & \cos\theta \end{smallmatrix}) \in K_\infty^1$ and $u_0 = (\begin{smallmatrix} a & b \\ c & d \end{smallmatrix}) \in K_0(\mathfrak{c})$ so that

$$f(t\gamma g u_\infty^1 u_0) = \chi(t) f(g) \mu(u_\infty^1) \chi_0(u_0)$$

7.4. Atkin-Lehner operator   117

where $\mu(u_\infty^1) = e^{i\pi(k+2)\theta}$. Note that $\overline{\chi}(t) = \chi^{-1}(t)|t|_{\mathbb{A}_L}^{-2[k+2m]}$ for $t \in \mathbb{A}_L^\times$ by equation (C.3.2). Hence

$$\begin{aligned}
f_c(t\gamma g u_\infty^1 u_0) &= \overline{f}(t\gamma g u_\infty^1 u_0 w) \\
&= \overline{\chi}(t)\overline{\mu}(w^{-1}u_\infty^1 w)\overline{\chi}_0(u_0)\overline{f}(gw) \quad \text{(since w commutes with u_0)} \\
&= \overline{\chi}(t)\overline{\mu}\left(\left(\begin{smallmatrix}\cos\theta & -\sin\theta \\ \sin\theta & \cos\theta\end{smallmatrix}\right)\right)\overline{\chi}_0(u_0)\overline{f}(gw) \\
&= \overline{\chi}(t)\mu(u_\infty^1)\overline{\chi}_0(u_0) f_c(g)
\end{aligned}$$

which shows that $f_c \in S_\kappa(K_0(\mathfrak{c}), \chi^{-1}|\cdot|^{-2[k+2m]})$. The argument for S_κ^{coh} is similar. □

The following lemma tells us that notions (2) and (4) of complex conjugation coincide:

Lemma 7.3. *Let $f \in S_\kappa(K_0(\mathfrak{c}), \chi)$. For $y \in \mathbb{A}_L^\times$ such that $y_\sigma > 0$ for all $\sigma \in \Sigma(L)$ one has*

$$f_c\left(\left(\begin{smallmatrix}y & x \\ 0 & 1\end{smallmatrix}\right)\right) = |y|_{\mathbb{A}_L} \sum_{\xi \gg 0} \overline{a(\xi y \mathcal{D}_{L/\mathbb{Q}}, f)} q_\kappa(\xi x, \xi y).$$

Proof. We have

$$f_c\left(\left(\begin{smallmatrix}y & x \\ 0 & 1\end{smallmatrix}\right)\right) = \overline{f}\left(\left(\begin{smallmatrix}(-y)_\infty y_0 & x \\ 0 & 1\end{smallmatrix}\right)\right) = \overline{f}\left(\left(\begin{smallmatrix}-1 & 0 \\ 0 & 1\end{smallmatrix}\right)\left(\begin{smallmatrix}(-y)_\infty y_0 & x \\ 0 & 1\end{smallmatrix}\right)\right) = \overline{f}\left(\left(\begin{smallmatrix}y_\infty(-y_0) & -x \\ 0 & 1\end{smallmatrix}\right)\right)$$

by the left $G(\mathbb{Q})$-invariance of f. On the other hand

$$\overline{f}\left(\left(\begin{smallmatrix}y_\infty(-y_0) & -x \\ 0 & 1\end{smallmatrix}\right)\right) = |y|_{\mathbb{A}_L} \sum_{\xi \gg 0} \overline{a(\xi y \mathcal{D}_{L/\mathbb{Q}}, f)} q_\kappa(\xi x, \xi y).$$

since $(\xi y_0 \mathcal{D}_{L/\mathbb{Q}}) = (\xi(-y_0)\mathcal{D}_{L/\mathbb{Q}})$ as ideals of $\mathcal{O}_L$. □

7.4 Atkin-Lehner operator

In this section we describe the Atkin-Lehner operator, which acts on the modular variety $Y_0(\mathfrak{c})$ and on its compactification $X_0(\mathfrak{c})$. This action preserves newforms and it takes a newform f to its complex conjugate f_c (times a factor). Since it acts on the space $Y_0(\mathfrak{c})$ it also acts on cycles in the space, and therefore gives us a cycle-level "lift" of complex conjugation $f \mapsto f_c$. The Atkin-Lehner operator is also needed in order to compare the Petersson inner product on modular forms with the intersection pairing (see Theorem 7.11). The intersection pairing involves two different groups, $IH^n(X_0(\mathfrak{c}); \mathcal{L}(\kappa, \chi_0))$ and $IH^n(X_0(\mathfrak{c}); \mathcal{L}^\vee(\kappa, \chi_0))$. The Atkin-Lehner operator converts the local system $\mathcal{L}^\vee(\kappa, \chi_0)$ into its dual, $\mathcal{L}(\kappa, \chi_0)$, see Lemma 7.4, so the intersection pairing becomes an inner product.

More precisely we will construct a commutative diagram

$$
\begin{array}{ccccc}
S_\kappa(K_0(\mathfrak{c}),\chi) & \xrightarrow{-\iota} & S_\kappa^{\mathrm{coh}}(K_0(\mathfrak{c}),\chi) & \xrightarrow{\omega_J} & \Omega^n(Y_0(\mathfrak{c}),\mathcal{L}(\kappa,\chi_0)) \\
c \Big\downarrow \Big\downarrow W_\mathfrak{c}^* & & \vdots W_\mathfrak{c}^{*\,\mathrm{coh}} & & \Big\downarrow W_\mathfrak{c}^* \\
S_\kappa(K_0(\mathfrak{c}),\psi) & \xrightarrow[-\iota]{} & S_\kappa^{\mathrm{coh}}(K_0(\mathfrak{c}),\psi) & \xrightarrow[\omega_J]{} & \Omega^n(Y_0(\mathfrak{c}),\mathcal{L}(\kappa,\psi_0))
\end{array}
\qquad (7.4.1)
$$

where $J \subset \Sigma(L)$, where $\psi = \overline{\chi} = \chi^{-1}|\cdot|^{-2[k+2m]}$, where c denotes complex conjugation, and where the mapping $W_\mathfrak{c}^*$ in the first column differs from c by a scalar multiple, see Proposition 7.7 and equation (7.4.9).

Assume as in the previous sections that $\mathfrak{c} \subset \mathcal{O}_L$ is a fixed ideal. Let $\widetilde{c} \in \mathbb{A}_L^\times$ be an idèle such that $\widetilde{c}_v = 1$ if $v|\infty$ and such that

$$[\widetilde{c}_0] = \mathfrak{c}$$

where $\widetilde{c}_0$ is the finite part of $\widetilde{c}$, and $[\widetilde{c}_0]$ is its associated fractional ideal. Then the Atkin-Lehner matrix is

$$W_\mathfrak{c} := \begin{pmatrix} 0 & -1 \\ \widetilde{c} & 0 \end{pmatrix} \in G(\mathbb{A}). \qquad (7.4.2)$$

Right multiplication by the element $W_\mathfrak{c}$ (or by $W_\mathfrak{c}^\iota = -W_\mathfrak{c}$) induces a well-defined mapping on the Hilbert modular variety $Y_0(\mathfrak{c}) = G(\mathbb{Q})\backslash G(\mathbb{A})/K_0(\mathfrak{c})K_\infty$, for if $g \in G(\mathbb{A})$, if $u_0 \in K_0(\mathfrak{c})$ and if $u_\infty \in K_\infty$ then

$$G(\mathbb{Q})gW_\mathfrak{c}K_0(\mathfrak{c})K_\infty = G(\mathbb{Q})gu_0u_\infty W_\mathfrak{c}K_0(\mathfrak{c})K_\infty \in Y_0(\mathfrak{c})$$

because $W_\mathfrak{c}$ commutes with K_∞ and because $W_\mathfrak{c}^{-1}u_0W_\mathfrak{c} \in K_0(\mathfrak{c})$. Moreover this mapping is invertible: in fact,

$$(W_\mathfrak{c})^2 = \begin{pmatrix} -\widetilde{c} & 0 \\ 0 & -\widetilde{c} \end{pmatrix}$$

which is obviously invertible. In fact its action on $Y_0(\mathfrak{c})$ has finite order because $W_\mathfrak{c}^2$ acts through the idèle class group.

The action of the component group $\pi_0(G(\mathbb{R}))$ (see Section 6.11) commutes with the Atkin-Lehner operator. The characters $\chi_0, \chi_0^\vee$ of $K_0(\mathfrak{c})$ are also related via the Atkin-Lehner operator as follows:

$$\chi_0^\vee(u_0) = \chi_0(W_\mathfrak{c}^{-1}u_0W_\mathfrak{c}) \quad \text{for all} \quad u_0 \in K_0(\mathfrak{c}). \qquad (7.4.3)$$

Let $E(\chi_0) \cong \mathbb{C}$ be the one-dimensional vector space on which $K_0(\mathfrak{c})$ acts by χ_0 and let

$$\mathcal{E}(\chi_0) = G(\mathbb{Q})\backslash D \times G(\mathbb{A}_f) \times_{K_0(\mathfrak{c})} E(\chi_0)$$

be the corresponding line bundle on $Y_0(\mathfrak{c})$.

7.4. Atkin-Lehner operator

Lemma 7.4. *The Atkin-Lehner operator $W_{\mathfrak{c}} : Y_0(\mathfrak{c}) \to Y_0(\mathfrak{c})$ determines natural isomorphisms*

$$\mathcal{E}(\chi_0) \cong W_{\mathfrak{c}}^*(\mathcal{E}(\chi_0^\vee)) \text{ and } \mathcal{E}(\chi_0^\vee) \cong W_{\mathfrak{c}}^*(\mathcal{E}(\chi_0))$$

and hence also, an isomorphism

$$\mathcal{L}^\vee(\kappa, \chi_0) \cong W_{\mathfrak{c}}^* \mathcal{L}(\kappa, \chi_0).$$

Proof. Since $W_{\mathfrak{c}}$ commutes with K_∞ it suffices to show that the map

$$\mathcal{E}(\chi_0^\vee) = G(\mathbb{A}_f) \times_{K_0(\mathfrak{c})} E(\chi_0^\vee) \to G(\mathbb{A}_f) \times_{K_0(\mathfrak{c})} E(\chi_0) = \mathcal{E}(\chi_0)$$

given by

$$[x_0, v] \mapsto [x_0 W_{\mathfrak{c}}, v]$$

is well defined (for then it will obviously be an isomorphism). If $[x_0 k, v] \in \mathcal{E}(\chi_0^\vee)$ with $x_0 \in G(\mathbb{A}_f)$ and $k \in K_0(\mathfrak{c})$, then the following diagram commutes:

$$\begin{array}{ccc}
\mathcal{E}(\chi_0^\vee) & \longrightarrow & \mathcal{E}(\chi_0) \\
[x_0 k, v] & \longrightarrow & [x_0 k W_{\mathfrak{c}}, v] = [x_0 W_{\mathfrak{c}}, \chi_0(W_{\mathfrak{c}}^{-1} k W_{\mathfrak{c}})v] \\
\| & & \| \\
[x_0, \chi_0^\vee(k) v] & \longrightarrow & [x_0 W_{\mathfrak{c}}, \chi_0^\vee(k) v]
\end{array}$$

which proves that the mapping is well defined. $\square$

In order to obtain a mapping $W_{\mathfrak{c}}^*$ as in diagram (7.4.1), we want to land in the space of modular forms with character

$$\psi(t) = \overline{\chi}(t) = \chi^{-1}(t) |t|_{\mathbb{A}_L}^{-2[k+2m]}$$

rather than $\chi^\vee$ so we need to modify the above isomorphism slightly by multiplying by a complex number of norm 1:

Definition 7.5. The mapping $W_{\mathfrak{c}}^* : S_\kappa(K_0(\mathfrak{c}), \chi) \to S_\kappa(K_0(\mathfrak{c}), \psi)$ is defined by

$$W_{\mathfrak{c}}^*(f)(\alpha) := \frac{|\chi(\det \alpha)|}{\chi(\det \alpha)} f(\alpha W_{\mathfrak{c}}). \tag{7.4.4}$$

(See also [Hid3] Section 7, p. 354.) We similarly define an isomorphism

$$\mathcal{L}(\kappa, \psi_0) \cong W_{\mathfrak{c}}^*(\mathcal{L}(\kappa, \chi_0))$$

by the mapping

$$G(\mathbb{A}) \times_{K_\infty K_0(\mathfrak{c})} L(\kappa) \otimes E(\chi_0^\vee) \to G(\mathbb{A}) \times_{K_\infty K_0(\mathfrak{c})} L(\kappa) \otimes E(\psi_0^\vee)$$

given by

$$[\alpha, v] \mapsto \left[\alpha W_{\mathfrak{c}}, \frac{\chi(\det \alpha)}{|\chi(\det \alpha)|} \chi(\det W_{\mathfrak{c}}) v\right] \tag{7.4.5}$$

for $\alpha \in G(\mathbb{A})$ and $v \in L(\kappa) \otimes E(\chi_0^\vee)$.

Proposition 7.6. *Definition (7.4.4) does indeed map $S_\kappa(K_0(\mathfrak{c}),\chi)$ to $S_\kappa(K_0(\mathfrak{c}),\psi)$. The mapping (7.4.5) is well defined. It induces a mapping on differential forms*

$$W_\mathfrak{c}^* : \Omega^n(Y_0(\mathfrak{c}), \mathcal{L}(\kappa,\chi_0)) \to \Omega^n(Y_0(\mathfrak{c}), \mathcal{L}(\kappa,\psi_0))$$

which is given by

$$W_\mathfrak{c}^*\omega(z,x_0) = \frac{\chi(\det \alpha_z x_0)}{|\chi(\det \alpha_z x_0)|} \chi(\det W_\mathfrak{c}) \omega(z, x_0 W_\mathfrak{c}) \qquad (7.4.6)$$

where $\alpha_z \cdot \mathbf{i} = z \in D = G(\mathbb{R})/K_\infty$ and $x_0 \in G(\mathbb{A}_f)$. This mapping commutes with the action ι_J of the component group.

Proof. First let us check that $f \in S_\kappa(K_0(\mathfrak{c}),\chi) \implies W_\mathfrak{c}^* f \in S_\kappa(K_0(\mathfrak{c}),\psi)$. Notice the following preliminary properties of χ. We may write $\chi(t) = \chi_1(t)|t|_{\mathbb{A}_L}^s$ for some unitary Hecke character χ_1 and $s = -[k+2m] = -k_\sigma - 2m_\sigma$ for any $\sigma \in \Sigma(L)$. If $u_\infty = \begin{pmatrix} u_1 & u_2 \\ -u_2 & u_1 \end{pmatrix} \in K_\infty$ then

$$\chi(\det u_\infty) = |\chi(\det u_\infty)| = (u_1^2 + u_2^2)^s.$$

Moreover, $\chi_{1,v}(t_v) = 1$ if $v \nmid \mathfrak{c}$ and $t_v \in \mathcal{O}_{F_v}^\times$. For any $u \in G(\widehat{\mathbb{Z}})$ and any $v < \infty$, we have that $(\det u)_v \in \mathcal{O}_{L_v}^\times$ so $|\det u_v|_v = 1$. Consequently, if $u_0 = \begin{pmatrix} a & b \\ c & d \end{pmatrix} \in K_0(\mathfrak{c})$ then:

$$\chi(\det u_0) = \prod_{\mathfrak{p}_v | \mathfrak{c}} \chi_v(a_v d_v) \quad \text{and} \quad |\chi(\det u_0)| = 1. \qquad (7.4.7)$$

Now let $t \in \mathbb{A}_L^\times$, $\gamma \in G(\mathbb{Q})$, $g \in G(\mathbb{A})$, $u_\infty^1 = \begin{pmatrix} \cos\theta & \sin\theta \\ -\sin\theta & \cos\theta \end{pmatrix} \in K_\infty^1$. Set $\mu(u_\infty^1) = e^{i(k+2)\theta}$. Then

$$\begin{aligned}
W_\mathfrak{c}^* f(t\gamma\alpha u_\infty^1 u_0) &= \frac{|t|_\mathbb{A}^{2s}}{\chi(t^2)} \frac{|\chi(\det \alpha u_0)|}{\chi(\det \alpha u_0)} f(t\gamma\alpha u_\infty^1 u_0 W_\mathfrak{c}) \\
&= \chi^{-1}(t)|t|_\mathbb{A}^{2s} \frac{|\chi(\det \alpha u_0)|}{\chi(\det \alpha u_0)} f(\alpha W_\mathfrak{c})\mu(u_\infty^1)\chi_0(W_\mathfrak{c}^{-1} u_0 W_\mathfrak{c}) \\
&= \psi(t) \frac{|\chi(\det \alpha)|}{\chi(\det \alpha)} f(\alpha W_\mathfrak{c})\mu(u_\infty^1) \prod_{\mathfrak{p}_v|\mathfrak{c}} \chi_v(a_v d_v)^{-1} \prod_{\mathfrak{p}_v|\mathfrak{c}} \chi_v(a_v) \\
&= \psi(t) \frac{|\chi(\det \alpha)|}{\chi(\det \alpha)} f(\alpha W_\mathfrak{c})\mu(u_\infty^1)\psi_0(u_0) \\
&= \psi(t) W_\mathfrak{c}^* f(\alpha)\mu(u_\infty^1)\psi_0(u_0)
\end{aligned}$$

A similar calculation shows that equation (7.4.5) is well defined. Equation (7.4.6) is obtained from (7.4.5). □

To complete the construction of diagram (7.4.1) we need to define the action of Atkin-Lehner on S_κ^{coh} and check that the diagram does indeed commute. There are no choices involved in this definition if we insist that the diagram commute, and we obtain

7.4. Atkin-Lehner operator

Proposition 7.7. *The mapping* $W_{\mathfrak{c}}^{*\,\mathrm{coh}} : S_\kappa^{\mathrm{coh}}(K_0(\mathfrak{c}), \chi) \to S_\kappa^{\mathrm{coh}}(K_0(\mathfrak{c}), \psi)$ *given by* $W_{\mathfrak{c}}^{*\,\mathrm{coh}} f := (W_{\mathfrak{c}}^* f^{-\iota})^{-\iota}$ *may be expressed as follows:*

$$W_{\mathfrak{c}*}^{\mathrm{coh}} f(\alpha) := \frac{\chi(\det \alpha)}{|\chi(\det \alpha)|} \chi(\det W_{\mathfrak{c}}) f(\alpha W_{\mathfrak{c}})$$

for $\alpha \in G(\mathbb{A})$. *In particular the diagram (7.4.1) is commutative so that*

$$W_{\mathfrak{c}}^* \omega_J(f^{-\iota}) = \omega_J(W_{\mathfrak{c}}^*(f)^{-\iota}) \tag{7.4.8}$$

for all $f \in S_\kappa^{\mathrm{coh}}(K_0(\mathfrak{c}), \chi)$ *and all* $J \subset \Sigma(L)$. *The complex conjugate of* f *is given by* $f_c = (-1)^{\{k\}}((f^{-\iota})_c)^{-\iota}$ *where* $\{k\} = \sum_{\sigma \in \Sigma(L)} k_\sigma$. □

We have the following result from Shimura [Shim4] and [Hid5, (4.10b)] which relates the complex conjugate f_c and the Atkin-Lehner operator.

Lemma 7.8. *If* $f \in S_\kappa(K_0(\mathfrak{c}), \chi)$ *is a newform then*

$$W_{\mathfrak{c}}^*(f) = W(f) \mathrm{N}_{L/\mathbb{Q}}(\mathfrak{c})^{[k+2m]/2} f_c, \tag{7.4.9}$$

where $W(f)$ *is a complex number of norm 1. If* $f \in S_\kappa^{\mathrm{coh}}(K_0(\mathfrak{c}), \chi)$ *is a newform it then follows that*

$$W_{\mathfrak{c}}^{*\,\mathrm{coh}}(f) = (-1)^{\{k\}} W(f^{-\iota}) \mathrm{N}_{L/\mathbb{Q}}(\mathfrak{c})^{[k+2m]/2} f_c \tag{7.4.10}$$

where $\{k\} = \sum_{\sigma \in \Sigma(L)} k_\sigma$.

Proof. Let $f \in S_\kappa(K_0(\mathfrak{c}), \chi)$ and let $\mathfrak{m} \subset \mathcal{O}_L$ be an ideal with $\mathfrak{m} + \mathfrak{c} = \mathcal{O}_L$. It is immediate from the definition of $W_{\mathfrak{c}}^*$ that

$$W_{\mathfrak{c}}^*(f)|T_{\mathfrak{c}}(\mathfrak{m}) = \frac{\chi(\mathfrak{m})}{|\chi(\mathfrak{m})|} W_{\mathfrak{c}}^*(f|T_{\mathfrak{c}}(\mathfrak{m})).$$

On the other hand, if $f \in S_\kappa(K_0(\mathfrak{c}), \chi)$ then

$$(f|T_{\mathfrak{c}}(\mathfrak{m}), f)_P = \chi| \cdot |^{[k+2m]}(\mathfrak{m})(f, f|T_{\mathfrak{c}}(\mathfrak{m}))_P$$

by Lemma 5.6. In other words, the eigenvalue of $T_{\mathfrak{c}}(\mathfrak{m})$ acting on $W_{\mathfrak{c}}^*(f)$ is the complex conjugate of the eigenvalue of $T_{\mathfrak{c}}(\mathfrak{m})$ acting on f, at least for $\mathfrak{m} + \mathfrak{c} = \mathcal{O}_L$.

It follows that if f is a simultaneous eigenform for all Hecke operators then the Hecke eigenvalues attached to ideals $\mathfrak{m} \subseteq \mathcal{O}_L$ with $\mathfrak{m} + \mathfrak{c} = \mathcal{O}_L$ of $W_{\mathfrak{c}}^*(f)$ and f_c agree. On the other hand, complex conjugation preserves newforms, and $W_{\mathfrak{c}}^*$ preserves newforms since it preserves the kernels of the trace maps recalled in Section 5.8. In particular by strong multiplicity one, $W_{\mathfrak{c}}^*$ maps f to a nonzero complex multiple of f_c, that is, if $f \in S_\kappa(K_0(\mathfrak{c}), \chi)$ is a newform then

$$W_{\mathfrak{c}}^*(f) = W(f) \mathrm{N}(\mathfrak{c})^{[k+2m]/2} f_c$$

for some nonzero complex number $W(f)$. To complete the proof we need to check that $|W(f)| = 1$. Let $f, g \in S_\kappa(K_0(\mathfrak{c}), \chi)$. One has

$$(W_\mathfrak{c}^*(f), W_\mathfrak{c}^*(g))_P = \int_{Y_0(\mathfrak{c})} \overline{g}(\alpha W_\mathfrak{c}) \frac{|\chi(\det \alpha)|^2}{\overline{\chi}(\det \alpha)\chi(\det \alpha)} f(\alpha W_\mathfrak{c}) |\det(\alpha)|_{\mathbb{A}_L}^{[k+2m]} d\mu_\mathfrak{c}(\alpha)$$

$$= \int_{Y_0(\mathfrak{c})} \overline{g}(\alpha W_\mathfrak{c}) f(\alpha W_\mathfrak{c}) |\det(\alpha)|_{\mathbb{A}_L}^{[k+2m]} d\mu_\mathfrak{c}(\alpha).$$

Since $W_\mathfrak{c}$ normalizes $K_0(\mathfrak{c})$ and $d\mu_\mathfrak{c}$ is constructed from a Haar measure, this last quantity is $\mathrm{N}(\mathfrak{c})^{-[k+2m]}(f, g)_P$. Thus

$$(W_\mathfrak{c}^*(f), W_\mathfrak{c}^*(g)) = \mathrm{N}(\mathfrak{c})^{[k+2m]}(f, g)_P.$$

On the other hand,

$$(W_\mathfrak{c}^*(f), W_\mathfrak{c}^*(f)) = |W(f)\mathrm{N}(\mathfrak{c})^{[k+2m]/2}|^2 (f_c, f_c)_P = |W(f)\mathrm{N}(\mathfrak{c})^{[k+2m]/2}|^2 (f, f)_P$$

which implies that $|W(f)|^2 = 1$. $\square$

7.5 Pairings of vector bundles

In order to have a product on the intersection cohomology of $X_0(\mathfrak{c})$ it is first necessary to have an inner product on the local coefficient systems. There is a natural pairing on $\mathrm{Sym}^k(V)$ (see below) which, when extended to the local system, gives a pairing between $\mathcal{L}(\kappa, \chi_0)$ and $\mathcal{L}^\vee(\kappa, \chi_0)$. Then we use use the Atkin-Lehner operator to convert this into a product on $\mathcal{L}(\kappa, \chi_0)$.

For each $k \in (\mathbb{Z}^{\geq 0})^{\Sigma(L)}$ we have the representation of $G(\mathbb{R}) \cong \mathrm{GL}_2(\mathbb{R})^{\Sigma(L)}$ on the vector space

$$\mathrm{Sym}^k(V^\vee) = \otimes_{\sigma \in \Sigma(L)} \mathrm{Sym}^{k_\sigma}(V_\sigma^\vee)$$

of polynomials that are homogeneous of degree k_σ on V_σ. Following [Shim1] Section 8.2 and [Hid6] Section 6.2, there is a natural $G(\mathbb{R})$-equivariant pairing

$$\mathrm{Sym}^k(V^\vee) \times \mathrm{Sym}^k(V^\vee) \to \det^{-k}, \qquad (7.5.1)$$

see Definition 7.9, which is constructed as follows.

Fix $\sigma \in \Sigma(L)$ and by abuse of notation write k rather than k_σ in this paragraph only. Use the standard basis of $\mathbb{C}^2$ to identify $\mathrm{Sym}^k(V_\sigma)$ with the vector space of homogeneous degree k polynomials in two variables, which in turn has the basis $\{x^k, x^{k-1}y, \ldots, xy^{k-1}, y^k\}$. The $k+1 \times k+1$ matrix

$$\Psi_{ij} := \begin{cases} (-1)^{j-1} \binom{k}{j-1} & \text{if } i-1+j-1 = k \\ 0 & \text{otherwise} \end{cases} \qquad (7.5.2)$$

7.5. Pairings of vector bundles

defines a bilinear form $\Psi(A, B) := {}^t A \Psi B$ on $\mathrm{Sym}^k(V_\sigma)$ such that

$$\Psi\left((u^k, u^{k-1}v, \ldots, v^k), (x^k, x^{k-1}y, \ldots, y^k)\right) = (uy - vx)^k = \det\begin{pmatrix} u & x \\ v & y \end{pmatrix}^k.$$

If $\gamma \in \mathrm{GL}(V_\sigma)$ then $\mathrm{Sym}^k(\gamma)$ acts on $\mathrm{Sym}^k(V_\sigma)$ (with action we denote by $\gamma \cdot$) and we have:

$$\Psi(\gamma \cdot A, \gamma \cdot B) = \det(\gamma)^k \Psi(A, B) \qquad (7.5.3)$$

for any $A, B \in \mathrm{Sym}^k(V_\sigma)$. Using the dual basis $\{X^k, X^{k-1}Y, \ldots, XY^{k-1}, Y^k\}$, and noting equation (6.6.3), this gives the following inner product on $\mathrm{Sym}^k(V_\sigma^\vee)$:

$$\left\langle \sum_{j=0}^{k} u_j X^{k-j} Y^j, \sum_{j=0}^{k} w_j X^{k-j} Y^j \right\rangle = \sum_{j=0}^{k} (-1)^j \binom{k}{j}^{-1} u_j w_{k-j}. \qquad (7.5.4)$$

In particular, we obtain the following, which will be needed in the proof of Theorem 7.11 and Proposition 9.2:

$$\left\langle (-X + zY)^k, (-X + wY)^k \right\rangle = (z - w)^k. \qquad (7.5.5)$$

Multiplying the infinite places together gives the inner product (7.5.1):

Definition 7.9. Let $k = (k_\sigma) \in (\mathbb{Z}^{\geq 0})^{\Sigma(L)}$. If $p = \otimes_{\sigma \in \Sigma(L)} p_\sigma$ and $p' = \otimes_{\sigma \in \Sigma(L)} p'_\sigma$ are elements of $\mathrm{Sym}^k(V^\vee)$, set

$$\langle p, p' \rangle := \prod_{\sigma \in \Sigma(L)} \langle p_\sigma, p'_\sigma \rangle.$$

This product has the following equivariance property:

$$\langle \gamma.p, \gamma.p' \rangle = \det(\gamma)^{-k} \langle p, p' \rangle \qquad (7.5.6)$$

for all $\gamma \in G(\mathbb{R})$, where as usual, $\det(\gamma)^{-k} := \prod_{\sigma \in \Sigma(L)} \det(\gamma_\sigma)^{-k_\sigma}$.

Let $\kappa = (k, m) \in \mathcal{X}(L)$ be an allowable weight and let $\chi = \chi_0 \chi_\infty : \mathbb{A}_L^\times \to \mathbb{C}^\times$ be a quasi character such that $\chi_\infty(b_\infty) = b_\infty^{-k-2m}$. Let $\mathcal{L}(\kappa, \chi_0)$ and $\mathcal{L}^\vee(\kappa, \chi_0)$ be the resulting local systems. Let $\chi \circ \det$ denote the quasi-character of $G(\mathbb{A})$ given by $\gamma \mapsto \chi(\det(\gamma))$.

Proposition 7.10. *The $G(\mathbb{R})$-equivariant pairing* (7.5.1),

$$\mathrm{Sym}^k(V^\vee) \times \mathrm{Sym}^k(V^\vee) \to \det_\infty^{-k}$$

extends canonically to a $G(\mathbb{R})K_0(\mathfrak{c})$-equivariant nondegenerate pairing

$$\langle \cdot, \cdot \rangle : L(\kappa, \chi_0) \times L^\vee(\kappa, \chi_0) \to \chi \circ \det \to \mathbb{C}.$$

Proof. Extend the pairing (7.5.1) to $L(\kappa, \chi_0)$ at the infinite places, using multiplication, which gives $\det_\infty^{-m} \otimes \det_\infty^{-m} \to \det_\infty^{-2m}$. Since $\chi_\infty \circ \det_\infty = \det_\infty^{-k-2m}$ this defines the pairing at the infinite places. At the finite places, define

$$E(\chi_0^\vee) \otimes E(\chi_0) \to \chi_0 \circ \det{}_0 \qquad (7.5.7)$$

(where $\det_0 : G(\mathbb{A}_f) \to \mathbb{A}_f^\times$ and $\chi_0 : \mathbb{A}_f^\times \to \mathbb{C}^\times$ denote the restrictions to the finite places) also using multiplication as follows. If $u_0 = \begin{pmatrix} a & b \\ c & d \end{pmatrix} \in K_0(\mathfrak{c})$ then

$$\chi_0(u_0)^\vee \chi_0(u_0) = \prod_{\mathfrak{p}_v | \mathfrak{c}} d_v a_v = \chi(\det(u_0))$$

so this defines the pairing to the one-dimensional vector space $E(\chi \circ \det)$ at the finite places.

The one-dimensional local system $\mathcal{E}(\chi \circ \det)$ is trivial because $\chi \circ \det$ is trivial on $G(\mathbb{Q})$, cf. Section 6.2. In fact, a vector in $\mathcal{E}(\chi \circ \det)$ is an equivalence class of pairs $[g, v]$ where $g \in G(\mathbb{A})$ and $v \in E(\chi \circ \det)$, with $[\gamma g k, v] \sim [g, \chi(\det(k)).v]$ for all $\gamma \in G(\mathbb{Q})$ and all $k \in K_0(\mathfrak{c})$. A canonical trivialization is given by the mapping

$$[g, v] \mapsto \chi(\det(g)).v$$

Thus we obtain a complex-valued non-degenerate pairing by composing this trivialization with the pairing (7.5.1)$\otimes$(7.5.7). To summarize, the pairing

$$\langle \cdot, \cdot, \rangle : \mathcal{L}(\kappa, \chi_0) \times \mathcal{L}^\vee(\kappa, \chi_0) \to \mathbb{C}$$

is given as follows. Let $p = (p_\sigma)$ and $p' = (p'_\sigma)$ ($\sigma \in \Sigma(L)$) be elements of $\mathrm{Sym}^k(V)$ and let $v \in E(\chi_0^\vee) = \mathbb{C}$ and let $v' \in E(\chi_0) = \mathbb{C}$. Let $g \in G(\mathbb{A})$. Then $[g, p \otimes v]$ represents a vector in the bundle $\mathcal{L}(\kappa, \chi_0)$ and $[g, p' \otimes v']$ represents a vector in $\mathcal{L}^\vee(\kappa, \chi_0)$, and

$$\langle [g, p \otimes v], [g, p' \otimes v'] \rangle = \chi(\det(g)) \prod_{\sigma \in \Sigma(L)} \langle p_\sigma, p'_\sigma \rangle v v' \in \mathbb{C}. \qquad (7.5.8)$$

$\square$

Since the above pairing $\langle \cdot, \cdot \rangle$ is nondegenerate, and since the line bundle corresponding to the representation $\det_\infty$ is trivial, it follows from Theorem 3.4 that the resulting pairing on middle intersection homology and cohomology

$$IH_{2n-j}(X_0(\mathfrak{c}), \mathcal{L}) \times IH_j(X_0(\mathfrak{c}), \mathcal{L}^\vee) \xrightarrow{\langle \cdot, \cdot \rangle_{IH_*}} IH_0(X_0(\mathfrak{c}), \mathbb{C}) \to \mathbb{C}$$

$$IH^{2n-j}(X_0(\mathfrak{c}), \mathcal{L}) \times IH^j(X_0(\mathfrak{c}), \mathcal{L}^\vee) \xrightarrow{\langle \cdot, \cdot \rangle_{IH^*}} IH^{2n}(X_0(\mathfrak{c}), \mathbb{C}) \to \mathbb{C}$$

are also nondegenerate, where $\mathcal{L} = \mathcal{L}(\kappa, \chi_0)$, $\mathcal{L}^\vee = \mathcal{L}^\vee(\kappa, \chi_0)$ and $n = [L : \mathbb{Q}] = \dim_\mathbb{C} X_0(\mathfrak{c})$.

7.5. Pairings of vector bundles

The Atkin-Lehner operator $W_{\mathfrak{c}} : Y_0(\mathfrak{c}) \to Y_0(\mathfrak{c})$ has a unique extension to an operator on $X_0(\mathfrak{c})$ and by Lemma 7.4 it induces isomorphisms

$$W_{\mathfrak{c}}^* : IH_j(X_0(\mathfrak{c}), \mathcal{L}) \to IH_j(X_0(\mathfrak{c}), \mathcal{L}^\vee)$$
$$W_{\mathfrak{c}}^* : IH^j(X_0(\mathfrak{c}), \mathcal{L}) \to IH^j(X_0(\mathfrak{c}), \mathcal{L}^\vee).$$

Consequently we obtain nondegenerate pairings on middle intersection (co)homology,

$$[\cdot, \cdot]_{IH_*} : IH_{2n-i}(X_0(\mathfrak{c}), \mathcal{L}) \times IH_i(X_0(\mathfrak{c}), \mathcal{L}) \longrightarrow \mathbb{C}$$
$$[\cdot, \cdot]_{IH^*} : IH^{2n-i}(X_0(\mathfrak{c}), \mathcal{L}) \times IH^i(X_0(\mathfrak{c}), \mathcal{L}) \longrightarrow \mathbb{C}$$

by defining

$$[a,b]_{IH_*} := \langle a, W_{\mathfrak{c}}^* b \rangle_{IH_*} \qquad (7.5.9)$$
$$[a,b]_{IH^*} := \langle a, W_{\mathfrak{c}}^* b \rangle_{IH^*}$$

where $\mathcal{L} = \mathcal{L}(\kappa, \chi_0)$ and $\mathcal{L}^\vee = \mathcal{L}^\vee(\kappa, \chi_0)$.

The main result of this section is the following theorem, which will be used in Chapter 8. It says that the Petersson product on automorphic forms coincides (up to a factor) with the topological product on intersection cohomology. Let $\kappa = (k,m) \in \mathcal{X}(L)$ be a weight. Write $[k+2m] := k_\sigma + 2m_\sigma$ (which is independent of σ) and $\{k\} := \sum_{\sigma \in \Sigma(L)} k_\sigma$. Let

$$T(g,J) := (-1)^{|J|}(-2i)^n (2i)^{\{k\}} N_{L/\mathbb{Q}}(\mathfrak{c})^{[k+2m]/2} W(g) \in \mathbb{C} \qquad (7.5.10)$$

where $n = [L:\mathbb{Q}]$ as above.

Theorem 7.11. *Let $f, g \in S_\kappa^{\text{new}}(K_0(\mathfrak{c}), \chi)$ be newforms. Let $J \subset \Sigma(L)$. Then*

$$[\omega_J(f^{-\iota}), \omega_{\Sigma(L)-J}(g^{-\iota})]_{IH^*} = T(g,J)(f,g)_P,$$

Proof. From Section 6.11 we have

$$\iota_J^*(dz) \wedge \iota_{\Sigma(L)-J}^*(dz) = (-1)^{|J|} dz \wedge d\bar{z}$$

so it suffices to consider the case when $J = \emptyset$. Using the definition of $(\cdot,\cdot)_P^{\text{coh}}$ (see above Lemma 5.5), it suffices to show that

$$[\omega(f), \omega_{\Sigma(L)}(g)]_{IH^*} = T(g^{-\iota}, \emptyset)(f,g)_P^{\text{coh}}$$

whenever $f,g \in S_\kappa^{\text{coh}}(K_0(\mathfrak{c},\chi))$ and $f^{-\iota}, g^{-\iota}$ are newforms.

According to Proposition 4.3 the intersection product $\langle \cdot, \cdot \rangle_{IH^*}$ on intersection cohomology corresponds to the exterior product on differential forms under the mapping $\mathcal{Z}$. Hence

$$[\omega(f), \omega_{\Sigma(L)}(g)]_{IH^*} = \langle \mathcal{Z}(\omega(f)), W_{\mathfrak{c}}^*(\mathcal{Z}(\iota_{\Sigma(L)}^* \omega(g))) \rangle_{IH^*}$$
$$= \epsilon \int_{Y_0(\mathfrak{c})} \omega(f) \wedge W_{\mathfrak{c}}^* \iota_{\Sigma(L)}^* \omega(g)$$

where (as in Proposition 4.3) the values of $\omega(f)$ and $W_{\mathfrak{c}}^* \iota_{\Sigma(L)}^* \omega(g)$ are multiplied using the pairing $\langle \cdot, \cdot \rangle$ of Definition 7.9, and where ϵ denotes the augmentation which adds the values of these integrals over the individual components of $Y_0(\mathfrak{c})$. Recall that

$$\omega(f)(z, h_0) := \det(\alpha_z)^{-m-k-1} j(\alpha_z, \mathbf{i})^{k+2} f(\alpha_z h_0) (-X + zY)^k \, dz$$

By equation (7.4.10) and using the fact that the diagram (7.4.1) commutes, we find

$$W_{\mathfrak{c}}^* \iota_{\Sigma(L)}^* \omega(g) = \iota_{\Sigma(L)}^* W_{\mathfrak{c}}^* \omega(g) = \iota_{\Sigma(L)}^* \omega \left(W_{\mathfrak{c}}^{* \, \mathrm{coh}}(g) \right)$$
$$= \mathrm{N}_{L/\mathbb{Q}}(\mathfrak{c})^{[k+2m]/2} W(g^{-\iota})(-1)^{\{k\}} \iota_{\Sigma(L)}^* \omega \left(g_c \right)$$

which, evaluated at $(z, h_0) \in D \times G(\mathbb{A}_f)$ (see Proposition 7.7) is:

$$\mathrm{N}_{L/\mathbb{Q}}(\mathfrak{c})^{[k+2m]/2} W(g^{-\iota})(-1)^{\{k\}} \det(\alpha_{\bar{z}})^{-m-k-1} j(\alpha_{\bar{z}}, \mathbf{i})^{k+2} g_c(\alpha_z h_0)(-X + \bar{z}Y)^k d\bar{z}$$
$$= \mathrm{N}_{L/\mathbb{Q}}(\mathfrak{c})^{[k+2m]/2} W(g^{-\iota})(-1)^{\{k\}} \det(\alpha_{\bar{z}})^{-m-k-1} j(\alpha_{\bar{z}}, \mathbf{i})^{k+2} \overline{g}(\alpha_z h_0)(-X + \bar{z}Y)^k d\bar{z}$$
$$= \mathrm{N}_{L/\mathbb{Q}}(\mathfrak{c})^{[k+2m]/2} W(g^{-\iota})(-1)^{\{k\}} (-y_\infty)^{-m-k-1} \overline{g}(\alpha_z h_0)(-X + \bar{z}Y)^k d\bar{z}$$

where $z = x_\infty + \mathbf{i} y_\infty \in D$, and $\alpha_z = \left(\begin{smallmatrix} y_\infty & x_\infty \\ 0 & 1 \end{smallmatrix}\right)$. Pairing $\omega(f)$ with $W_{\mathfrak{c}}^* i_{\Sigma(L)}^* \omega(g)$ and using equation (7.5.5) we find

$$\omega(f) \wedge W_{\mathfrak{c}}^* i_{\Sigma(L)}^* \omega(g)(z, h_0)$$
$$= \mathrm{N}_{L/\mathbb{Q}}(\mathfrak{c})^{[k+2m]/2} W(g^{-\iota})(-1)^{\{k\}} y_\infty^{-2m-2k-2} (z - \bar{z})^k \overline{g}(\alpha_z h_0) f(\alpha_z h_0) dz \wedge d\bar{z}$$
$$= T(g^{-\iota}, \emptyset)(-2i)^{-n} y_\infty^{-2m-k} \overline{g}(\alpha_z h_0) f(\alpha_z h_0) \frac{dz \wedge d\bar{z}}{y_\infty^2}.$$

The result now follows from equation (5.7.3) and the fact that

$$dz \wedge d\bar{z} = (-2i)^n dx \wedge dy. \qquad \square$$

7.6 Generalities on Hecke correspondences

The action of the Hecke algebra on modular forms given in equation (5.6.5) comes from an action of a Hecke correspondence on the associated differential forms. See Section 3.5 for a general introduction to correspondences. In the next few sections we explain the translation between Hecke operators and Hecke correspondences and we prove that the mapping $f \mapsto \omega_f$ (which associates a differential form ω_f to an automorphic form f) is Hecke-equivariant. We also show that the action of the Hecke algebra commutes with that of the component group of G and that the Atkin-Lehner operator is Hecke-covariant.

It is notationally simpler to describe the Hecke correspondences in a slightly more general setting. So, for the purposes of this section only, let G be a group, $K \subset G$ a subgroup and let $y \in G$. Suppose that

$$H = H(y) = K \cap yKy^{-1}$$

7.6. Generalities on Hecke correspondences

has finite index in K. The resulting Hecke correspondence

$$(c(I), c(y)) : G/H \rightrightarrows G/K$$

is given by $c(I)(gH) = gK$ and $c(y)(gH) = gyK$. The isomorphism class of this correspondence depends only on the double coset KyK. Indeed, if $y' = yk$ for some $k \in K$ then $H(y') = H(y)$ and $c(y') = c(y)$ so we obtain the same correspondence. If $y' = ky$ then $H(y') = kH(y)k^{-1}$ and we have a well-defined mapping $\phi : G/H(y') \to G/H(y)$ given by $\phi(gH(y')) = gkH(y)$. Then the following diagram commutes and defines an isomorphism of correspondences:

$$\begin{array}{ccc} G/H(y') & \xrightarrow{\phi} & G/H(y) \\ c(I) \downarrow \;\; \downarrow c(y') & & c(I) \downarrow \;\; \downarrow c(y) \\ G/K & \longrightarrow & G/K . \end{array}$$

Let $R \subseteq G$ be a semigroup that contains K and y. As in [Shim1, p. 51] the decomposition into left cosets $K = \coprod_{i=1}^{r} a_i H$ corresponds to the decomposition of KyK into left cosets

$$KyK = \coprod_{i=1}^{r} (a_i y) K$$

(with $a_i y \in R$), so it defines a one to one correspondence

$$K/H \leftrightarrow (KyK)/K. \tag{7.6.1}$$

We now check that this is indeed a bijection. First, the mapping $a_i H \mapsto a_i y K$ is well defined since any $h \in H$ may be expressed as $h = yky^{-1}$ with $k \in K$, so

$$a_i h H \mapsto a_i h y K = a_i y k y^{-1} y K = a_i y K.$$

The mapping (7.6.1) is one to one for if $a_i y K = a_j y K$ then $(a_j)^{-1} a_i \in yKy^{-1}$, and since $a_i, a_j \in K$ we have: $a_j^{-1} a_i \in H$. The mapping (7.6.1) is surjective for if $bK \subset KyK$ then $b = ay$ for some $a \in K$.

Similarly there is a canonical one to one correspondence $H\backslash K \leftrightarrow K\backslash(KyK)$: the decomposition into cosets $K = \coprod_{j=1}^{s} Hb_j$ gives a decomposition $KyK = \coprod_{j=1}^{s} Kyb_j$ and moreover, $yb_j \in R$.

The *Hecke algebra* of R consists of formal $\mathbb{Z}$-linear combinations of double cosets $[KyK]$ with $y \in R$. Multiplication is defined as follows. Suppose $KyK = \coprod_{i=1}^{r} Ka_i$ and $KwK = \coprod_{j=1}^{s} Kb_j$. Then

$$[KyK].[KwK] = \sum_{i=1}^{r} \sum_{j=1}^{s} [Ka_i b_j K].$$

It can be shown that this product arises from the composition (see equation (4.5.2)) of the correspondences $(c(I), c(y))$ and $(c(I), c(x))$.

Action on a vector bundle. Suppose that a representation $\psi : R \to GL(E)$ is given, where $R \subset G$ is a semigroup containing K, y, and y^{-1}. The representation ψ defines a vector bundle (or, if $R = G$, a local system) on G/K,

$$\mathcal{E}_K = E_K(\psi) = G \times_K E$$

consisting of equivalence classes of pairs $[g, v]$ ($g \in G$; $v \in E$) where $[gk, v] \sim [g, \psi(k)v]$ for all $k \in K$. A section $s : G/K \to \mathcal{E}_K$ of $\mathcal{E}_K$ may be expressed as $s(gK) = [g, f(g)]$ where $f : G \to E$ and $f(gk) = \psi(k)^{-1} f(g)$ for any $k \in K$. The following proposition describes the action of the Hecke correspondence on sections of the vector bundle $\mathcal{E}_K$.

Lemma 7.12. *Let $y \in R$ and let $H = K \cap yKy^{-1}$. The Hecke correspondence $(c(I), c(y)) : G/H \rightrightarrows G/K$ has a natural vector bundle lift, $c(I)^* \mathcal{E}_K \to c(y)^* \mathcal{E}_K$. Let $a_1, \ldots, a_r \in K$ be a complete set of representatives for the quotient $K/y^{-1}Hy$. Let $s(gK) = [g, f(g)]$ be a section of $\mathcal{E}_K$. Then $s' = c(y)_* c(I)^* s$ is the section $s'(gK) = [g, f'(g)]$ where*

$$f'(g) = \sum_{i=1}^{r} \psi(a_i y^{-1}) f(g a_i y^{-1}). \tag{7.6.2}$$

Proof. Let $c(I)^*(\mathcal{E}_K) = \mathcal{E}_H = G \times_H E$ be the corresponding local system on G/H. Then $c(I)^*(s)$ is a section of $c(I)^*(\mathcal{E}_K)$ which is given by the same function f, that is, $c(I)^*(s)(gH) = [g, f(g)]$. The mapping $c(y)$ is covered by a morphism of vector bundles (that we denote with the same symbol), $c(y) : \mathcal{E}_H \to \mathcal{E}_K$ defined by

$$c(y)([g, v]) = [gy, \psi(y)^{-1} v] \tag{7.6.3}$$

which is well defined because for any $h \in H$,

$$c(y)([gh, v]) = [ghy, \psi(y)^{-1} v] = [gy, \psi(y^{-1} hy) \psi(y^{-1}) v] = c(y)([g, \psi(h) v]).$$

The Hecke correspondence acts on the section s to give $c(y)_* c(I)^* s$, where $c(y)_*$ denotes summing over the fibers of $c(y)$, which we determine as follows. Suppose $c(y)(gbH) = c(y)(gH)$. Then $gbyK = gyK$ so $b \in yKy^{-1}$. Therefore $c(y)^{-1}(gyK)$ consists of the points $gb_1 H, \ldots, gb_r H$ where $b_1, \ldots, b_r$ form a complete set of representatives for the quotient yKy^{-1}/H. Equivalently, the elements $a_i = y^{-1} b_i y$ ($1 \leq i \leq r$) form a complete set of representatives for $K/y^{-1} Hy$, and $c(y)^{-1}(gyK) = \{gya_i y^{-1} H : 1 \leq i \leq r\}$. Replacing gy with g we conclude that $c(y)^{-1}(gK) = \{ga_i y^{-1} H : 1 \leq i \leq r\}$. At such a point, the section $c(I)^*(s)$ has the value

$$c(I)^*(s)(ga_i y^{-1} H) = [ga_i y^{-1}, f(ga_i y^{-1})].$$

So applying $c(y)$ and using equation (7.6.3) gives

$$c(y)\left(c(I)^* s(ga_i K)\right) = [ga_i, \psi(y^{-1}) f(ga_i y^{-1})] = [g, \psi(a_i y^{-1}) f(ga_i y^{-1})].$$

Summing over these points in the fiber $c(y)^{-1}(gK)$ gives equation (7.6.2). $\square$

7.7 Hecke correspondences in the Hilbert modular case

Let us return to the Hilbert modular case with $\mathfrak{c} \in \mathcal{O}_L$, and $Y_0(\mathfrak{c}) = G(\mathbb{Q})\backslash D \times G(\mathbb{A}_f)/K_0(\mathfrak{c})$ where $D = G(\mathbb{R})/K_\infty$. As in Section 5.6 let $R(\mathfrak{c})$ be the semigroup of matrices

$$G(\mathbb{A}_f) \cap \left\{ x \in M_2(\widehat{\mathcal{O}}_L) : x = \begin{pmatrix} a & b \\ c & d \end{pmatrix} \text{ with } c \in \mathfrak{c}\widehat{\mathcal{O}}_L \text{ and } d_v \in \mathcal{O}_v^\times \text{ whenever } \mathfrak{p}_v | \mathfrak{c} \right\}.$$

and let $R^\vee(\mathfrak{c})$ be the semigroup

$$G(\mathbb{A}_f) \cap \left\{ x \in M_2(\widehat{\mathcal{O}}_L) : x = \begin{pmatrix} a & b \\ c & d \end{pmatrix} \text{ with } c \in \mathfrak{c}\widehat{\mathcal{O}}_L \text{ and } a_v \in \mathcal{O}_v^\times \text{ whenever } \mathfrak{p}_v | \mathfrak{c} \right\}.$$

Then $x \in R(\mathfrak{c}) \iff x^\iota \in R^\vee(\mathfrak{c}) \iff W_\mathfrak{c} x W_\mathfrak{c}^{-1} \in R^\vee(\mathfrak{c})$. Let $x \in R(\mathfrak{c})$. To obtain agreement with the "cohomological" normalization (see Section 5.5) which is used in the definition of the differential form ω_f, it is necessary to use the Hecke correspondence $(c(I), c(x^\iota))$ rather than $(c(I), c(x))$, in other words, we associate to the double coset $K_0(\mathfrak{c}) x K_0(\mathfrak{c})$ the Hecke correspondence

$$(c(I), c(x^\iota)) : Y_H \rightrightarrows Y_0(\mathfrak{c})$$

where $Y_H = G(\mathbb{Q})\backslash D \times G(\mathbb{A}_f)/H$ with $H = H(x^\iota) = K_0(\mathfrak{c}) \cap x^\iota K_0(\mathfrak{c}) x^{-\iota}$, and

$$c(I)(G(\mathbb{Q})(z, gH)) = G(\mathbb{Q})(z, gK_0(\mathfrak{c}))$$

and

$$c(x^\iota)(G(\mathbb{Q})(z, gH)) = G(\mathbb{Q})(z, gx^\iota K_0(\mathfrak{c}))$$

for any $z \in D$ and $g \in G(\mathbb{A}_f)$. This action depends only on the finite adèlic part $G(\mathbb{A}_f)/K_0(\mathfrak{c})$ so we may write

$$c(I)(gH) = gK_0(\mathfrak{c}) \quad \text{and} \quad c(x^\iota)(gH) = gx^\iota K_0(\mathfrak{c})$$

which places us in the framework of Section 7.6 where $G = G(\mathbb{A}_f)$, $y = x^\iota$, $K = K_0(\mathfrak{c})$, $R = R(\mathfrak{c})$ and $\psi = \chi_0^\vee$ (as defined in Section 7.1).

Recall from Section 5.4 and Proposition 6.3, for any $\gamma = \begin{pmatrix} a & b \\ c & d \end{pmatrix} \in G(\mathbb{A}_f)$ that $\chi_0(\gamma) = \prod_v (\chi_v(\gamma_v))$ and $\chi_0^\vee(\gamma) = \prod_v (\chi_v^\vee(\gamma_v))$ where

$$\chi_v\left(\begin{pmatrix} a_v & b_v \\ c_v & d_v \end{pmatrix}\right) = \begin{cases} \chi_v(d_v) & \text{if } \mathfrak{p}_v | \mathfrak{c} \\ 1 & \text{otherwise} \end{cases}$$

and

$$\chi_v^\vee\left(\begin{pmatrix} a_v & b_v \\ c_v & d_v \end{pmatrix}\right) = \begin{cases} \chi_v(a_v) & \text{if } \mathfrak{p}_v | \mathfrak{c} \\ 1 & \text{otherwise} \end{cases}$$

Recall also from Section 7.1 that the local system

$$\mathcal{L}(\kappa, \chi_0) = \mathcal{L}(\kappa) \otimes \mathcal{E}(\chi_0^\vee)$$

is the vector bundle on $Y_0(\mathfrak{c})$ corresponding to the representation $L(\kappa)$ of K_∞ and the character $\chi_0^\vee$ of $K_0(\mathfrak{c})$.

Proposition 7.13. *As in Section 7.1, fix $\kappa = (k, m) \in \mathcal{X}(L)$, and fix $\chi : \mathbb{A}_L^\times \to \mathbb{C}^\times$, a Hecke character whose conductor divides $\mathfrak{c}$. Let $J \subset \Sigma(L)$. Then the mapping* (7.2.3)

$$\omega_J : S_\kappa^{\mathrm{coh}}(K_0(\mathfrak{c}), \chi) \to I^m H^n(X_0(\mathfrak{c}), \mathcal{L}(\kappa, \chi_0))$$
$$f \mapsto \mathcal{Z}([\omega_J(f)]).$$

is Hecke-equivariant in the sense that

$$\omega_J(f|K_0(\mathfrak{c})xK_0(\mathfrak{c})) = c(x^\iota)_* c(I)^*(\omega_J(f))$$

for all $x \in R(\mathfrak{c})$.

Proof. Several steps are involved in translating Lemma 7.12 into a proof of this Proposition. The action of the component group $\pi_0(G(\mathbb{R}))$ (see Section 6.11) commutes with the action of any Hecke operator (since one involves the finite places and the other involves the infinite places), so it suffices to take $J = \emptyset$. Let $f \in \mathcal{M}_\kappa^{\mathrm{coh}}(K_0(\mathfrak{c}), \chi)$ be an automorphic form. By equation (6.10.2), it determines a differential form $\omega(f) = \omega_f \in \Omega^n(Y_0(\mathfrak{c}), \mathcal{L}(\kappa, \chi_0))$. Let $x \in R(\mathfrak{c})$. It determines a Hecke operator $K_0(\mathfrak{c})xK_0(\mathfrak{c})$ which acts on f by equation (5.6.5) to give an automorphic form $f|K_0(\mathfrak{c})xK_0(\mathfrak{c}) \in M_\kappa^{\mathrm{coh}}(K_0(\mathfrak{c}), \chi)$ namely

$$(f|K_0(\mathfrak{c})xK_0(\mathfrak{c}))(\alpha) := \sum_i \chi_0(x_i)^{-1} f(\alpha x_i^{-\iota})$$

where $K_0(\mathfrak{c})xK_0(\mathfrak{c}) = \coprod_{i=1}^r x_i K_0(\mathfrak{c})$ with $x_i \in R(\mathfrak{c})$ $(1 \le i \le r)$. The element x also determines a Hecke correspondence

$$(c(I), c(x^\iota)) : Y_{H(x^\iota)} \rightrightarrows Y_0(\mathfrak{c})$$

with a lift to the local system $\mathcal{L}(\kappa, \chi_0)$, as defined in Lemma 7.12. Then we must show that

$$c(x^\iota)_* c(I)^*(\omega_f) = \omega_{f|T_x}.$$

The action of the Hecke correspondence only involves the finite places, so we may suppress the dependence of f on $G(\mathbb{R})$ and consider it to be a function only of $g \in G(\mathbb{A}_f)$. A similar remark applies to ω_f, see equation (6.10.2).

For brevity, write $K = K_0(\mathfrak{c})$. According to Lemma 7.12, using $\psi = \chi_0^\vee$, $G = G(\mathbb{A}_f)$, $K = K_0(\mathfrak{c})$, $y = x^\iota$, we find:

$$(c(x^\iota)_* c(I)^* \omega_f)(\alpha) = \sum_{i=1}^r \chi_0^\vee(a_i x^{-\iota}) \omega_f(\alpha a_i x^{-\iota})$$

where $a_1, \ldots, a_r$ form a complete set of representatives for the quotient K/H' where $H' = K \cap (x^\iota)^{-1} K x^\iota = K \cap xKx^{-1}$. Since $\chi_0^\vee(a_i) = \chi(\det(a_i))\chi_0(a_i)^{-1}$ and

7.8. Atkin-Lehner-Hecke compatibility

$\chi_0^\vee(x^{-\iota}) = \chi_0(x)^{-1}$ we find

$$c(x^\iota)_* c(I)^* \omega_f(\alpha) = \sum_{i=1}^r \chi_0(a_i x)^{-1} \chi(\det(a_i)) \omega_f(\alpha a_i x^{-\iota})$$

$$= \sum_{i=1}^r \chi_0(x_i)^{-1} \omega_f(\alpha x_i^{-\iota}) = \omega_{f|T_x}(\alpha)$$

where, as in Section 7.6 we take $x_i = a_i x \in R(\mathfrak{c})$ in the decomposition $KxK = \coprod_{i=1}^r x_i K$. □

Action of the Hecke algebra. In the previous paragraph we have defined, for any double coset $K_0(\mathfrak{c}) x K_0(\mathfrak{c})$ a Hecke correspondence $Y_{H(x^\iota)} \rightrightarrows Y_0(\mathfrak{c})$ and an induced mapping on differential forms and on intersection cohomology (for any perversity p)

$$K_0(\mathfrak{c}) x K_0(\mathfrak{c})_* := c(x^\iota)_* c(I)^* : I^\mathsf{p} H_*(X_0(\mathfrak{c}), \mathcal{L}(\kappa, \chi_0)) \longrightarrow I^\mathsf{p} H_*(X_0(\mathfrak{c}), \mathcal{L}(\kappa, \chi_0))$$
$$I^\mathsf{p} H^*(X_0(\mathfrak{c}), \mathcal{L}(\kappa, \chi_0)) \longrightarrow I^\mathsf{p} H^*(X_0(\mathfrak{c}), \mathcal{L}(\kappa, \chi_0))$$
(7.7.1)

which is compatible with the action of $K_0(\mathfrak{c}) x K_0(\mathfrak{c})$ on modular forms in $S_\kappa^{\mathrm{coh}}(K_0(\mathfrak{c}), \chi)$. As in (5.6.6), for any ideal $\mathfrak{m} \subset \mathcal{O}_L$ define the following element of the Hecke algebra,

$$T_\mathfrak{c}(\mathfrak{m})_* := \sum K_0(\mathfrak{c}) y K_0(\mathfrak{c})_*, \qquad (7.7.2)$$

where the sum is over double cosets $K_0(\mathfrak{c}) y K_0(\mathfrak{c})$ with $y \in R(\mathfrak{c})$ such that $[\det(y)] = \mathfrak{m}$.

7.8 Atkin-Lehner-Hecke compatibility

The Atkin-Lehner operator does not commute with the action of the Hecke algebra. Rather, it converts the action of $\mathbb{T}_\mathfrak{c}$ into the action of the "opposite" algebra $\mathbb{T}_\mathfrak{c}^\vee$ as we will now explain. Let $R^\vee(\mathfrak{c})$ denote the semigroup

$$G(\mathbb{A}_f) \cap \left\{ y \in M_2(\widehat{\mathcal{O}}_L) : y = \begin{pmatrix} a & b \\ c & d \end{pmatrix} \text{ with } c \in \mathfrak{c}\widehat{\mathcal{O}}_L \text{ and } a_v \in \mathcal{O}_v^\times \text{ whenever } \mathfrak{p}_v | \mathfrak{c} \right\}.$$

Elements of $R^\vee(\mathfrak{c})$ determine Hecke correspondences exactly as in Section 7.6. Moreover,

$$y \in R(\mathfrak{c}) \iff W_\mathfrak{c} y W_\mathfrak{c}^{-1} \in R^\vee(\mathfrak{c}).$$

7.9 Integral coefficients

By inverting finitely many primes it is possible to produce the pairings and mappings described in this chapter with coefficients in certain subalgebras $A \subset \mathbb{C}$. In this section we outline the requirements on such an algebra. Throughout this section we fix a weight $\kappa = (k, m) \in \mathcal{X}(L)$ and a quasicharacter χ such that $\chi_\infty(b_\infty) = b_\infty^{-k-2m}$.

Fix a "large" number field M, Galois over $\mathbb{Q}$, such that

(1) All Galois conjugates of L are in M.

(2) All Galois conjugates of the number field generated by the values of the finite-order character $\chi| \cdot |_{\mathbb{A}_L}^{k+2m}$ on $\mathbb{A}_L^\times$ are in M.

Choose, once and for all, an embedding $\mathcal{O}_M \hookrightarrow \mathbb{C}$. We now view $\mathbb{C}$ as a $\mathcal{O}_M$-algebra via this embedding. In this section, $A \subseteq \mathbb{C}$ will be an $\mathcal{O}_M$-subalgebra satisfying the following conditions:

(1A) The algebra A contains 1 and is regular, Noetherian, and of finite cohomological dimension.

(2A) The algebra A contains the values of $\chi|_{\prod_{\mathfrak{p}|\mathfrak{c}} \mathcal{O}_{v(\mathfrak{p})}^\times}$.

(3A) The integers j with $1 \leq j \leq k$ are invertible in A.

(4A) We can choose a torsion-free finite-index normal subgroup $K^{\mathrm{fr}} \trianglelefteq K_0(\mathfrak{c})$ normalized by $W_\mathfrak{c}$ such that every rational integer dividing $|K_0(\mathfrak{c})/K^{\mathrm{fr}}|$ is invertible in A.

(5A) For any $1 \leq i \leq h = h(K_0(\mathfrak{c}))$ and $\sigma \in \Sigma(L)$, the ideal

$$(\det(\sigma(t_i(\mathfrak{c}))) \cap M)A$$

is principal and invertible in A.

In the last condition, with notation as in Section 5.2, $h = h(K_0(\mathfrak{c}))$ is the narrow class number of L, and $t_i(\mathfrak{c})$ are representatives for the narrow class group. In the display, we extended the notation slightly, denoting $\sigma : \widehat{\mathcal{O}}_L \hookrightarrow \widehat{\mathcal{O}}_M$ the embedding induced by $\sigma : L \hookrightarrow \mathbb{R}$. Notice that all of these conditions are automatically satisfied if $A = \mathbb{C}$ or A is a sufficiently large field. In particular, for condition (4A), we may use the K^{fr} given in the following remarks.

Remarks.

(1) For each rational prime $p \in \mathbb{Z}$, let $\mathbb{Q}(\mu_p)^+$ be the maximal totally real subfield of the pth cyclotomic field $\mathbb{Q}(\mu_p)$. If $\mathfrak{p} \subset \mathcal{O}_M$ is a prime ideal outside of the set of prime ideals dividing

$$\{p : \mathbb{Q}(\mu_p)^+ \subseteq L\},$$

7.9. Integral coefficients

then using the argument in Lemma 1 of [Gh, Section 3.1] we have that $\mathcal{O}_{M,\mathfrak{p}}$ satisfies (4A) with

$$K^{\mathrm{fr}} = K_0(\mathfrak{c}) \cap \left\{ \begin{pmatrix} a & b \\ c & d \end{pmatrix} \in G(\widehat{\mathbb{Z}}) : \begin{pmatrix} a & b \\ c & d \end{pmatrix} \equiv \begin{pmatrix} 1 & 0 \\ 0 & 1 \end{pmatrix} \pmod{\mathfrak{q}^n} \right\}$$

for some auxiliary prime $\mathfrak{q} \subset \mathcal{O}_L$ coprime to $\mathfrak{p}$ and sufficiently large n.

(2) If $A = \mathcal{O}_{M,\mathfrak{p}}$ for some prime $\mathfrak{p} \subset \mathcal{O}_M$, we can always choose the $t_i(\mathfrak{c})$ so that condition (5A) is satisfied. This condition is needed for the construction of local systems (as outlined below).

Using the results recalled in Section B.5, one can prove the following lemma:

Lemma 7.14. *If $A \subseteq \mathbb{C}$ is an algebra satisfying the requirements* (1A)–(5A) *above, then the Hilbert modular variety $Y_0(\mathfrak{c})$ is an A-homology manifold and this determines an A-orbifold stratification of the Baily-Borel compactification $X_0(\mathfrak{c})$.*

Proof. Recall from (5.1.7) that we have an identification

$$Y_0(\mathfrak{c}) \longleftrightarrow \bigcup_{i=1}^{h(K_0(\mathfrak{c}))} Y_0^i(\mathfrak{c}) = \bigcup_{i=1}^{h(K_0(\mathfrak{c}))} \Gamma_0^i(\mathfrak{c}) \backslash \mathfrak{h}^{\Sigma(L)}.$$

Let $K^{\mathrm{fr}} \triangleleft K_0(\mathfrak{c})$ be the torsion-free, finite index normal subgroup normalized by $W_{\mathfrak{c}}$ that appeared in the set of assumptions on A above (any torsion-free, finite index normal subgroup will do if A is a field). We obtain torsion free subgroups

$$\Gamma_0^{i,\mathrm{fr}} := G(\mathbb{Q}) \cap t_i(\mathfrak{c}) G(\mathbb{R})^0 K^{\mathrm{fr}} t_i(\mathfrak{c})^{-1} \triangleleft \Gamma_0^i(\mathfrak{c}).$$

Let

$$Y_0^{i,\mathrm{fr}} := \Gamma_0^{i,\mathrm{fr}} \backslash \mathfrak{h}^{\Sigma(L)} \quad \text{and} \quad \Gamma_i^{\mathrm{qu}} := \Gamma_0^i(\mathfrak{c}) / \Gamma_0^{i,\mathrm{fr}}.$$

Then Γ_i^{qu} acts on $Y_0^{i,\mathrm{fr}}$ in a manner such that the quotient $\Gamma_i^{\mathrm{qu}} \backslash Y_0^{i,\mathrm{fr}}$ is naturally isomorphic to $Y_0^i(\mathfrak{c})$. This action extends by continuity to an action on the Bailey-Borel compactification $X_0^{i,\mathrm{fr}}$ of $Y_0^{i,\mathrm{fr}}$. Then the key point in the proof of Lemma 7.14 is that the order of any subgroup of $\Gamma_0^{i,\mathrm{qu}}$ is invertible in A (which follows from the various assumptions on A). $\square$

In particular this gives rise to an A-orbifold structure on $Y_0(\mathfrak{c})$, see Section B.2, and hence also an orbifold stratification of $X_0(\mathfrak{c})$ (where the singular strata are the cusps), as described in Section 5.2. The lemma implies, in particular, that we may use the orbifold stratification of $X_0(\mathfrak{c})$ to compute the intersection homology and cohomology groups with respect to any local system of A-modules (see Section 3.3).

Local Systems. Let $A \subseteq \mathbb{C}$ be an $\mathcal{O}_M$-module as described above. Using the weight $\kappa = (k, m) \in \mathcal{X}(L)$ and the quasicharacter χ we have a representation $L(\kappa, \chi_0)$

as described in Section 7.1. For an integral model, we replace the representation $L(\kappa) = \mathrm{Sym}^k(V^\vee) \otimes \det^{-m}$ of $G(\mathbb{R})$ with the representation $L(\kappa, A)$ of $\mathrm{GL}_2(A)$, consisting of polynomial functions on $V(A) \cong A^2$ with coefficients in the algebra A. The construction of Section 7.1 cannot be used to create a local system from this data because the group $G(\mathbb{R})$ does not preserve the A-valued polynomials. Instead, we use the fact that $G(\mathbb{R})/K_\infty$ is a disjoint union of finitely many contractible manifolds (each diffeomorphic to a product of upper half-planes). So we may start with the trivial bundle $L(\kappa, A)$ on $G(\mathbb{R})/K_\infty$ and then divide by $\Gamma_0^i(\mathfrak{c})$ for $1 \subseteq i \subseteq h$. Here we embed $\Gamma_0^i(\mathfrak{c}) \hookrightarrow \mathrm{GL}_2(A)$ via a choice of embedding $\mathcal{O}_L \hookrightarrow \mathcal{O}_M$. The group $\Gamma_0^i(\mathfrak{c})$ does preserve the A-valued polynomials because of the assumption that A is an $\mathcal{O}_M$-module, the assumption that all Galois conjugates of L are contained in M, and assumption (5A). Working component by component, this construction gives us a local system $\mathcal{L}(\kappa)$ over $Y_0(\mathfrak{c})$. Similarly, by hypothesis (2A), the one-dimensional representation $E(\chi_0^\vee)$ (of complex vector spaces) of $K_0(\mathfrak{c})$ may be replaced by a one-dimensional representation $E(\chi_0^\vee, A)$ on A, which gives rise to a one-dimensional local system $\mathcal{E}(\chi_0^\vee, A)$ of A-modules. In summary we obtain a local system

$$\mathcal{L}(\kappa, \chi, A) = \mathcal{L}(\kappa, A) \otimes_A \mathcal{E}(\chi_0^\vee, A)$$

on the orbifold $Y_0(\mathfrak{c})$. The resulting pairings

$$IH_i(X_0(\mathfrak{c}), \mathcal{L}(\kappa, \chi_0, A)) \times IH_{n-i}(X_0(\mathfrak{c}), \mathcal{L}(\kappa, \chi_0, A)) \to A$$
$$IH^i(X_0(\mathfrak{c}), \mathcal{L}(\kappa, \chi_0, A)) \times IH^{n-i}(X_0(\mathfrak{c}), \mathcal{L}(\kappa, \chi_0, A)) \to A$$

take values in A. These pairings are not necessarily perfect over A. However, if the order of the torsion in the middle degree cohomology of the links of the singular points of $X_0(\mathfrak{c})$ is invertible in A (see [Gre8]) then these pairings will be perfect.

Chapter 8

Hilbert Modular Forms with Coefficients in a Hecke Module

The goal of this chapter is to prove Theorems 8.4 and 8.5, the full versions of Theorems 1.1 and 1.2 given in the introduction. We consider a quadratic extension of totally real number fields L/E. Let

$$\iota : G_E := \mathrm{Res}_{E/\mathbb{Q}}(\mathrm{GL}_2) \longrightarrow G_L := \mathrm{Res}_{L/\mathbb{Q}}(\mathrm{GL}_2)$$

be the diagonal embedding. In Section 8.2, we define the Hecke operators $\widehat{T}(\mathfrak{m})$ (on L): they are "lifts" of the operators $T(\mathfrak{m})$ on E. We prove that these operators have the "key property" (1.5.3) mentioned in the introduction: if f is a newform on G_E with Hecke eigenvalues $\lambda_f(\mathfrak{m})$ and if its base change $\widehat{f}$ is a newform on G_L then $\widehat{f}$ is an eigenform for $\widehat{T}(\mathfrak{m})$ with the same eigenvalue, $\lambda_f(\mathfrak{m})$.

Let $\mathfrak{c} \subset \mathcal{O}_L$ be an ideal. Let $\chi = \chi_E \circ \mathrm{N}_{L/E}$ be a Hecke character for L whose conductor divides $\mathfrak{c}$. Let $\mathcal{M}$ be a module over the Hecke algebra $\mathbb{T}_\mathfrak{c}$ for G_L. Let $\gamma \in \mathcal{M}$ be an appropriately chosen simultaneous eigenvector of all Hecke operators. The operators $\widehat{T}(\mathfrak{m})$ are used to form a family of Hecke translates, $\gamma(\mathfrak{m}) = \gamma_{\chi_E}(\mathfrak{m})$ of γ. Theorem 8.3 then states that for any $\Lambda \in (\mathcal{M}^{\chi_E})^\vee$ there exists a Hilbert modular form for G_E whose Fourier coefficients are equal to the products $\langle \Lambda, \gamma_{\chi_E}(\mathfrak{m}) \rangle$, at least if $\mathfrak{m}$ is a norm from $\mathcal{O}_L$. This is a purely algebraic version of the phenomena described by Hirzebruch and Zagier [Hirz] and it is a formal consequence the existence of a section of the base change mapping on the Hecke algebra.

In Section 8.3 we consider the case where the Hecke module $\mathcal{M}$ is the intersection homology group, $I^m H_*(X_0(\mathfrak{c}), \mathcal{L}(\kappa, \chi_0))$, and γ is an intersection homology class. Then the linear mapping Λ may be taken to be the intersection product with any other intersection homology class. So we may conclude that the intersection numbers $\langle \Lambda, \gamma_{\chi_E}(\mathfrak{m}) \rangle_{IH}$ are the Fourier coefficients of a modular form.

In Theorem 8.5 we express these Fourier coefficients in terms of f-isotypical components under the Hecke algebra. The resulting formula involves the pairing between the two natural sources of intersection (co-)homology classes on $X_0(\mathfrak{c})$.

One source is the "diagonally embedded" Hilbert modular variety Z_0 attached to E and its "twists" Z_θ, see Chapter 9. A second source comes from the L^2 differential form $\omega_J(f)$ associated to a cusp form f for G_L. According to Proposition 4.8 the intersection product between such classes may be identified with the integral $\int_{Z_\theta} \omega_J(f)$.

8.1 Notation

Let L/E be a quadratic extension of totally real number fields. To simplify notation, write
$$n := [L:\mathbb{Q}], \quad d = d_{L/E}, \quad \mathcal{D} = \mathcal{D}_{L/E}$$
for the degree of L, the discriminant of L/E and the different of L/E respectively. If $\mathfrak{c} \subset \mathcal{O}_L$ is an ideal, write $\mathfrak{c}_E = \mathfrak{c} \cap \mathcal{O}_E$. We use class field theory to identify the group of Hecke characters $\eta : E^\times \backslash \mathbb{A}_E^\times \to \mathbb{C}^\times$ trivial on the image of the norm $\mathrm{N}_{L/E} : \mathbb{A}_L^\times \to \mathbb{A}_E^\times$ with $\mathrm{Gal}(L/E)^\wedge$. In other words, the nontrivial element η of $\mathrm{Gal}(L/E)^\wedge$ is trivial at infinity, assigns the value 1 to finite primes in E that split in L and assigns the value -1 to finite primes that are inert in L. We may use the construction of Section 5.11 to define twists of Hilbert modular forms by η. As above, let $\Sigma(E) := \{\sigma_{01}, \ldots, \sigma_{0[E:\mathbb{Q}]}\}$ denote the set of embeddings $E \hookrightarrow \mathbb{R}$ and let $\Sigma(L) := \{\sigma_1, \ldots, \sigma_n\}$ denote the set of embeddings $L \hookrightarrow \mathbb{R}$. For $\kappa = (k,m) \in \mathcal{X}(E)$, define $\widehat{\kappa} = (\widehat{k}, \widehat{m}) \in \mathbb{Z}^{\Sigma(L)}$ by stipulating that $\widehat{k}_\sigma = k_{\sigma_0}$ and $\widehat{m}_\sigma = m_{\sigma_0}$ if σ extends σ_0. Fix an ideal $\mathfrak{c} \subset \mathcal{O}_L$, and let χ be a quasicharacter of $\mathbb{A}_L^\times$ with conductor dividing $\mathfrak{c}$. We assume that χ is of the form $\chi = \chi_E \circ \mathrm{N}_{L/E}$ for some quasicharacter $\chi_E : E^\times \backslash \mathbb{A}_E^\times \to \mathbb{C}^\times$ satisfying $\chi_{E\infty}(b_\infty) = b_\infty^{-k-2m}$ for $b_\infty \in \mathbb{A}_E^{\infty,\times}$.

If $\mathcal{M}$ is a module over a Hecke algebra $\mathbb{T}$ and if $\gamma \in \mathcal{M}$ and $t \in \mathbb{T}$ we write $t(\gamma) = \gamma | t = t.\gamma$ for the action of $\mathbb{T}$ on $\mathcal{M}$.

If $\mathfrak{n} \subset \mathcal{O}_L$ is coprime to $\mathfrak{c}$, we often abuse notation and set $\chi(\mathfrak{n}) = \chi(n)$, where $n \in \mathbb{A}_L^\times$ is any idéle trivial at the infinite places and at the places dividing $\mathfrak{c}$ such that $[n] = \mathfrak{n}$. We will commit the same abuse of notation in dealing with $\chi_E(\mathfrak{n}')$ for $\mathfrak{n}'$ coprime to $d\mathfrak{c}_E = d_{L/E}(\mathfrak{c} \cap \mathcal{O}_E)$. In this section we will use the theory of prime degree base change for GL_2 developed by Langlands [Lan]. This theory is recalled in Appendix E.

8.2 Base change for the Hecke algebra

Let $\mathfrak{c} \subset \mathcal{O}_L$, be an ideal and set $\mathfrak{c}_E = \mathfrak{c} \cap \mathcal{O}_E$. From Section 5.6 we have Hecke algebras $\mathbb{T}_\mathfrak{c}$ and $\mathbb{T}_{\mathfrak{c}_E}$. Define Hecke subalgebras

$$\mathbb{T}^{\mathfrak{c}\mathcal{D}} := \mathbb{Z}[\{K_0(\mathfrak{c})xK_0(\mathfrak{c}) \in \mathbb{T}_\mathfrak{c} : x_{v(\mathfrak{p})} = \begin{pmatrix} 1 & 0 \\ 0 & 1 \end{pmatrix} \text{ whenever } \mathfrak{P}|\mathfrak{c}\mathcal{D}\}] \subseteq \mathbb{T}_\mathfrak{c} \quad (8.2.1)$$

$$\mathbb{T}^{\mathfrak{c}_E d} := \mathbb{Z}[\{K_0(\mathfrak{c}_E)xK_0(\mathfrak{c}_E) \in \mathbb{T}_{\mathfrak{c}_E} : x_{v(\mathfrak{p})} = \begin{pmatrix} 1 & 0 \\ 0 & 1 \end{pmatrix} \text{ whenever } \mathfrak{p}|\mathfrak{c}_E d\}] \subseteq \mathbb{T}_{\mathfrak{c}_E}.$$

8.2. Base change for the Hecke algebra

Thus $\mathbb{T}^{\mathfrak{c}_E d} \subseteq \mathbb{T}_{\mathfrak{c}_E}$ and $\mathbb{T}^{\mathfrak{c}\mathcal{D}} \subseteq \mathbb{T}_{\mathfrak{c}}$ are the subalgebras consisting of operators trivial at the ramified places. There is an algebra homomorphism

$$b : \mathbb{T}^{\mathfrak{c}\mathcal{D}} \longrightarrow \mathbb{T}^{\mathfrak{c}_E d} \qquad (8.2.2)$$

induced by the L-map

$$b : {}^L G_E \longrightarrow {}^L G_L$$

given on the connected components of the L-groups by the diagonal embedding, as explained in Section E.4 below. Explicitly, b is defined by requiring that for $\mathfrak{P} \nmid \mathfrak{c}\mathcal{D}$ we have

$$b(T_{\mathfrak{c}}(\mathfrak{P})) = \begin{cases} T_{\mathfrak{c}_E d}(\mathfrak{p}) & \text{for } \mathfrak{p} = \mathfrak{P}\bar{\mathfrak{P}} \text{ split in } L/E \\ T_{\mathfrak{c}_E d}(\mathfrak{p}^2) - N_{E/\mathbb{Q}}(\mathfrak{p}) T_{\mathfrak{c}_E d}(\mathfrak{p},\mathfrak{p}) & \text{for } \mathfrak{p} = \mathfrak{P} \text{ inert in } L/E \end{cases} \qquad (8.2.3)$$

$$b(T_{\mathfrak{c}}(\mathfrak{P},\mathfrak{P})) = \begin{cases} T_{\mathfrak{c}_E d}(\mathfrak{p},\mathfrak{p}) & \text{for } \mathfrak{p} = \mathfrak{P}\bar{\mathfrak{P}} \text{ split in } L/E \\ T_{\mathfrak{c}_E d}(\mathfrak{p}^2,\mathfrak{p}^2) & \text{for } \mathfrak{p} = \mathfrak{P} \text{ inert in } L/E \end{cases}.$$

(see Lemma E.5). It follows from [Bu, Proposition 4.6.5] that the operators $T_{\mathfrak{c}}(\mathfrak{P})$, $T_{\mathfrak{c}}(\mathfrak{P},\mathfrak{P})$ for $\mathfrak{P} \nmid \mathfrak{c}\mathcal{D}$ generate the algebra $\mathbb{T}^{\mathfrak{c}\mathcal{D}}$, so these equations suffice to define b.

Let $\mathbb{Z}[\chi]$ (resp. $\mathbb{Z}[\chi_E]$) be the subalgebra of $\mathbb{C}$ generated by $\chi(\mathfrak{m})$ (resp. $\chi_E(\mathfrak{n})$) as $\mathfrak{m}$ (resp. $\mathfrak{n}$) ranges over the ideals of $\mathcal{O}_L$ coprime to $\mathfrak{c}\mathcal{D}$ (resp. $\mathcal{O}_E$ coprime to $\mathfrak{c}_E d$). We note that $\mathbb{Z}[\chi] \subseteq \mathbb{Z}[\chi_E]$. We then have ideals

$$I(\chi) := \langle T_{\mathfrak{c}}(\mathfrak{m},\mathfrak{m}) - \chi(\mathfrak{m}) \rangle_{\mathfrak{m}+\mathfrak{c}\mathcal{D}=\mathcal{O}_L} \subset \mathbb{T}^{\mathfrak{c}\mathcal{D}} \otimes_{\mathbb{Z}} \mathbb{Z}[\chi] \qquad (8.2.4)$$

$$I(\chi_E) := \langle T_{\mathfrak{c}_E}(\mathfrak{n},\mathfrak{n}) - \chi_E(\mathfrak{n}) \rangle_{\mathfrak{n}+\mathfrak{c}_E d=\mathcal{O}_E} \subset \mathbb{T}^{\mathfrak{c}_E d} \otimes_{\mathbb{Z}} \mathbb{Z}[\chi_E].$$

It follows from (8.2.3) that b induces a morphism

$$b : \mathbb{T}^{\mathfrak{c}\mathcal{D}} \otimes_{\mathbb{Z}} \mathbb{Z}[\chi_E]/I(\chi) \longrightarrow \mathbb{T}^{\mathfrak{c}_E d} \otimes_{\mathbb{Z}} \mathbb{Z}[\chi_E]/I(\chi_E).$$

Using the $\mathfrak{p}$-adic Cartan decomposition and (5.6.8), one checks that its image is generated as a $\mathbb{Z}[\chi_E]$-algebra by $T_{\mathfrak{c}}(N_{L/E}(\mathfrak{P}))$ as $\mathfrak{P}$ ranges over the prime ideals of $\mathcal{O}_L$ coprime to $\mathfrak{c}\mathcal{D}$, and as a $\mathbb{Z}[\chi_E]$-module by $T_{\mathfrak{c}}(N_{L/E}(\mathfrak{m}))$ as $\mathfrak{m}$ ranges over the ideals of $\mathcal{O}_L$ coprime to $\mathfrak{c}\mathcal{D}$ (compare [Bu, Proposition 4.6.2 and Proposition 4.6.4]). In other words, we have "killed" the Hecke operators $T_{\mathfrak{c}}(\mathfrak{m},\mathfrak{m})$ and $T_{\mathfrak{c}_E}(\mathfrak{n},\mathfrak{n})$.

We are now going to define a linear map

$$\widehat{} : \mathbb{T}^{\mathfrak{c}_E d} \otimes_{\mathbb{Z}} \mathbb{Z}[\chi_E]/I(\chi_E) \longrightarrow \mathbb{T}^{\mathfrak{c}\mathcal{D}} \otimes_{\mathbb{Z}} \mathbb{Z}[\chi_E]/I(\chi). \qquad (8.2.5)$$

that is a section of the base change map b. It is defined on basis elements

$$T(\mathfrak{m}) = T_{\mathfrak{c}_E d}(\mathfrak{m}) \in \mathbb{T}^{\mathfrak{c}_E d}$$

and extended $\mathbb{Z}[\chi_E]$-linearly. The Hecke operator $\widehat{T}(\mathfrak{m}) \in \mathbb{T}_{\mathfrak{c}} \otimes \mathbb{C}$ is defined as follows. If $\mathfrak{m} \subset \mathcal{O}_E$ is not a norm from $\mathcal{O}_L$ set $\widehat{T}(\mathfrak{m}) := 0$. The definition of the operator $\widehat{T}(\mathfrak{p}^n)$ involves a choice: if $\mathfrak{p} \subset \mathcal{O}_E$ is a prime ideal that splits in $\mathcal{O}_L$ (say, $\mathfrak{p}\mathcal{O}_L = \mathfrak{P}\bar{\mathfrak{P}}$) then choose a prime $\mathfrak{P}$ lying above $\mathfrak{p}$. Otherwise let $\mathfrak{P} = \mathfrak{p}\mathcal{O}_L$. Then

$$\widehat{T}(\mathfrak{p}^r) := \begin{cases} \mathrm{Id} & \text{if } r = 0 \\ T_{\mathfrak{c}}(\mathfrak{P}^r) & \text{if } \mathfrak{p} \text{ splits in } L \\ T_{\mathfrak{c}}(\mathfrak{P}^{r/2}) + \chi_E(\mathfrak{p}) N_{E/\mathbb{Q}}(\mathfrak{p}) T_{\mathfrak{c}}(\mathfrak{P}^{r/2-2}) & \text{if } \mathfrak{p} \text{ is inert and } r \text{ is even} \\ 0 & \text{otherwise} \end{cases}$$

where "otherwise" means that either $\mathfrak{p}$ ramifies in L or $\mathfrak{p}$ is inert and r is odd. Finally, if $\mathfrak{m} = \prod_{\mathfrak{p} \in I} \mathfrak{p}^{r_\mathfrak{p}}$ for some set I of primes in $\mathcal{O}_E$ and some integers $r_\mathfrak{p} \geq 0$ then set

$$\widehat{T}(\mathfrak{m}) := \prod_{\mathfrak{p} \in I} \widehat{T}(\mathfrak{p}^{r_\mathfrak{p}}).$$

Lemma 8.1. *Let $\mathfrak{c} \subset \mathcal{O}_L$ be an ideal, set $\mathfrak{c}_E = \mathfrak{c} \cap \mathcal{O}_L$ and let χ_E be a quasicharacter as in Section 8.1. The restriction of the map*

$$\widehat{} : \mathbb{T}^{\mathfrak{c}_E d} \otimes_{\mathbb{Z}} \mathbb{Z}[\chi_E]/I(\chi_E) \longrightarrow \mathbb{T}^{\mathfrak{c}\mathcal{D}} \otimes_{\mathbb{Z}} \mathbb{Z}[\chi_E]/I(\chi)$$

to $b(\mathbb{T}^{\mathfrak{c}\mathcal{D}})$ is a $\mathbb{Z}[\chi_E]$-algebra morphism that is a section of b.

Proof. The definition of $\widehat{}$ is multiplicative in the sense that

$$\widehat{T}(\mathfrak{m})\widehat{T}(\mathfrak{n}) = (T_{\mathfrak{c}'_E}(\mathfrak{m}) T_{\mathfrak{c}'_E}(\mathfrak{n}))^{\widehat{}}$$

whenever $\mathfrak{m} + \mathfrak{n} = \mathcal{O}_E$. Since the Hecke algebras $\mathbb{T}^{\mathfrak{c}\mathcal{D}}$ and $\mathbb{T}^{\mathfrak{c}_E d}$ are multiplicative as well, to prove that $\widehat{}$ is an algebra morphism it suffices to check that

$$\widehat{} : b(\mathbb{T}^{\mathfrak{c}\mathcal{D}} \otimes_{\mathbb{Z}} \mathbb{Z}[\chi_E]/I(\chi_E)) \longrightarrow \mathbb{T}^{\mathfrak{c}\mathcal{D}} \otimes_{\mathbb{Z}} \mathbb{Z}[\chi_E]/I(\chi)$$

is a $\mathbb{Z}[\chi_E]$ algebra homomorphism when restricted to the subalgebra

$$A := \langle T_{\mathfrak{c}_E d}(\mathfrak{p}^r) \rangle_{r \geq 0} \cap b(\mathbb{T}^{\mathfrak{c}\mathcal{D}} \otimes_{\mathbb{Z}} \mathbb{Z}[\chi_E]/I(\chi_E)) = \langle T_{\mathfrak{c}_E d}(N(\mathfrak{P}^r)) \rangle_{r \in \mathbb{Z}^{\geq 0}}$$

for each prime $\mathfrak{p} \subset \mathcal{O}_E$ coprime to $\mathfrak{c}_E d$ with prime $\mathfrak{P} \subseteq \mathcal{O}_L$ the prime above $\mathfrak{p}$. Here the $\langle \cdot \rangle$ denotes the span as a $\mathbb{Z}[\chi_E]$-module, and we are using the $\mathfrak{p}$-adic Cartan decomposition together with (5.6.8) to deduce the equality (compare [Bu, Proposition 4.6.2 and Proposition 4.6.4]). If $\mathfrak{p}$ is split in L/E this is trivial, so we henceforth assume that $\mathfrak{p}$ is inert in L/E and write $\mathfrak{P}$ for the prime above $\mathfrak{p}$. In this case, the image of A under $\widehat{}$ is contained in the $\mathbb{Z}[\chi_E]$-subalgebra

$$B := \langle T_{\mathfrak{c}}(\mathfrak{P})^r \rangle_{r \geq 0} \subseteq \mathbb{T}^{\mathfrak{c}\mathcal{D}} \otimes_{\mathbb{Z}} \mathbb{Z}[\chi_E]/I(\chi).$$

8.2. Base change for the Hecke algebra

Let
$$\mathbb{T}(\mathfrak{P}) := \langle T_{\mathfrak{c}\mathcal{D}}(\mathfrak{P}^r)\rangle_{r\geq 0} \subseteq \mathbb{T}^{\mathfrak{c}\mathcal{D}} \otimes_{\mathbb{Z}} \mathbb{Z}[\chi_E]$$
$$\mathbb{T}(\mathfrak{p}) := \langle T_{\mathfrak{c}_E d}(\mathfrak{p}^{2r})\rangle_{r\geq 0} \subseteq \mathbb{T}^{\mathfrak{c}_E d} \otimes_{\mathbb{Z}} \mathbb{Z}[\chi_E].$$

Thus $A = \mathbb{T}(\mathfrak{p}) \otimes_{\mathbb{Z}} \mathbb{Z}[\chi_E]/I(\chi_E)$ and $B = \mathbb{T}(\mathfrak{P}) \otimes_{\mathbb{Z}} \mathbb{Z}[\chi_E]/I(\chi)$. It follows from [Shim1, Theorem 3.20] that $T(\mathfrak{P})$ and $T(\mathfrak{p})$ are free $\mathbb{Z}[\chi_E]$-algebras on two generators:

$$T(\mathfrak{P}) = \mathbb{Z}[\chi_E][T_{\mathfrak{c}\mathcal{D}}(\mathfrak{P}), T_{\mathfrak{c}\mathcal{D}}(\mathfrak{P},\mathfrak{P})]$$
$$T(\mathfrak{p}) = \mathbb{Z}[\chi_E][T_{\mathfrak{c}_E d}(\mathfrak{p}^2), T_{\mathfrak{c}_E d}(\mathfrak{p}^2,\mathfrak{p}^2)]$$

This implies that A and B are the free $\mathbb{Z}[\chi_E]$-algebras on one generator given by

$$B = \mathbb{Z}[\chi_E][T_{\mathfrak{c}}(\mathfrak{P})] \tag{8.2.6}$$
$$A = \mathbb{Z}[\chi_E][T_{\mathfrak{c}_E d}(\mathfrak{p}^2)]$$

In view of (8.2.6) we have an algebra morphism $s: A \to B$ defined by stipulating that
$$s(T_{\mathfrak{c}_E d}(\mathfrak{p}^2)) = T_{\mathfrak{c}}(\mathfrak{P}) + \chi_E(\mathfrak{p})N_{E/\mathbb{Q}}(\mathfrak{p}),$$

and it is easy to see that s is a section of b. Thus to prove the lemma it suffices to show that $s = \widehat{}$. For this it suffices to show

$$s(T_{\mathfrak{c}_E d}(\mathfrak{p}^{2r})) = \widehat{T}(\mathfrak{p}^{2r}).$$

We proceed by induction on r. The statement is obviously true for $r = 0, 1$. Assume it is true for $r - 1$. Applying (5.6.8), we have that $s(T_{\mathfrak{c}_E}(\mathfrak{p}^{2r}))$ is equal to

$$s(T_{\mathfrak{c}_E d}(\mathfrak{p}^{2r-2})T_{\mathfrak{c}_E d}(\mathfrak{p}^2) - \chi_E N_{E/\mathbb{Q}}(\mathfrak{p})T_{\mathfrak{c}_E d}(\mathfrak{p}^{2r-2}) - \chi_E N_{E/\mathbb{Q}}(\mathfrak{p}^2)T_{\mathfrak{c}_E d}(\mathfrak{p}^{2r-4})) \tag{8.2.7}$$

$$= \widehat{T}(\mathfrak{p}^{2r-2})\widehat{T}(\mathfrak{p}^2) - \chi_E N_{E/\mathbb{Q}}(\mathfrak{p})\widehat{T}(\mathfrak{p}^{2r-2}) - \chi_E N_{E/\mathbb{Q}}(\mathfrak{p}^2)\widehat{T}(\mathfrak{p}^{2r-4})$$

Here we have written (and will continue to write) $\chi_E N_{E/\mathbb{Q}}(\mathfrak{m}) = \chi_E(\mathfrak{m})N_{E/\mathbb{Q}}(\mathfrak{m})$ for ideals $\mathfrak{m} \subset \mathcal{O}_E$. Substituting

$$\widehat{T}(\mathfrak{p}^{2r-2})\widehat{T}(\mathfrak{p}^2) = (T_{\mathfrak{c}}(\mathfrak{P}^{r-1}) + \chi_E N_{E/\mathbb{Q}}(\mathfrak{p})T_{\mathfrak{c}}(\mathfrak{P}^{r-2}))(T_{\mathfrak{c}}(\mathfrak{P}) + \chi_E N_{E/\mathbb{Q}}(\mathfrak{p}))$$
$$\widehat{T}(\mathfrak{p}^{2r-2}) = T_{\mathfrak{c}}(\mathfrak{P}^{r-1}) + \chi_E N_{E/\mathbb{Q}}(\mathfrak{p})T_{\mathfrak{c}}(\mathfrak{P}^{r-2})$$
$$\widehat{T}(\mathfrak{p}^{2r-4}) = T_{\mathfrak{c}}(\mathfrak{P}^{r-2}) + \chi_E N_{E/\mathbb{Q}}(\mathfrak{p})T_{\mathfrak{c}}(\mathfrak{P}^{r-3})$$

into (8.2.7), we have that $s(T_{\mathfrak{c}_E d}(\mathfrak{p}^{2r}))$ is equal to

$$T_{\mathfrak{c}}(\mathfrak{P}^{r-1})T_{\mathfrak{c}}(\mathfrak{P}) + \chi_E N_{E/\mathbb{Q}}(\mathfrak{p})T_{\mathfrak{c}}(\mathfrak{P}^{r-2})T_{\mathfrak{c}}(\mathfrak{P}) + \chi_E N_{E/\mathbb{Q}}(\mathfrak{p})T_{\mathfrak{c}}(\mathfrak{P}^{r-1})$$
$$+ \chi_E N_{E/\mathbb{Q}}(\mathfrak{p}^2)T_{\mathfrak{c}}(\mathfrak{P}^{r-2}) - \chi_E N_{E/\mathbb{Q}}(\mathfrak{p})(T_{\mathfrak{c}}(\mathfrak{P}^{r-1}) + \chi_E N_{E/\mathbb{Q}}(\mathfrak{p})T_{\mathfrak{c}}(\mathfrak{P}^{r-2}))$$
$$- \chi_E N_{E/\mathbb{Q}}(\mathfrak{p}^2)(T(\mathfrak{P}^{r-2}) + \chi_E N_{E/\mathbb{Q}}(\mathfrak{p})T_{\mathfrak{c}}(\mathfrak{P}^{r-3}))$$

$$\begin{aligned}
&= T_{\mathfrak{c}}(\mathfrak{P}^{r-1})T_{\mathfrak{c}}(\mathfrak{P}) + \chi_E N_{E/\mathbb{Q}}(\mathfrak{p})T_{\mathfrak{c}}(\mathfrak{P}^{r-2})T_{\mathfrak{c}}(\mathfrak{P}) \\
&\quad - \chi_E N_{E/\mathbb{Q}}(\mathfrak{p}^2)(T_{\mathfrak{c}}(\mathfrak{P}^{r-2}) + \chi_E N_{E/\mathbb{Q}}(\mathfrak{p})T_{\mathfrak{c}}(\mathfrak{P}^{r-3})) \\
&= T_{\mathfrak{c}}(\mathfrak{P}^r) + \chi_E N_{E/\mathbb{Q}}(\mathfrak{p}^2)T_{\mathfrak{c}}(\mathfrak{P}^{r-2}) \\
&\quad + \chi_E N_{E/\mathbb{Q}}(\mathfrak{p})(T_{\mathfrak{c}}(\mathfrak{P}^{r-1}) + \chi_E N_{E/\mathbb{Q}}(\mathfrak{p}^2)T_{\mathfrak{c}}(\mathfrak{P}^{r-3})) \\
&\quad - \chi_E N_{E/\mathbb{Q}}(\mathfrak{p}^2)(T_{\mathfrak{c}}(\mathfrak{P}^{r-2}) + \chi_E N_{E/\mathbb{Q}}(\mathfrak{p})T_{\mathfrak{c}}(\mathfrak{P}^{r-3})) \\
&= T_{\mathfrak{c}}(\mathfrak{P}^r) + \chi_E N_{E/\mathbb{Q}}(\mathfrak{p})T_{\mathfrak{c}}(\mathfrak{P}^{r-1}) = \widehat{T}(\mathfrak{p}^{2r}).
\end{aligned}$$

This completes the proof of the lemma. □

An immediate consequence of the fact that $\widehat{}$ is a section of b over its image is the crucial Proposition 8.2 below. In order to state it, let $\mathfrak{c}'_E \subseteq \mathcal{O}_E$ be an ideal. For any newform $f \in S_\kappa^{\text{new}}(K_0(\mathfrak{c}'_E), \chi_E)$, there exists a base change $\widehat{f} \in S_{\widehat{\kappa}}^{\text{new}}(K_0(\mathfrak{c}'), \chi)$ for some $\mathfrak{c}' \subset \mathcal{O}_L$ with $\mathfrak{c}' \cap \mathcal{O}_E$ equal to $\mathfrak{c}'_E$ up to powers of primes dividing $2d$. This $\widehat{f}$ is uniquely characterized as the newform generating the automorphic representation $\widehat{\pi}$ that is the base change of π in the sense of Section E.4 below. The following diagram may help in keeping track of the base change mappings.

$$\begin{array}{ccc}
L & \mathbb{T}^{\mathfrak{c}\mathcal{D}} & S_{\widehat{\kappa}}^{\text{new}}(K_0(\mathfrak{c}'), \chi) \\
\uparrow & b \downarrow \uparrow \widehat{} & \uparrow \widehat{} \\
E & \mathbb{T}^{\mathfrak{c}_E d} & S_\kappa^{\text{new}}(K_0(\mathfrak{c}'_E), \chi)
\end{array}$$

Proposition 8.2. *Let $\mathfrak{m} \subset \mathcal{O}_E$ be an ideal coprime to $d\mathfrak{c}_E$. If $\mathfrak{m}$ is a norm from $\mathcal{O}_L$, then the Hecke operator $\widehat{T}(\mathfrak{m})$ satisfies*

$$\widehat{f} | \widehat{T}(\mathfrak{m}) = \lambda_f(\mathfrak{m}) \widehat{f}$$

for all newforms $f \in S_\kappa^{\text{new}}(K_0(\mathfrak{c}'_E), \chi_E)$ such that $\widehat{f} \in S_{\widehat{\kappa}}^{\text{new}}(K_0(\mathfrak{c}'), \chi)$.

Proof. Let H be an unramified connected reductive group over F_v, where v is a finite place of the number field F, and let $K_H \subseteq H(F_v)$ be a hyperspecial subgroup. Recall that an unramified irreducible representation of $H(F_v)$ has a unique vector fixed by K_H [Cart, p. 152]. We apply this fact in the case $H = \mathrm{GL}_2$ and $K_H = \mathrm{GL}_2(\mathcal{O}_{F_v})$ to conclude that the Hecke eigenvalues of $\widehat{f}$ (resp. f) appearing in the proposition are the same as the Hecke eigenvalues of the automorphic representation $\widehat{\pi}$ (resp. π) generated by $\widehat{\pi}$ (resp. π) (compare Section E.1). Thus by definition of the base change $\widehat{\pi}$ of π (see Section E.4 and (E.8.2)) one has

$$\lambda_{\widehat{f}}(t) = \lambda_f(b(t))$$

for all $t \in \mathbb{T}^{\mathfrak{c}\mathcal{D}}$. Since $\widehat{}$ is a section of b, the proposition follows. □

Remark. The simple shape of the formula given above for the $\widehat{T}(\mathfrak{m})$ depends on the fact that L/E is quadratic. The root of this is the "$\mathfrak{a}^{-2}$" in the Hecke algebra

identity
$$T_{\mathfrak{c}}(\mathfrak{m})T_{\mathfrak{c}}(\mathfrak{n}) = \sum_{\mathfrak{m}+\mathfrak{n} \subseteq \mathfrak{a}} N_{L/\mathbb{Q}}(\mathfrak{a}) T_{\mathfrak{c}}(\mathfrak{a},\mathfrak{a}) T_{\mathfrak{c}}(\mathfrak{a}^{-2}\mathfrak{mn}),$$

(see (5.6.8)). However, if L/E is an arbitrary prime degree extension of totally real fields and $\mathfrak{p}$ is a prime of $\mathcal{O}_E$ that totally splits as $\mathfrak{p} = \mathfrak{P}_1 \ldots \mathfrak{P}_n$ in $\mathcal{O}_L$, then an obvious extension of the argument given above proves that

$$T_{\mathfrak{c}}(\mathfrak{P}_1)\widehat{f} = \lambda_f(\mathfrak{p})\widehat{f}$$

if the newform $\widehat{f} \in S_{\widehat{\kappa}}^{\mathrm{new}}(K_0(\mathfrak{c}), \chi)$ is the base change of a Hilbert modular form f on E and $\mathfrak{p}$ is coprime to $\mathfrak{c}$.

8.3 Hilbert modular forms with coefficients in a Hecke module

The main goal of this section is to prove Theorem 8.3, which uses the operators $\widehat{T}(\mathfrak{m})$ of the previous section to produce Hilbert modular forms on E with coefficients in $\mathcal{M}$. As in the previous paragraphs, we fix ideals $\mathfrak{c} \subset \mathcal{O}_L$ and $\mathfrak{c}_E = \mathfrak{c} \cap E$, we fix Hecke characters $\chi = \chi_E \circ N_{L/E}$ such that $\chi_{E\infty}(b_\infty) = b_\infty^{-k-2m}$ for all $b_\infty \in \mathbb{A}_E^{\infty,\times}$ and we assume the conductor of χ divides $\mathfrak{c}$.

Let $\mathcal{M}$ be a finite-dimensional, left $\mathbb{T}_{\mathfrak{c}} \otimes \mathbb{C}$-module (for example, the intersection cohomology of an appropriate Hilbert modular variety). Then it is also a module over the abelian subalgebra $\mathbb{T}^{\mathfrak{c}\mathcal{D}}$ so it decomposes into Hecke eigen subspaces which correspond to newforms in the following way. If $f \in S_{\widehat{\kappa}}^{\mathrm{new}}(K_0(\mathfrak{c}), \chi)$ is a newform (and therefore a simultaneous eigenform for all Hecke operators), let

$$\lambda_f : \mathbb{T}^{\mathfrak{c}\mathcal{D}} \longrightarrow \mathbb{C}$$

be the linear functional defined by $\lambda_f(t)f := f|t$ as above. Let $\mathcal{M}(f)$ denote the f-isotypical component of $\mathcal{M}$, that is,

$$\mathcal{M}(f) := \{B \in \mathcal{M} : tB = \lambda_f(t)B \text{ for all } t \in \mathbb{T}^{\mathfrak{c}\mathcal{D}}\}.$$

Define the submodule $\mathcal{M}^E$ to be the sum of isotypical components

$$\mathcal{M}^E := \bigoplus_f \mathcal{M}(f) \subseteq \mathcal{M}, \qquad (8.3.1)$$

where the sum is over all (normalized) *newforms* $f \in S_{\widehat{\kappa}}^{\mathrm{new}}(K_0(\mathfrak{c}), \chi)$ such that for almost all primes $\mathfrak{P} \subset \mathcal{O}_L$ we have $\lambda_f(\mathfrak{P}) = \lambda_f(\mathfrak{P}^\sigma)$ for all $\sigma \in \mathrm{Gal}(L/E)$. By the theory of quadratic base change (see Corollary E.12) there is a canonical decomposition

$$\mathcal{M}^E = \mathcal{M}^{\chi_E} \oplus \mathcal{M}^{\chi_E \eta},$$

where η is the generator of $\mathrm{Gal}(L/E)^\wedge$, and

$$\mathcal{M}^{\chi_E} := \bigoplus_f \mathcal{M}(f) \subseteq \mathcal{M}^E \tag{8.3.2}$$

is the submodule spanned by the f-isotypical components such that $f = \widehat{g}$ is a base change of a newform $g \in S_\kappa^{\mathrm{new}}(K_0(\mathfrak{c}'_E), \chi_E)$, where $\mathfrak{c}'_E \subseteq \mathcal{O}_E$ is an ideal equal to $\mathfrak{c}_E$ up to primes dividing $2d_{L/E}$.

The dual space $(\mathcal{M}^E)^\vee := \mathrm{Hom}_\mathbb{C}(\mathcal{M}^E, \mathbb{C})$ is tautologically a $\mathbb{T}^{\mathfrak{c}\mathcal{D}} \otimes \mathbb{C}$-module. Denote the endomorphism of $(\mathcal{M}^E)^\vee$ induced by $t \in \mathbb{T}_\mathfrak{c}$ by

$$t^* : (\mathcal{M}^E)^\vee \longrightarrow (\mathcal{M}^E)^\vee.$$

Suppose that $\gamma \in \mathcal{M}^{\chi_E}$. In analogy with (1.5.2), for each ideal $\mathfrak{m} \subset \mathcal{O}_E$ define the following Hecke translate of γ:

$$\gamma(\mathfrak{m}) = \gamma_{\chi_E}(\mathfrak{m}) := \begin{cases} \widehat{T}(\mathfrak{m})\gamma & \text{if } \mathfrak{m} \text{ is a norm from } \mathcal{O}_L \text{ and } \mathfrak{m} + \mathfrak{c}_E d = \mathcal{O}_E \\ 0 & \text{otherwise.} \end{cases} \tag{8.3.3}$$

In Section 8.5 we will give a proof of Theorem 8.3 below, which interprets the generating series associated to $\widehat{T}(\mathfrak{m})(\gamma)$ (as $\mathfrak{m}$ varies) as a modular form with coefficients in $\mathcal{M}^{\chi_E}$. To state it, note that from any $\Phi \in \mathcal{M}^{\chi_E} \otimes S_\kappa(K_0(\mathcal{N}(\mathfrak{c})), \chi_E)$ we obtain a map

$$\langle \cdot, \Phi \rangle : (\mathcal{M}^{\chi_E})^\vee \longrightarrow S_\kappa(K_0(\mathcal{N}(\mathfrak{c})), \chi_E)$$
$$\Lambda \longmapsto \langle \Lambda, \Phi \rangle$$

Theorem 8.3. *If $\gamma \in \mathcal{M}^{\chi_E}$, then there exists an ideal $\mathcal{N}(\mathfrak{c})$ and a unique*

$$\Phi_{\gamma, \chi_E} \in \mathcal{M}^{\chi_E} \otimes S_\kappa(K_0(\mathcal{N}(\mathfrak{c})), \chi_E).$$

such that

(1) *The map $\langle \cdot, \Phi_{\gamma, \chi_E} \rangle$ is Hecke-equivariant in the sense that for all $\Lambda \in (\mathcal{M}^{\chi_E})^\vee$ and $t \in \mathbb{T}^{\mathfrak{c}\mathcal{D}} \otimes \mathbb{C}$ we have*

$$\langle t^*\Lambda, \Phi_{\gamma, \chi_E} \rangle = \langle \Lambda, \Phi_{\gamma, \chi_E} \rangle | b(t)$$

(2) *If $\mathfrak{m} \subset \mathcal{O}_E$ is a norm from $\mathcal{O}_L$, $\mathfrak{m} + (\mathfrak{c} \cap \mathcal{O}_E)d_{L/E} \neq \mathcal{O}_E$ or $\mathfrak{m} + d_{L/E} = \mathcal{O}_E$ and $\eta(\mathfrak{m}) = -1$ then the $\mathfrak{m}$th Fourier coefficient of Φ_{γ, χ_E} is $\gamma_{\chi_E}(\mathfrak{m})$.*

Remark. In fact the ideal $\mathcal{N}(\mathfrak{c})$ may be taken to be of the following form,

$$\mathcal{N}(\mathfrak{c}) = \mathfrak{m}_2 \mathfrak{b}_{L/E} \mathfrak{c}_E \prod_{\mathfrak{p} | \mathfrak{c}_E} \mathfrak{p}, \tag{8.3.4}$$

where $\mathfrak{m}_2 \subset \mathcal{O}_E$ is an ideal divisible only by dyadic primes, which we take to be $\mathcal{O}_E$ if $\mathfrak{c}+2\mathcal{O}_L = \mathcal{O}_L$, and $\mathfrak{b}_{L/E}$ is an ideal divisible only by the primes ramifying in L/E. If we choose $\mathcal{N}(\mathfrak{c})$ as in (8.3.4), the modular form Φ_{γ,χ_E} actually lies in the subspace

$$\mathcal{M}^{\chi_E} \otimes S_\kappa^+(\mathcal{N}(\mathfrak{c}), \chi_E) \tag{8.3.5}$$

where

$$S_\kappa^+(\mathcal{N}(\mathfrak{c}), \chi_E) := \{f \in S_\kappa(K_0(\mathcal{N}(\mathfrak{c})), \chi_E) : a(\mathfrak{m}, f) = 0$$
$$\text{if } \eta(\mathfrak{m}) = -1 \text{ or } \mathfrak{m} + \mathcal{N}(\mathfrak{c}) \neq \mathcal{O}_E\}.$$

Although the "plus" subspace $S_\kappa^+(\mathfrak{c}_E d, \chi_E)$ is not preserved by $\mathbb{T}^{\mathfrak{c}_E d}$, it is preserved by $b(\mathbb{T}^{\mathfrak{c}\mathcal{D}})$. This may be checked using the explicit description of b given in (8.2.3) above.

8.4 Hilbert modular forms with coefficients in intersection homology

Recall from Section 7.6 that $I^m H_n(X_0(\mathfrak{c}), \mathcal{L}(\widehat{\kappa}, \chi_0))$ is a $\mathbb{T}_\mathfrak{c} \otimes \mathbb{C}$-module. Thus it makes sense to define the *isotypical part*,

$$I^m H_n^E(X_0(\mathfrak{c}), \mathcal{L}(\widehat{\kappa}, \chi_0)) := I^m H_n(X_0(\mathfrak{c}), \mathcal{L}(\widehat{\kappa}, \chi_0))^E$$
$$I^m H_n^{\chi_E}(X_0(\mathfrak{c}), \mathcal{L}(\widehat{\kappa}, \chi_0)) := I^m H_n(X_0(\mathfrak{c}), \mathcal{L}(\widehat{\kappa}, \chi_0))^{\chi_E}$$

as in (8.3.1)[1]. With this notation, Theorem 8.3 immediately implies the following generalization of Theorem 1.1 to nontrivial local coefficient systems:

Theorem 8.4. *If $\gamma \in I^m H_n^{\chi_E}(X_0(\mathfrak{c}), \mathcal{L}(\widehat{\kappa}, \chi_0, \mathbb{C}))$, then there exists an ideal $\mathcal{N}(\mathfrak{c}) \subseteq \mathcal{O}_L$ and a unique*

$$\Phi_{\gamma, \chi_E} \in I^m H_n^{\chi_E}(X_0(\mathfrak{c}), \mathcal{L}(\widehat{\kappa}, \chi_0)) \otimes S_\kappa(\mathcal{N}(\mathfrak{c}), \chi_E).$$

such that

(1) *The map $\langle \cdot, \Phi_{\gamma, \chi_E} \rangle_{IH}$ is Hecke-equivariant: if $\Lambda \in I^m H_n^{\chi_E}(X_0(\mathfrak{c}), \mathcal{L}(\widehat{\kappa}, \chi_0))^\vee$ and $t \in \mathbb{T}^{\mathfrak{c}\mathcal{D}}$, then*

$$\langle t^*\Lambda, \Phi_{\gamma, \chi_E} \rangle = \langle \Lambda, \Phi_{\gamma, \chi_E} \rangle | b(t).$$

(2) *If $\mathfrak{m} \subset \mathcal{O}_E$ is a norm from $\mathcal{O}_L$, $\mathfrak{m} + (\mathfrak{c} \cap \mathcal{O}_E) d_{L/E} \neq \mathcal{O}_E$ or $\mathfrak{m} + d_{L/E} = \mathcal{O}_E$ and $\eta(\mathfrak{m}) = -1$ then the $\mathfrak{m}$th Fourier coefficient of Φ_{γ, χ_E} is $\gamma_{\chi_E}(\mathfrak{m})$.* □

Remark. The definition of $I^m H_n^{\chi_E}(X_0(\mathfrak{c}), \mathcal{L}(\widehat{\kappa}, \chi_0))$ implicitly depends on a choice of quasicharacter $\chi : L^\times \backslash \mathbb{A}_L^\times \to \mathbb{C}^\times$ whose restriction is χ_0. There may be more than one choice of such a character, a fact which also played a role in the statement of Theorem 7.1 above.

---
[1] Notice that each of these groups are Hecke submodules of $I^m H_{\text{cusp}}(X_0(\mathfrak{c}), \mathcal{L}(\widehat{\kappa}, \chi_0))$.

8.5 Proof of Theorem 8.3

As in equations (8.3.1) and (8.3.2), using Corollary E.12 we have a canonical decomposition of the following subspace,

$$S_{\widehat{\kappa}}^{\mathrm{new},\chi_E}(\mathfrak{c},\chi) \oplus S_{\widehat{\kappa}}^{\mathrm{new},\chi_E\eta}(\mathfrak{c},\chi) = S_{\widehat{\kappa}}^{\mathrm{new},E}(\mathfrak{c},\chi) \subset S_{\widehat{\kappa}}^{\mathrm{new}}(K_0(\mathfrak{c}),\chi)$$

where $S_{\widehat{\kappa}}^{\mathrm{new},E}(\mathfrak{c},\chi)$ is the subspace spanned by those newforms f such that, for almost all prime ideals $\mathfrak{P} \subset \mathcal{O}_L$, we have $\lambda_f(\mathfrak{P}) = \lambda_f(\mathfrak{P}^\sigma)$ for all $\sigma \in \mathrm{Gal}(L/E)$, and where $S_{\widehat{\kappa}}^{\mathrm{new},\chi_E}(\mathfrak{c},\chi)$ is the subspace spanned by those newforms f such that $f = \widehat{g}$ for some g of nebentypus χ_E (and where η is the nontrivial element in $\mathrm{Gal}(L/E)^\wedge$).

Given $\gamma \in \mathcal{M}^{\chi_E}$ we may therefore write

$$\gamma = \sum_{\widehat{g}} \gamma(\widehat{g}), \tag{8.5.1}$$

where $\widehat{g} \in S_{\widehat{\kappa}}^{\mathrm{new},\chi_E}(\mathfrak{c},\chi)$ is a newform and $\gamma(\widehat{g}) \in \mathcal{M}^{\chi_E}$ is $\widehat{g}$-isotypical under the action of $\mathbb{T}_\mathfrak{c}$. If $\mathfrak{m} + d\mathfrak{c}_E \neq \mathcal{O}_E$ or if $\mathfrak{m}$ is not a norm from $\mathcal{O}_L$, then by definition $\gamma_{\chi_E}(\mathfrak{m}) = 0$. Otherwise, we have

$$\gamma_{\chi_E}(\mathfrak{m}) = \sum_{\widehat{g}} \widehat{T}(\mathfrak{m})\gamma(\widehat{g}) \tag{8.5.2}$$

where the sum is over newforms $\widehat{g} \in S_{\widehat{\kappa}}^{\mathrm{new},\chi_E}(\mathfrak{c},\chi)$. Applying Proposition 8.2, we obtain

$$\gamma_{\chi_E}(\mathfrak{m}) = \frac{1}{2}\sum_{g} \lambda_g(\mathfrak{m})\gamma(\widehat{g}) \tag{8.5.3}$$

where the sum is over newforms g such that $\widehat{g} \in S_{\widehat{\kappa}}^{\mathrm{new},\chi_E}(\mathfrak{c},\chi)$. Here the $\frac{1}{2}$ factor appears because there are exactly two newforms g contributing the same summand. To see this, note that it suffices by Theorem E.11 to check that $\pi(g) \not\cong \pi(g \otimes \eta)$, where η is the nontrivial element of $\mathrm{Gal}(L/E)^\wedge$ and $\pi(g)$ (resp. $\pi(g \otimes \eta)$) is the automorphic representation generated by g (resp. $g \otimes \eta$). If $\pi(g) \cong \pi(g \otimes \eta)$, then $\widehat{\pi(g)}$ will not be cuspidal (see [Ger, Theorem 2 and Appendix C]) which is a contradiction. It follows that

$$\Phi_{\gamma,\chi_E} := \frac{1}{2}\sum_{g} \gamma(\widehat{g}) \otimes g^{\mathfrak{c}_E d}$$

satisfies requirement (2) of the theorem. Here the sum is over newforms g such that $\widehat{g} \in S_{\widehat{\kappa}}^{\mathrm{new},\chi_E}(\mathfrak{c},\chi)$ and where $g^{\mathfrak{c}_E d}$ is the cusp form obtained from g by deleting the Fourier coefficients having a prime in common with $\mathfrak{c}_E d$ (and $\mathfrak{c}_E = \mathfrak{c} \cap \mathcal{O}_E$) as in Lemma 5.9. By Corollary E.12 in conjunction with Lemma 5.9, for each g in the sum we have that $g^{\mathfrak{c}_E d} \in S_\kappa(K_0(\mathfrak{c}'_E),\chi_E)$ for some ideal

$$\mathfrak{c}'_E \subseteq \mathfrak{m}_2\mathfrak{b}_{L/E}\mathfrak{c}_E \prod_{\mathfrak{p}|\mathfrak{c}_E} \mathfrak{p}$$

8.5. Proof of Theorem 8.3

where $\mathfrak{m}_2$ is an ideal divisible only by dyadic primes which we may take to be $\mathcal{O}_E$ if $\mathfrak{c} + 2\mathcal{O}_L = \mathcal{O}_L$ and $\mathfrak{b}_{L/E}$ is an ideal divisible only by those primes dividing d.

For the Hecke equivariance statement in the theorem, suppose that $\widehat{g}$ contributes to (8.5.1) and let $t \in \mathbb{T}^{\mathfrak{c}\mathcal{D}}$. Note that g and $\widehat{g}$ are unramified outside of $\mathfrak{c}\mathcal{D}$ and $\mathfrak{c}_E d$, respectively. Therefore, the eigenvalue of t on $\widehat{g}$ is the same as the eigenvalue of t on the automorphic representation generated by $\widehat{g}$, and similarly for the eigenvalue of $b(t)$ on $\widehat{g}$ (compare the proof of Proposition 8.2). Thus the eigenvalue of t on $\widehat{g}$ is equal to the eigenvalue of $b(t)$ on $g^{\mathfrak{c}_E d}$ by the definition of base change (see Section E.4). This implies the Hecke equivariance statement in Theorem 8.3.

As for the uniqueness of Φ_{γ, χ_E}, suppose that

$$\Phi_0 = \sum_h \sum_{g_h} \beta(h) \otimes g_h \in \mathcal{M}^{\chi_E} \otimes S_\kappa(K_0(\mathcal{N}(\mathfrak{c})), \chi_E)$$

satisfies property (2) in the theorem. Here the sum on h is over a basis of newforms for $S_{\widehat{\kappa}}^{\mathrm{new}, \chi_E}(\mathfrak{c}, \chi)$, $\beta(h)$ is h-isotypical under the action of $\mathbb{T}^{\mathfrak{c}\mathcal{D}_E}$, and for each h the collection $\{g_h\}$ is a subset of $S_\kappa(K_0(\mathcal{N}(\mathfrak{c})), \chi_E)$. If $\Phi_0 \neq \Phi_{\gamma, \chi_E}$ then we can find a linear functional $\Lambda \in (\mathcal{M}^{\chi_E})^\vee$ such that

$$\langle \Lambda, \Phi_0 - \Phi_{\gamma, \chi_E} \rangle \neq 0.$$

Decomposing $(\mathcal{M}^{\chi_E})^\vee$ into eigenspaces under the action of $\mathbb{T}^{\mathfrak{c}\mathcal{D}_E}$, we see that we may assume that Λ is a projector onto the h_0-isotypical component of $\mathcal{M}^{\chi_E}$ for some newform $h_0 \in S_{\widehat{\kappa}}^{\mathrm{new}}(K_0(\mathfrak{c}), \chi)$ in the sense that $\Lambda(\beta(h_0)) = 0$ if $h \neq ch_0$ for some $c \in \mathbb{C}^\times$ (this step uses strong multiplicity one for newforms, see Section 5.8). By definition of $\mathcal{M}^{\chi_E}$ we have $h_0 = \widehat{g}_0$ for some newform $g_0 \in S_\kappa(K_0(\mathcal{N}(\mathfrak{c})), \chi_E)$ (though it may not be new for the level $\mathcal{N}(\mathfrak{c})$). Thus if $\Phi_0 \neq \Phi_{\gamma, \chi_E}$ then the following pairing is not zero:

$$\langle \Lambda, \Phi_0 - \Phi_{\gamma, \chi_E} \rangle \qquad (8.5.4)$$
$$= \left(\langle \Lambda, \beta(\widehat{g}_0) \rangle \sum_{g_{h_0}} g_{h_0} \right) - \langle \Lambda, \gamma(\widehat{g}_0) \rangle \otimes \tfrac{1}{2}(g_0^{\mathfrak{c}d} + \underline{g_0 \otimes \eta}^{\mathfrak{c}d})$$

Here $\underline{g_0 \otimes \eta}$ is the newform generating the automorphic representation $\pi(g_0 \otimes \eta)$. We will use the assumption that (8.5.4) is not zero to derive a contradiction.

Upon renormalizing $\beta(\widehat{g}_0)$ if necessary, we may assume that $\sum_{g_{h_0}} g_{h_0}$ has $\mathcal{O}_E$-Fourier coefficient equal to 1. Thus, since the $\mathcal{O}_E$-Fourier coefficients of Φ_0 and Φ_{γ, χ_E} agree by requirement (2), we have

$$\langle \Lambda, \beta(\widehat{g}_0) \rangle = \langle \Lambda, \gamma(\widehat{g}_0) \rangle.$$

We also know that for all $\mathfrak{m}$ satisfying the conditions in (2) the $\mathfrak{m}$th Fourier coefficients of Φ_0 and Φ_{γ, χ_E} agree; thus for such $\mathfrak{m}$ we have

$$\sum_{g_{h_0}} a(\mathfrak{m}, g_{h_0}) = \tfrac{1}{2}\left(a(\mathfrak{m}, g_0^{\mathfrak{c}d}) + a(\mathfrak{m}, \underline{g_0 \otimes \eta}^{\mathfrak{c}d}) \right). \qquad (8.5.5)$$

In particular (8.5.5) is valid for $\mathfrak{m}$ a norm from $\mathcal{O}_L$. In view of the section s of the base change map b constructed above, this implies that all of the g_{h_0} contributing to the left-hand side have the property that the base change of $\pi(g_{h_0})$ to L is $\pi(h_0)$. As noted previously in the proof, there are precisely two nonisomorphic cuspidal automorphic representations that base change to $\pi(h_0)$, namely $\pi(g_0)$ and $\pi(g_0 \otimes \eta)$. Thus the left-hand side of (8.5.5) is actually a sum over g_{h_0} such that $\pi(g_{h_0})$ is isomorphic to either $\pi(g_0)$ or $\pi(g_0 \otimes \eta)$. Decompose

$$\{g_{h_0}\} = \{g_{1i}\}_{i=1}^r \cup \{g_{2j}\}_{j=1}^s$$

where the automorphic representation generated by g_{1i} (resp. g_{2j}) is $\pi(g_0)$ (resp. $\pi(g_0 \otimes \eta)$). We claim that

$$\sum_i g_{1i} = \tfrac{1}{2} \underline{g_0}^{cd} \quad \text{and} \quad \sum_j g_{2j} = \tfrac{1}{2} \underline{g_0 \otimes \eta}^{cd}.$$

The claim implies that (8.5.4) is in fact zero and this contradiction completes the proof of the theorem.

To prove the claim, let $V = V_1 \oplus V_2 \subseteq S_\kappa(K_0(\mathcal{N}(\mathfrak{c})), \chi_E)$, where V_1 (resp. V_2) is the subspace spanned by forms in the automorphic representation $\pi(g_0)$ (resp. $\pi(g_0 \otimes \eta)$). Newform theory provides a basis for the space of modular forms generating a given fixed automorphic representation with bounded level (compare Section 5.8). In particular, it implies that any element $f \in V$ is determined by the Fourier coefficients $a(\mathfrak{p}^n, f)$ with $\mathfrak{p} \nmid \mathcal{N}(\mathfrak{c})$ and $n \geq 0$. This implies the claim. □

8.6 The Fourier coefficients of $[\gamma(\mathfrak{m}), \Phi_{\gamma, \chi_E}]_{IH_*}$

The goal. By Theorem 8.4, the fact that Φ_{γ, χ_E} for $\gamma \in I^m H_n^{\chi_E}(X_0(\mathfrak{c}), \mathcal{L}(\widehat{\kappa}, \chi_0))$ is a modular form with coefficients in $I^m H_n^{\chi_E}(X_0(\mathfrak{c}), \mathcal{L}(\widehat{\kappa}, \chi_0))$ is a formal consequence of base change and the existence of a section of the base change morphism on the Hecke algebra. If we start with a cycle Z and a flat section s that represents a class $[Z] = [Z, s] \in I^m H_n(X_0(\mathfrak{c}), \mathcal{L}^\vee(\widehat{\kappa}, \chi_0))$ then it can be used to construct modular forms as follows: first apply[2] the Atkin-Lehner operator $W_\mathfrak{c}^{*-1}$ to obtain an intersection homology class with coefficients in $\mathcal{L}(\widehat{\kappa}, \chi_0)$, and project the cycle to obtain a class

$$\gamma = Q(W_\mathfrak{c}^{*-1}[Z]) \in I^m H_n^{\chi_E}(X_0(\mathfrak{c}), \mathcal{L}(\widehat{\kappa}, \chi_0))$$

where

$$Q = Q_{\chi_E} : I^m H_n(X_0(\mathfrak{c}), \mathcal{L}(\widehat{\kappa}, \chi_0)) \longrightarrow I^m H_n^{\chi_E}(X_0(\mathfrak{c}), \mathcal{L}(\widehat{\kappa}, \chi_0))$$

---

[2]This step, togther with all occurrences of $W_\mathfrak{c}^{*-1}$ on this page could be eliminated by starting with a flat section of $\mathcal{L}$ rather than of $\mathcal{L}^\vee$; but then $W_\mathfrak{c}^*(Z)$ would occur in the integrals (8.6.1). We choose to apply $W_\mathfrak{c}^{*-1}$ here, rather than later, so that the equations in this section will match those in Chapters 9 and 10.

8.6. The Fourier coefficients of $[\gamma(\mathfrak{m}), \Phi_{\gamma,\chi_E}]_{IH_*}$

is the orthogonal projection (with respect to the Petersson product or equivalently, with respect to the decomposition into isotypical components under the action of the Hecke algebra). Then feed this into the formal generating series Φ_{γ,χ_E} of Theorem 8.3 to obtain a modular form with coefficients in $I^m H_n(X_0(\mathfrak{c}), \mathcal{L}(\widehat{\kappa}, \chi_0))$. Finally, any linear functional on $I^m H_n(X_0(\mathfrak{c}), \mathcal{L}(\widehat{\kappa}, \chi_0))$ will give a modular form for G_E. Such a linear functional is given by the intersection pairing with other intersection homology classes, for example, with Hecke translates of γ such as

$$\gamma(\mathfrak{m}) = \gamma_{\chi_E}(\mathfrak{m}) = \begin{cases} \widehat{T}(\mathfrak{m})\gamma & \text{if } \mathfrak{m} \text{ is a norm from } \mathcal{O}_L \text{ and } \mathfrak{m} + \mathfrak{c}_E d = \mathcal{O}_E \\ 0 & \text{otherwise.} \end{cases}$$

In this section we express the Fourier coefficients of the resulting modular form,

$$[\gamma(\mathfrak{m}), \Phi_{\gamma,\chi_E}]_{IH_*} = [Q(W_\mathfrak{c}^{*-1}[Z])(\mathfrak{m}), \Phi_{Q(W_\mathfrak{c}^{*-1}[Z]),\chi_E}]_{IH_*}$$

in terms of certain periods, that is, in terms of integrals over the cycle Z when Z is the "diagonal" cycle that comes from the diagonal embedding

$$\iota : G_E = \operatorname{Res}_{E/\mathbb{Q}} \mathrm{GL}_2 \longrightarrow \operatorname{Res}_{L/\mathbb{Q}} \mathrm{GL}_2 = G_L.$$

In this situation a flat section s of $\mathcal{L}^\vee(\widehat{\kappa}, \chi_0)$ can be constructed canonically over Z. Here is a brief description of this procedure in the case that $\chi_0 = \chi_{\mathrm{triv}}$ is the trivial character. In Section 7.5 we constructed a bilinear form

$$\langle \cdot, \cdot \rangle : L(\kappa, \chi_{\mathrm{triv}}) \times L^\vee(\kappa, \chi_{\mathrm{triv}}) \to \mathbb{C}$$

invariant under the action of G_E. This implies the existence of an element in $L(\kappa, \chi_{\mathrm{triv}}) \otimes L^\vee(\kappa, \chi_{\mathrm{triv}})$ invariant under the action of G_E, which therefore passes to a flat section of

$$\mathcal{L}(\kappa, \chi_{\mathrm{triv}}) \otimes \mathcal{L}^\vee(\kappa, \chi_{\mathrm{triv}})|Z \cong \mathcal{L}^\vee(\widehat{\kappa}, \chi_{\mathrm{triv}})|Z.$$

The general case is discussed in Section 9.2. For more details in the classical Hirzebruch-Zagier setting, see [Ton].

The setup. Let $\mathfrak{c} \subset \mathcal{O}_L$ be an ideal and let $\mathfrak{c}_E = \mathfrak{c} \cap \mathcal{O}_E$. Then we obtain a (non-compact) Shimura subvariety

$$Z = Z_0(\mathfrak{c}_E) := \pi(\iota(G_E(\mathbb{A}))) \cong G_E(\mathbb{Q}) \backslash G_E(\mathbb{A}) / K_{E\infty} K_0(\mathfrak{c}_E)$$

of $Y_0(\mathfrak{c})$, where

$$\pi : G_L(\mathbb{A}) \to Y_0(\mathfrak{c}) = G_L(\mathbb{Q}) \backslash G_L(\mathbb{A}) / K_\infty K_0(\mathfrak{c})$$

is the canonical projection and $K_{E\infty} \subset G_E(\mathbb{R})$, $K_0(\mathfrak{c}_E) \subset G_E(\mathbb{A}_f)$ are defined as in Sections 5.1 and 5.2 with L replaced by E.

Let s be a flat section of $\mathcal{L}^\vee(\widehat{\kappa},\chi_0)$ over Z, or over an open subset of Z that is the complement of a proper subvariety. As in Section 2.7 the pair (Z,s) determines a Borel-Moore homology class which, by Theorem 4.6 has a canonical lift to intersection homology,

$$[Z] \in I^m H_n(X_0(\mathfrak{c}); \mathcal{L}^\vee(\widehat{\kappa},\chi_0)).$$

Let $f \in S_{\widehat{\kappa}}(K_0(\mathfrak{c}),\chi)$, let $J \subset \Sigma(L)$ and let $\omega_J(f^{-\iota}) \in \Omega^n(Y_0(\mathfrak{c}),\mathcal{L}(\widehat{\kappa},\chi_0))$ be the resulting differential form (cf. (7.2.3)). Using the pairing $\mathcal{L}(\widehat{\kappa},\chi_0) \times \mathcal{L}^\vee(\widehat{\kappa},\chi_0) \to \mathbb{C}$ of Section 7.5 the section s can be paired with the differential form, so the integral $\int_{(Z,s)} \omega_J(f^{-\iota})$ is defined. Assume, as in Theorem 4.8 that there exists a character θ of $G_E(\mathbb{A})$ so that, viewed as a function on $G_E(\mathbb{A})/K_{E\infty}K_0(\mathfrak{c}_E)$, the section s satisfies $|s(gx)| \le |\theta(g)||s(x)|$ for all $g \in G_E(\mathbb{A})$. Then by Theorem 4.8, the integral

$$\int_{(Z,s)} \omega_J(f^{-\iota})$$

is finite and it equals the Kronecker product $\langle \mathcal{Z}(\omega_J(f^{-\iota})), [Z,s]\rangle_K$.

By Lemma 7.4 the operator $W_\mathfrak{c} : Y_0(\mathfrak{c}) \to Y_0(\mathfrak{c})$ converts $\mathcal{L}(\widehat{\kappa},\chi_0)$ to $\mathcal{L}^\vee(\widehat{\kappa},\chi_0)$ so

$$[W_\mathfrak{c}^{*-1}(Z,s)] = W_\mathfrak{c}^{*-1}([Z,s]) \in I^m H_n(Y_0(\mathfrak{c}),\mathcal{L}(\widehat{\kappa},\chi_0)).$$

Theorem 8.5. *Let $\gamma = Q(W_\mathfrak{c}^{*-1}[Z])$. If $\mathfrak{m} + N_{L/E}(\mathfrak{c})\mathfrak{d} = \mathfrak{n} + N_{L/E}(\mathfrak{c})\mathfrak{d} = \mathcal{O}_E$ and $\mathfrak{m}, \mathfrak{n}$ are both norms from $\mathcal{O}_L$, then the $\mathfrak{m}$th Fourier coefficient of*

$$[\gamma(\mathfrak{n}), \Phi_{\gamma,\chi_E}]_{IH_*}$$

is

$$\frac{1}{4} \sum_{J \subset \Sigma(L)} \sum_f \frac{\int_Z \omega_J(\widehat{f}^{-\iota}) \int_Z \omega_{\Sigma(L)-J}(\widehat{f}^{-\iota})}{T(\widehat{f},\Sigma(L)-J)(\widehat{f},\widehat{f})_P} \lambda_f(\mathfrak{n})\overline{\lambda_f(\mathfrak{m})}, \qquad (8.6.1)$$

where σ is any element of $\Sigma(L)$, the sum is over the normalized newforms f whose base change $\widehat{f}$ is an element of $S_{\widehat{\kappa}}^{\mathrm{new},\chi_E}(\mathfrak{c},\chi)$, and $T(J,\widehat{f})$ is defined as in (7.5.10).

We remark that Theorem 1.2 is just Theorem 8.5 in the special case that $L(\kappa,\chi_0)$ is the trivial representation $\mathbb{C}$.

Proof. By Theorem 7.1 and Corollary E.12 any class in $I^m H_n^{\chi_E}(X_0(\mathfrak{c}),\mathcal{L}(\widehat{\kappa},\chi_0))$ can be expressed as a linear combination of classes represented by differential forms $\omega_J(\widehat{f}^{-\iota})$, where $\widehat{f}^{-\iota}(x) := \widehat{f}(x^{-\iota})$ as in (5.5.1) (thus replacing $\widehat{f} \in S_{\widehat{\kappa}}(K_0(\mathfrak{c}),\chi)$ with $\widehat{f}^{-\iota} \in S_{\widehat{\kappa}}^{\mathrm{coh}}(K_0(\mathfrak{c}),\chi)$). More precisely, we may write

$$\gamma = Q(W_\mathfrak{c}^{*-1}[Z]) = \frac{1}{2} \sum_{J \subset \Sigma(L)} \sum_f a_{J,f} \mathcal{P}[\omega_J(\widehat{f}^{-\iota})] \qquad (8.6.2)$$

8.6. The Fourier coefficients of $[\gamma(\mathfrak{m}), \Phi_{\gamma, \chi_E}]_{IH_*}$

for some $a_{J,f} \in \mathbb{C}$, where the sum is over the same set of f as in the theorem. Here $\mathcal{P}$ is the Poincaré duality isomorphism of Chapter 3, and the $\frac{1}{2}$ appears for the same reason explained in the proof of Theorem 8.3. We assume, without loss of generality, that $a_{J,f} = a_{J,f \otimes \eta}$, where $f \otimes \eta$ is the newform generating the same automorphic representation as $f \otimes \eta$.

Note that $\gamma = \gamma(\mathcal{O}_L)$. Thus, if $\mathfrak{n} \subset \mathcal{O}_E$ is a norm from $\mathcal{O}_L$, we have:

$$2\gamma(\mathfrak{n}) = \widehat{T}(\mathfrak{n})_* \left(\sum_{J \subset \Sigma(L)} \sum_f a_{J,f} \mathcal{P}[\omega_J(\widehat{f}^{-\iota})] \right)$$

$$= \sum_{J \subset \Sigma(L)} \sum_f a_{J,f} \widehat{T}(\mathfrak{n})_* \mathcal{P}[\omega_J(\widehat{f}^{-\iota})]$$

$$= \sum_{J \subset \Sigma(L)} \sum_f a_{J,f} \mathcal{P}\widehat{T}(\mathfrak{n})_*[\omega_J(\widehat{f}^{-\iota})] \quad \text{(by Hecke-equivariance of } \mathcal{P})$$

$$= \sum_{J \subset \Sigma(L)} \sum_f a_{J,f} \mathcal{P}[\omega_J(\widehat{f}|\widehat{T}(\mathfrak{n})^{-\iota})] \quad \text{(by Hecke equivariance of } \omega_J)$$

$$= \sum_{J \subset \Sigma(L)} \sum_f \lambda_f(\mathfrak{n}) a_{J,f} \mathcal{P}[\omega_J(\widehat{f}^{-\iota})] \quad \text{(Proposition 8.2)}.$$

By the same argument, for $\mathfrak{m} \subset \mathcal{O}_E$ a norm from $\mathcal{O}_L$, using diagram (7.4.1) and Lemma 7.8 we obtain

$$2\left(W_c^* \gamma\right)(\mathfrak{m}) = \sum_{J \subset \Sigma(L)} \sum_f a_{J,f} \mathcal{P}[\omega_J(\widehat{T}(\mathfrak{m}) W_c^* \widehat{f}^{-\iota})]$$

$$= \sum_{J \subset \Sigma(L)} \sum_f \overline{\lambda_f(\mathfrak{m})} a_{J,f} \mathcal{P} W_c^* [\omega_J(\widehat{f}^{-\iota})]$$

$$= \sum_{J \subset \Sigma(L)} \sum_f \overline{\lambda_f(\mathfrak{m})} a_{\Sigma(L)-J, f} \mathcal{P} W_c^* [\omega_{\Sigma(L)-J}(\widehat{f}^{-\iota})].$$

Newforms with distinct Hecke eigenvalues are orthogonal with respect to the Petersson product, and differential n-forms of pure Hodge types are orthogonal unless the Hodge types are complementary. Therefore the product

$$[\gamma(\mathfrak{n}), \gamma(\mathfrak{m})]_{IH_*} = \langle \gamma(\mathfrak{n}), W_c^* \gamma(\mathfrak{m}) \rangle_{IH_*}$$

is given by

$$\frac{1}{4} \sum_{J \subset \Sigma(L)} \sum_f \lambda_f(\mathfrak{n}) \overline{\lambda_f(\mathfrak{m})} a_{J,f} a_{\Sigma(L)-J, f} \left\langle \mathcal{P}[\omega_J(\widehat{f}^{-\iota})], \mathcal{P} W_c^* [\omega_{\Sigma(L)-J}(\widehat{f}^{-\iota})] \right\rangle_{IH_*}$$

$$= \frac{1}{4} \sum_{J \subset \Sigma(L)} \sum_f \lambda_f(\mathfrak{n}) \overline{\lambda_f(\mathfrak{m})} a_{J,f} a_{\Sigma(L)-J, f} \left\langle [\omega_J(\widehat{f}^{-\iota})], W_c^* [\omega_{\Sigma(L)-J}(\widehat{f}^{-\iota})] \right\rangle_{IH^*}$$

(see Section 3.4). It remains to prove that

$$a_{J,f} a_{\Sigma(L)-J,f} \left[[\omega_J(\widehat{f}^{-\iota})], [\omega_{\Sigma(L)-J}(\widehat{f}^{-\iota})] \right]_{IH^*} \quad (8.6.3)$$

$$= \frac{\left(\int_Z \omega_J(\widehat{f}^{-\iota}) \right) \left(\int_Z \omega_{\Sigma(L)-J}(\widehat{f}^{-\iota}) \right)}{T(\widehat{f}, \Sigma(L)-J)(\widehat{f}, \widehat{f})_P}.$$

For this, recall that if $g \in S_{\widehat{\kappa}}(K_0(\mathfrak{c}), \chi)$, then $[\omega_J(g)]$ is the image $\alpha_*([\omega_0])$ of a compactly supported class $[\omega_0] \in H^i_c(Y_0(\mathfrak{c}), \mathcal{L}(\widehat{\kappa}, \chi_0))$. This follows from [Har, Section 3.1] (the compactly supported cohomology is contained in $\ker(r)$ in the notation of [Har, Section 3.1]) (see [Hid7, Proposition 3.1] as well). So the differential form $\omega_J(g)$ determines an intersection cohomology class in two (potentially) different ways: as the image $\alpha_*([\omega_0])$ of a compactly supported class, and as $\mathcal{Z}[\omega_J(g)]$; a push-forward of the image of a square integrable form. Of course, both procedures determine the same intersection cohomology class (see diagram (4.3.1)). Moreover, since Q is an orthogonal projection, the product of $(Z - Q[Z])$ with any class in IH^{χ_E} vanishes (whether the Petersson product, intersection product, or Kronecker product is used). Consequently, from Proposition 4.8 we have:

$$\int_Z \omega_J(g) = \left[[\omega_J(g)], W^{-1}_{\mathfrak{c}*}[Z] \right]_K = \left[\omega_J(g), Q(W^{-1}_{\mathfrak{c}*}[Z]) \right]_K < \infty, \quad (8.6.4)$$

where $[\cdot, \cdot]_K$ denotes the Kronecker pairing between intersection cohomology and intersection homology induced by $[\cdot, \cdot]_{IH^*}$. Restricting to the case $g = \widehat{f}^{-\iota}$ and using (3.4.4) (which says that the Kronecker pairing plus Poincaré duality equals the intersection pairing) along with (8.6.2), we have

$$\left[[\omega_J(\widehat{f}^{-\iota})], W^{*-1}_{\mathfrak{c}}[Z] \right]_K = \left[\mathcal{P}[\omega_J(\widehat{f}^{-\iota})], W^{*-1}_{\mathfrak{c}}[Z] \right]_{IH_*}$$

$$= a_{\Sigma(L)-J,f} \left[\mathcal{P}[\omega_J(\widehat{f}^{-\iota})], \mathcal{P}[\omega_{\Sigma(L)-J}(\widehat{f}^{-\iota})] \right]_{IH_*}$$

$$= a_{\Sigma(L)-J,f} \left[[\omega_J(\widehat{f}^{-\iota})], [\omega_{\Sigma(L)-J}(\widehat{f}^{-\iota})] \right]_{IH^*}.$$

Here we also used the commutativity of diagram (3.4.3), and our assumption from above that $a_{J,f} = a_{J,f \otimes \eta}$. The same argument is valid with J replaced by $\Sigma(L) - J$ and we obtain

$$a_{J,f} a_{\Sigma(L)-J,f} \left[\omega_{\Sigma(L)-J}(\widehat{f}^{-\iota})], [\omega_J(\widehat{f}^{-\iota})] \right]_{IH^*} \left[\omega_J(\widehat{f}^{-\iota})], [\omega_{\Sigma(L)-J}(\widehat{f}^{-\iota})] \right]_{IH^*}$$

$$= \int_Z \omega_J(g) \int_Z \omega_{\Sigma(L)-J}(g).$$

Combining this with Theorem 7.11 implies (8.6.3) and hence the theorem. $\square$

Chapter 9
Explicit Construction of Cycles

In this chapter we consider a quadratic extension L/E of totally real fields. The inclusion $E \to L$ gives rise to Hilbert modular subvarieties, known as Hirzebruch-Zagier cycles, $Z \subset Y$ with $\dim(Y) = 2\dim(Z)$. If the local system $\mathcal{L}$ on Y is chosen appropriately then it admits a canonical section over Z, which gives a Borel Moore homology class $[Z] \in H_*^{BM}(Y, \mathcal{L})$. We use the machinery in this book to find a canonical lift of this class to the (middle) intersection homology $[Z] \in I^m H_*(X, \mathcal{L})$ of the Baily-Borel compactification X of Y (see Proposition 9.2).

We also consider the "twist", Z_θ of the cycle Z by a Hecke character $\theta : L^\times \backslash \mathbb{A}_L^\times \to \mathbb{C}^\times$ with its resulting intersection homology class $[Z_\theta]$. These "twisted" Hirzebruch-Zagier cycles were described in [MurR, Ram3]; they provide examples of cycles whose class in intersection homology satisfies the hypotheses of Theorems 8.4 and 8.5. Using some notation from below, the significance of the cycles Z_θ is the following: If the characters θ and χ_E are related as in Section 10.1 (so that $\theta|\mathbb{A}_E^\times = \eta\chi_E$) then the Hecke module generated by (the Hecke translates of) the intersection cohomology class $P_{\text{new}}(W_\mathfrak{c}^{*-1}[Z_\theta])$ is $IH_n^{\chi_E}(X_0(\mathfrak{c}), \mathcal{L}(\widehat{\kappa}, \chi_0))$, i.e., the part of the intersection cohomology that is "new" and comes from the base changes of newforms of nebentypus χ_E. For a precise statement, see Theorem 10.1. As a byproduct of the Rankin-Selberg computations that underly the proof of Theorem 10.1, we explicitly compute the Fourier coefficients appearing in Theorem 8.5 in the case where $[Z] = P_{\text{new}}[Z_\theta]$ (the "new part" of Z_θ). The result of this computation is recorded in Theorem 10.2.

9.1 Notation for the quadratic extension L/E

Let L/E be a quadratic extension of totally real number fields. Set

$$n := [L : \mathbb{Q}], \quad d := d_{L/E},$$

for the degree and discriminant respectively. Let $\langle \varsigma \rangle = \text{Gal}(L/E)$ and $\langle \eta \rangle = \text{Gal}(L/E)^\wedge$. As in Chapter 8, by class field theory, η may be identified with a Hecke character $\eta : \mathbb{A}_E \to \mathbb{C}^\times$ that is trivial on the image of the norm from L, and hence also trivial at infinity (the fact that L/E is totally real implies $\mathbb{A}_{E\infty}^\times \subset N_{L/E}\mathbb{A}_L^\times$). For every $\sigma \in \Sigma(L)$, write $\sigma' := \sigma \circ \varsigma$. Borrowing terminology from the theory of CM extensions, we say that a subset $J \subset \Sigma(L)$ is a *type* for L/E if exactly one of each pair $\{\sigma, \sigma'\}$ is in J. Without loss of generality, we write $L = E(\sqrt{\Delta})$ for a totally positive $\Delta \in E$. There is a distinguished type $J_E = \{\sigma \in \Sigma(L) : \sigma(\sqrt{\Delta}) < 0\}$ for L/E. We assume that the set $\Sigma(L)$ is ordered so that for $z \in \mathfrak{h}^{\Sigma(L)}$ we have

$$dz = dz_{\sigma_1} \wedge dz_{\sigma_1'} \wedge \cdots \wedge dz_{\sigma_{[E:\mathbb{Q}]}} \wedge dz_{\sigma'_{[E:\mathbb{Q}]}}$$

where $J_E = \{\sigma_1, \ldots, \sigma_{[E:\mathbb{Q}]}\}$ and $J_E' := \{\sigma_1', \ldots, \sigma'_{[E:\mathbb{Q}]}\}$.

Choose a weight $\kappa = (k, m) \in \mathcal{X}(E)$ and fix a quasicharacter $\chi_E : E^\times \backslash \mathbb{A}_E^\times \to \mathbb{C}^\times$ satisfying $\chi_{E\infty}(b_\infty) = b_\infty^{-k-2m}$ for all $b \in \mathbb{A}_E^\times$. We obtain a weight $\widehat{\kappa}$ and a Hecke character $\chi = \chi_L$ for L by setting $\widehat{\kappa} = (\widehat{k}, \widehat{m}) \in \mathcal{X}(L)$ where $\widehat{k}_\sigma := k_{\sigma_0}$ and $\widehat{m}_\sigma = m_{\sigma_0}$ if the infinite place $\sigma \in \Sigma(L)$ extends $\sigma_0 \in \Sigma(E)$, and by setting

$$\chi = \chi_L = \chi_E \circ N_{L/E} : L^\times \backslash \mathbb{A}_L^\times \to \mathbb{C}^\times$$

(which is the base change of χ_E to L). We assume that the conductor of χ divides an ideal $\mathfrak{c} \subset \mathcal{O}_L$. Then the associated character $\chi_0 = \chi_{L,0}$ of $K_0(\mathfrak{c}) \subset G_L(\mathbb{A})$ is defined as in Section 7.1. Denote by h_L^+ the narrow class number of $\mathcal{O}_L$. Recall that $[k + 2m] = k_\sigma + 2m_\sigma$, an integer independent of the place σ. For our later convenience, write $[k + 2m]\mathbf{1} = k + 2m$ and similarly $[\widehat{k} + 2\widehat{m}]\mathbf{1} := \widehat{k} + 2\widehat{m}$.

9.2 Canonical section over the diagonal

In this chapter we construct flat sections of the local systems $\mathcal{L}(\widehat{\kappa}, \chi_0)$ and $\mathcal{L}^\vee(\widehat{\kappa}, \chi_0)$ over the "diagonal cycle" $Z_0(\mathfrak{c}_E) \subset Y_0(\mathfrak{c})$ (defined below) that arises from the quadratic extension $E \subset L$. Let $\mathfrak{c} \subset \mathcal{O}_L$ be an ideal and let $\mathfrak{c}_E = \mathfrak{c} \cap \mathcal{O}_E$. Let

$$\iota : G_E = \mathrm{Res}_{E/\mathbb{Q}} \mathrm{GL}_2 \longrightarrow \mathrm{Res}_{L/\mathbb{Q}} \mathrm{GL}_2 = G_L$$

be the diagonal embedding and let

$$\pi : G_L(\mathbb{A}) \to Y_0(\mathfrak{c}) = G_L(\mathbb{Q}) \backslash G_L(\mathbb{A}) / K_\infty K_0(\mathfrak{c})$$

be the canonical projection. Associated to the diagonal embedding is a (non-compact) Shimura subvariety

$$Z_0 = Z_0(\mathfrak{c}_E) := \pi(\iota(G_E(\mathbb{A}))) \cong G_E(\mathbb{Q}) \backslash G_E(\mathbb{A}) / K_{E\infty} K_0(\mathfrak{c}_E).$$

Here $K_{E\infty} \subset G_E(\mathbb{R})$ and $K_0(\mathfrak{c}_E) \subset G_E(\mathbb{A}_f)$ are defined as in Sections 5.1 and 5.2 with L replaced by E. We begin by constructing a section

$$s : Z_0 \longrightarrow \mathcal{L}^\vee(\widehat{\kappa}, \chi_0) = \mathcal{L}(\widehat{\kappa}) \otimes \mathcal{E}(\chi_0)$$

9.2. Canonical section over the diagonal

where $\mathcal{L}(\widehat{\kappa})$ is the local system associated to the representation $\operatorname{Sym}^{\widehat{k}}(V^{\vee}) \otimes \det^{-\widehat{m}}$ of K_∞ as in Proposition 6.3, and where $\chi = \chi_L$ is the character that we fixed in Section 9.1 and $\chi_0 = \chi_{L,0}$ is the resulting character of $K_0(\mathfrak{c})$ as in Section 5.6. This is accomplished by first constructing a section over the finite cover

$$Z_1(\mathfrak{c}_E) = G_E(\mathbb{Q}) \backslash G_E(\mathbb{A}) / K_{E\infty} K_1(\mathfrak{c}_E) \to Z_0(\mathfrak{c}_E)$$

where

$$K_1(\mathfrak{c}) := \left\{ \begin{pmatrix} a & b \\ c & d \end{pmatrix} \in K_0(\mathfrak{c}) : a \equiv 1 \pmod{\mathfrak{c}} \right\}$$

and $K_1(\mathfrak{c}_E) := K_1(\mathfrak{c}) \cap G_E(\mathbb{A})$ so that

$$K_1(\mathfrak{c}_E) = \left\{ \begin{pmatrix} a & b \\ c & d \end{pmatrix} \in K_0(\mathfrak{c}_E) : a \equiv 1 \pmod{\mathfrak{c}_E} \right\}.$$

Proposition 9.1. *The function* $s_1 : G_E(\mathbb{A}) \to \operatorname{Sym}^{\widehat{k}}(V^{\vee})$ *defined by*

$$s_1(g) := \chi_L(\det(g))^{-1} |\det g|_{\mathbb{A}_E}^{-[k+2m]} \cdot \Delta \tag{9.2.1}$$

where

$$\Delta := \prod_{\sigma \in \Sigma(E)} \sum_{j=0}^{k_\sigma} \binom{k_\sigma}{j} (-1)^{k_\sigma - j} X_\sigma^j Y_\sigma^{k_\sigma - j} X_{\sigma'}^{k_\sigma - j} Y_{\sigma'}^j$$

defines a flat section of the vector bundle $\mathcal{L}^{\vee}(\widehat{\kappa}, \chi_0)$ *over* $Z_1(\mathfrak{c}_E)$, *where the elements*

$$\left\{ X_\sigma^i Y_\sigma^{k-i} X_{\sigma'}^j Y_{\sigma'}^{k-j} \right\} \text{ for } 0 \le i, j \le k$$

form the standard basis of $\operatorname{Sym}^k(V_\sigma^{\vee}) \otimes \operatorname{Sym}^k(V_{\sigma'}^{\vee})$ *that arises from the standard basis of* V, *the standard representation of* GL_2.

Proof. First let us check that s_1 does indeed define a (continuous) section of the vector bundle $\mathcal{L}^{\vee}(\widehat{\kappa}, \chi_0)$ over $Z_1(\mathfrak{c}_E)$. As in equation (6.9.3) we need to check that s_1 has the correct equivariance properties, viz.

(A) $s_1(\gamma g) = s_1(g)$ for all $\gamma \in G_E(\mathbb{Q})$
(B) $s_1(g u_0) = \chi_0(u_0^{-1}) s_1(g)$ for all $u_0 \in K_1(\mathfrak{c}_E)$
(C) $s_1(g u_\infty) = \lambda_{\widehat{k}}^{\vee}(u_\infty^{-1}) \det^{-\widehat{m}}(u_\infty^{-1}) s_1(g)$ for all $u_\infty \in K_{E\infty} \subset G_E(\mathbb{R})$

where $\lambda_{\widehat{k}}^{\vee}$ is the irreducible representation of

$$G_L(\mathbb{R}) = \prod_{\sigma \in \Sigma(E)} \operatorname{GL}_2(\mathbb{R}_\sigma) \times \operatorname{GL}_2(\mathbb{R}_{\sigma'})$$

on the vector space

$$\mathcal{L}(\widehat{\kappa}) := \operatorname{Sym}^{\widehat{k}}(V^{\vee}) \otimes \det^{-\widehat{m}}$$
$$= \bigotimes_{\sigma \in \Sigma(E)} \operatorname{Sym}^{k_\sigma}(V_\sigma^{\vee}) \otimes \det^{-m_\sigma} \otimes \operatorname{Sym}^{k_{\sigma'}}(V_{\sigma'}^{\vee}) \otimes \det^{-m_{\sigma'}}$$

which we also identify with the corresponding vector bundle on $G_L(\mathbb{R})/K_\infty$.

For item (A), use the fact that $\det(\gamma) \in E^\times \subset L^\times$ for all $\gamma \in G_E(\mathbb{Q})$ so that $\chi_L(\det \gamma) = 1$, and $|\det \gamma|_{\mathbb{A}_E} = 1$. For item (B) let $u_0 = \begin{pmatrix} a & b \\ c & d \end{pmatrix} \in K_1(\mathfrak{c}_E)$. Then

$$\chi_L(\det(u_0))^{-1} = \prod_{\mathfrak{p}_v | \mathfrak{c}} \chi_v(a_v d_v)^{-1} = \prod_{\mathfrak{p}_v | \mathfrak{c}} \chi_v(d_v)^{-1} = \chi_0(u_0)^{-1}.$$

Moreover, $|\det u_0|_{\mathbb{A}_E} = 1$ as in equation (7.4.7). So it remains to verify condition (C). First we make a few comments about the general situation. Suppose a Lie group H acts on a finite-dimensional complex vector space W. The dual, or contragredient representation of H on $W^\vee = \mathrm{Hom}(W, \mathbb{C})$ is given by

$$(h \cdot \lambda)(w) := (\lambda \circ h^{-1})w = \lambda(h^{-1} \cdot w)$$

where $h \in H, \lambda \in W^\vee, w \in W$. We obtain a representation of $H \times H$ on $W \otimes W^\vee$. There is a canonical element $x_0 \in W \otimes W^\vee$ which may be described as the dual of the evaluation mapping $W^\vee \otimes W \to \mathbb{C}$ or equivalently as $x_0 = \sum_i e_i \otimes e_i^*$ where $\{e_1, e_2, \ldots, e_r\}$ is any basis of W and $\{e_1^*, \ldots, e_r^*\}$ is the dual basis. Since $\{h \cdot e_i\}$ and $\{h \cdot e_i^* = e_i^* \circ h^{-1}\}$ are also dual bases, it follows that the element x_0 is fixed under the action of the diagonal subgroup $H \subset H \times H$.

Suppose that $K \subset H$ is a Lie subgroup. Then we obtain an $H \times H$-homogeneous vector bundle

$$\widetilde{\mathcal{M}} = (H \times H)_{(K \times K)}(W \otimes W^\vee)$$

over the quotient $(H \times H)/(K \times K)$. The diagonal embedding $H \to H \times H$ gives an embedding of $\widetilde{Z} := H/K \to \widetilde{Z} \times \widetilde{Z}$ so the restriction $\widetilde{\mathcal{M}}|\widetilde{Z}$ becomes an H-homogeneous vector bundle over $\widetilde{Z}$. Therefore the element x_0 defines a canonical section of $\widetilde{\mathcal{M}}|\widetilde{Z}$ which is invariant under the action of H on $\widetilde{Z}$, and it is a flat section (with respect to the natural flat connection on $\widetilde{\mathcal{M}}$, see Section 6.2.3).

Now assume we are given a nondegenerate bilinear pairing $\Theta: W \times W \to \mathbb{C}$ and suppose there exists a character $\psi: H \to \mathbb{C}^\times$ such that

$$\Theta(h \cdot A, h \cdot B) = \psi(h)\Theta(A, B) \qquad (9.2.2)$$

for all $A, B \in W$ and all $h \in H$. We construct a canonical flat section of the vector bundle
$$\widetilde{\mathcal{L}} = \widetilde{\mathcal{L}}_1 \otimes \widetilde{\mathcal{L}}_2 := (H \times H)_{K \times K}(W \otimes W)$$

over the diagonal $\widetilde{Z}$ as follows. The pairing Θ determines vector space isomorphisms

$$\Theta^\sharp: W \to W^\vee \quad \text{and} \quad \Theta^\flat = (\Theta^\sharp)^{-1} : W^\vee \to W$$

by $\Theta^\sharp(A)(B) = \Theta(A, B)$, hence $\Theta(\Theta^\flat(\lambda), B) = \lambda(B)$ for all $\lambda \in W^\vee$. It follows that

$$h \cdot \Theta^\flat(\lambda) = \psi(h)\Theta^\flat(h \cdot \lambda)$$

9.2. Canonical section over the diagonal

for all $h \in H$. So we obtain an isomorphism of vector bundles

$$\widetilde{\mathcal{M}}|\widetilde{Z} = H \times_K (W \otimes W^\vee) \to H \times_K (W \otimes W) = \widetilde{\mathcal{L}}|\widetilde{Z}$$

by $[h, v \otimes \lambda] \mapsto [h, \psi(h)^{-1} v \otimes \Theta^\flat(\lambda)]$. (This mapping is not H-equivariant.)

If $\{e_1, \ldots, e_r\}$ and $\{e_1^*, \ldots, e_r^*\}$ are dual bases of W and $W^\vee$ respectively then

$$\Theta^\sharp(e_i) = \sum_j \Theta_{ij} e_j^* \text{ and } \Theta^\flat(e_i^*) = \sum_j \Xi_{ij} e_j$$

where $(\Theta_{ij}) = (\Theta(e_i, e_j))$ is the matrix for Θ and (Ξ_{ij}) is its inverse. Applying this to the constant section $x_0 = \sum_i e_i \otimes e_i^*$ of the vector bundle $\widetilde{M}$ gives a canonical section over $\widetilde{Z} = H/K$ of $\widetilde{\mathcal{L}}$:

$$hK \mapsto \left[h, \psi(h)^{-1} \sum_i \sum_j (e_i \otimes \Xi_{ij} e_j) \right] \quad (9.2.3)$$

Now consider the case that $H = \mathrm{GL}_2(\mathbb{R})$ and $W = \mathrm{Sym}^k(V^\vee)$ where $V \cong \mathbb{C}^2$ is the standard representation, and k is a positive integer. Let $K = K_\infty$ be the set of elements of the form $\begin{pmatrix} a & b \\ -b & a \end{pmatrix}$. The standard basis for V gives a basis $\{X_1^i Y_1^{k-i} X_2^j Y_2^{k-j}\}$ ($0 \le i, j \le k$) for $W \otimes W$. An inner product on W is given in equation (7.5.4), that is,

$$\Theta_{ij} = (-1)^{j-1} \binom{k}{j-1} \text{ if } i - 1 + j - 1 = k$$

(and $\Theta_{ij} = 0$ otherwise) with respect to this basis. It satisfies equation (7.5.6), viz

$$\Theta(g_\infty.A, g_\infty.B) = \det(g_\infty)^{-k} \Theta(A, B) \quad (9.2.4)$$

for all $g_\infty \in H = \mathrm{GL}_2(\mathbb{R})$ and $A, B \in W$. Substituting all of this into equation (9.2.3), with $h = g_\infty$ and $\psi(h) = \det(g_\infty)^{-k}$, and taking the product over all infinite places gives a flat section

$$G_E(\mathbb{R}) \to G_E(\mathbb{R}) \times_{K_{E\infty}} \mathrm{Sym}^{\widehat{k}}(V^\vee)$$
$$g_\infty \mapsto \det(g_\infty)^k . \Delta,$$

see Section 6.2.3 and 6.4.2. Multiplying by $\det(g_\infty)^{2m}$, we therefore obtain a flat section of $\mathcal{L}(\widehat{\kappa}) = \mathrm{Sym}^{\widehat{k}}(V^\vee) \otimes \det^{-\widehat{m}}$,

$$s_1'(g_\infty) = \det(g_\infty)^k \det(g_\infty)^{2m} . \Delta$$
$$= \chi_E(\det g_\infty)^{-1} . \Delta$$

over $G_E(\mathbb{R})/K_{E\infty}$, which is to say that the mapping $s_1' : G_E(\mathbb{R}) \to \operatorname{Sym}^{\widehat{k}}$ satisfies condition (C) above. Unfortunately this is not quite the same as the section $s_1(g)$; it may differ by a sign. But the factors

$$\chi_E(\det g_\infty)^{-1} \quad \text{and} \quad \chi_L(\det g_\infty)^{-1}|\det g_\infty|_{\mathbb{A}_E}^{-[k+2m]}$$

agree when $g_\infty = u_\infty \in K_{E\infty}$ so the function s_1 also satisfies condition (C) above, hence it is also a section of the same vector bundle $\mathcal{L}(\widehat{\kappa})$. This completes the verification that the section (9.2.1) exhibits the correct equivariance at the infinite places. $\square$

So far, we have a flat section of the vector bundle $\mathcal{L}^\vee(\widehat{\kappa}, \chi_0)$ over $Z_1(\mathfrak{c}_E)$, which is a finite cover of $Z_0(\mathfrak{c}_E)$. So we can obtain a flat section of the same vector bundle over $Z_0(\mathfrak{c}_E)$ by summing over the fibers of the projection $Z_1(\mathfrak{c}_E) \to Z_0(\mathfrak{c}_E)$ as described in equation (6.1.2). In other words, if $g_1 K_1(\mathfrak{c}_E), \ldots, g_t K_1(\mathfrak{c}_E)$ is a minimal complete set of coset representatives for $K_0(\mathfrak{c}_E)/K_1(\mathfrak{c}_E)$ then the function $s : G_E(\mathbb{A}) \to \operatorname{Sym}^{\widehat{k}}(V^\vee)$,

$$s(g) := \sum_{i=1}^{t} \chi_0(g_i) s_1(gg_i) \qquad (9.2.5)$$

defines a section of $\mathcal{L}^\vee(\widehat{\kappa}, \chi_0)$ over $Z_0(\mathfrak{c}_E)$. Finally, by replacing the group

$$K_1(\mathfrak{c}) := \left\{ \begin{pmatrix} a & b \\ c & d \end{pmatrix} \in K_0(\mathfrak{c}) : a \equiv 1 \pmod{\mathfrak{c}} \right\}$$

with the group

$$K_1^\vee(\mathfrak{c}) := \left\{ \begin{pmatrix} a & b \\ c & d \end{pmatrix} \in K_0(\mathfrak{c}) : d \equiv 1 \pmod{\mathfrak{c}} \right\}$$

in the definition of s_1, we will obtain a flat section of the vector bundle $\mathcal{L}(\widehat{\kappa}, \chi_0) = \mathcal{L}(\widehat{\kappa}) \otimes \mathcal{E}(\chi_0^\vee)$ over $Z_0(\mathfrak{c}_E)$.

In the next section we will pair the section s with a differential form constructed from an automorphic form. In order to do this we must first express the section s as a function

$$S : D_E \times G_E(\mathbb{A}_f) \to \operatorname{Sym}^{\widehat{k}}(V^\vee)$$

following the procedure described in equations (6.3.3) and (6.4.3), (that is, by multiplying by an automorphy factor), where

$$D_E := G_E(\mathbb{R})/K_{E,\infty}$$

is the symmetric space associated to G_E, and where $K_{E,\infty}$ is the subgroup obtained by replacing L with E in the definition of K_∞. The automorphy factor in this case is $\lambda_{\widehat{k}}^\vee \det^{-\widehat{m}}$ (and is independent of z), so the function S is simply

$$S(z, g_0) = \lambda_{\widehat{k}}^\vee(\alpha_z) \det(\alpha_z)^{-\widehat{m}} s(\alpha_z g_0) \qquad (9.2.6)$$

where $\alpha_z \in G_E(\mathbb{R})$ is any element with $\alpha_z \mathbf{i} = z$. Equations (9.2.2) and (9.2.4) in the proof Proposition 9.1 above implies that the element Δ satisfies

$$\lambda_{\hat{k}}^{\vee}(g_\infty)\Delta = \det(g_\infty)^{-k}\Delta$$

for $g_\infty \in G_E(\mathbb{R})$ (which is the key requirement in order that it define a flat section, as in Section 6.2.3). Hence (9.2.6) becomes the following:

$$S(z, g_0) = \sum_{i=1}^{t} \chi_0(g_i)\det(\alpha_z)^{-\widehat{m}}\chi_L(\det(\alpha_z g_0 g_i))^{-1}|\det(\alpha_z g_0 g_i)|_{\mathbb{A}_E}^{-[k+2m]}\det(\alpha_z)^{-k}\Delta$$

$$= \sum_{i=1}^{t} \chi_0(g_i)\det(\alpha_z)^{-2m-k}\chi_L(\det(\alpha_z g_0 g_i))^{-1}|\det(\alpha_z g_0 g_i)|_{\mathbb{A}_E}^{-[k+2m]}\Delta$$

$$= \epsilon(z)\sum_{i=1}^{t} \chi_0^{\vee}(g_i)\chi_L(\det(g_0))^{-1}|\det(g_0)|_{\mathbb{A}_E}^{-[k+2m]}\Delta \qquad (9.2.7)$$

since $\det(g_i) \in \widehat{\mathcal{O}}_E^{\times}$ and $|\det(g_i)|_{\mathbb{A}_E} = 1$. Here, g_i are defined in (9.2.5) above and

$$\epsilon(z) = \prod_{\sigma \in \Sigma(E)} \left(\frac{\det(\alpha_z)_\sigma}{|\det(\alpha_z)_\sigma|}\right)^{k_\sigma + 2m_\sigma} = \pm 1$$

depends only on the signs of $\text{Im}(z_\sigma)$. If we choose the point z to lie in the product of *upper* half-planes, which we may do without loss of generality, then $\epsilon(z) = 1$.

9.3 Homological properties of $Z_0(\mathfrak{c}_E)$

Let $n := [L : \mathbb{Q}]$. Let $J \subset \Sigma(L)$. Let $f \in S_{\hat{\kappa}}^{\text{coh}}(K_0(\mathfrak{c}), \chi)$ be a modular form. In Proposition 6.5 and Definition 6.8 we constructed an associated differential form

$$\omega_J(f) = \iota_J^*(\omega_f) \in \Omega^n(Y_0(\mathfrak{c}), \mathcal{L}(\hat{\kappa}, \chi_0))$$

which was shown to be square integrable, and to determine an intersection cohomology class

$$[\omega_J(f)] \in I^m H^n(X_0(\mathfrak{c}), \mathcal{L}(\hat{\kappa}, \chi_0))$$

in the Baily-Borel compactification. On the other hand, in the previous section we have constructed a flat section S of the local system $\mathcal{L}^{\vee}(\hat{\kappa}, \chi_0)$ over the diagonal cycle $Z_0 := Z_0(\mathfrak{c}_E)$ in the Hilbert modular variety $Y_0(\mathfrak{c})$. By Theorem 4.6 the pair (Z_0, S) determine an intersection homology class

$$[Z_0] \in I^m H_n(X_0(\mathfrak{c}), \mathcal{L}^{\vee}(\hat{\kappa}, \chi_0)).$$

(Recall that first they determine a Borel-Moore homology class, which then lifts canonically to intersection homology by Saper's theorem.) In Proposition 7.10 we constructed a pairing

$$\langle \cdot, \cdot \rangle : \mathcal{L}(\widehat{\kappa}, \chi_0) \times \mathcal{L}^\vee(\widehat{\kappa}, \chi_0) \to \mathcal{E}(\chi \circ \det) \to \mathbb{C}.$$

Consequently the product

$$\langle \omega_J(f), S \rangle \in \Omega^n(Y_0(\mathfrak{c}), \mathbb{C})$$

is a differential form with coefficients in the trivial local system and so its integral

$$\int_{Z_0} \omega_J(f) := \int_{Z_0} \langle \omega_J(f), S \rangle = \int_{Z_0,S} \omega_f \in \mathbb{C}$$

is well defined. By Proposition 4.8 the integral is finite and equal to the Kronecker product

$$\int_{Z_0} \omega_J(f) = \langle [\omega_J(f)], [Z_0] \rangle_K \tag{9.3.1}$$

between the intersection cohomology class $[\omega_J(f)]$ and the class that is obtained from the section S and the subvariety Z_0.

Proposition 9.2. *If J is not a type for E then this integral vanishes. If J is a type for E then it equals*

$$\sum_{i=1}^t \chi_0(g_i)(2i)^{[L:\mathbb{Q}]}(-1)^{[E:\mathbb{Q}]} \int_{Z_0} \iota_J^* \left(\det(\alpha_z)^{-\widehat{m}-\widehat{k}-1} j(\alpha_z, i)^{\widehat{k}+2} f(g_0 g_i \alpha_z) \right)$$

$$\times \chi_L(\det(\alpha_z g_i))^{-1} |\det(\alpha_z g_0 g_i)|_{\mathbb{A}_E}^{-[k+2m]} \det(\alpha_z)^{-2m-k} \prod_{\sigma \in J_E} y_\sigma^{k_\sigma + 2} d\mu_{\mathfrak{c}_E},$$

where $d\mu_{\mathfrak{c}_E}$ is the canonical measure on $Z_0 = Y_0(\mathfrak{c}_E)$.

Proof. In order to evaluate this integral, first recall that the differential form ω_f of equation (6.10.2) was defined using a carefully constructed section P_z from equation (6.7.1) of a certain vector bundle. Then, as described in equation (6.11.2), for $(z, x_0) \in D \times G(\mathbb{A}_f)$,

$$\omega_J(f)(z, x_0) = \iota_J^* \omega_f(z, x_0)$$

$$= \iota_J^* \left(\det(\alpha_z)^{-\widehat{m}-\widehat{k}-1} j(\alpha_z, i)^{\widehat{k}+2} f(x_0 \alpha_z) dz \right) \prod_{\sigma \in \Sigma(L)} (-X_\sigma + w_\sigma Y_\sigma)^{\widehat{k}_\sigma}$$

$$\tag{9.3.2}$$

where $\alpha_z \in G(\mathbb{R})$, $\alpha_z \cdot \mathbf{i} = z$ and

$$w_\sigma := \begin{cases} z_\sigma & \text{if } \sigma \notin J \\ \bar{z}_\sigma & \text{otherwise} \end{cases}.$$

9.3. Homological properties of $Z_0(\mathfrak{c}_E)$

This must be paired with the canonical section S of equation (9.2.7) using the pairing of equation (7.5.4); in other words we must calculate the integral

$$\sum_{i=1}^{t}\chi_0(g_i)\int_{Z_0}\iota_J^*\left(\det(\alpha_z)^{-\widehat{m}-\widehat{k}-1}j(\alpha_z,i)^{\widehat{k}+2}f(g_0g_i\alpha_z)dz\right) \qquad (9.3.3)$$

$$\times \chi_L(\det(\alpha_zg_i))^{-1}|\det(\alpha_z)|_{\mathbb{A}_E}^{-[k+2m]}\det(\alpha_z)^{-2m-k}\left\langle \prod_{\sigma\in\Sigma(L)}(-X_\sigma+w_\sigma Y_\sigma)^{\widehat{k}_\sigma},\Delta\right\rangle$$

where the pairing is the pairing $L(\kappa)\times L(\kappa)\to\det_\infty^{-\widehat{k}}\to\mathbb{C}$ of Proposition 7.10. Notice that we have multiplied the naïve product of the two sections by $\chi_L(\det g_0)$ as in equation (7.5.8), because the naïve product takes values in the line bundle $\chi\circ\det$. The paired quantities give

$$\left\langle \prod_{\sigma\in\Sigma(L)}(-X_\sigma+w_\sigma Y_\sigma)^{\widehat{k}_\sigma}, \prod_{\sigma\in J_E}\sum_{j=0}^{k_\sigma}\binom{k_\sigma}{j}(-1)^{k_\sigma-j}X_\sigma^j Y_\sigma^{k_\sigma-j}X_{\sigma'}^{k_\sigma-j}Y_{\sigma'}^{j}\right\rangle$$

$$=\prod_{\sigma\in J_E}\sum_{j=0}^{k_\sigma}\frac{(-1)^{2j}}{\binom{k_\sigma}{j}^2}\binom{k_\sigma}{j}(-1)^j w_\sigma^{k_\sigma-j}\binom{k_\sigma}{j}(-1)^{k_\sigma-j}w_{\sigma'}^j\binom{k_\sigma}{j}(-1)^{k_\sigma-j}$$

$$=\prod_{\sigma\in J_E}\sum_{j=0}^{k_\sigma}\binom{k_\sigma}{j}w_\sigma^{k_\sigma-j}w_{\sigma'}^j(-1)^j=\prod_{\sigma\in J_E}(w_\sigma-w_{\sigma'})^{k_\sigma},$$

see equation (7.5.5). This vanishes unless J is a type for E, in which case it is

$$=(-1)^{|J\cap J_E|}(2i)^{[E:\mathbb{Q}]}\prod_{\sigma\in J_E}y_\sigma^{k_\sigma}.$$

Assume that J is a type for L/E. By the calculation we have that (9.3.3) is equal to

$$\sum_{i=1}^{t}(-1)^{|J\cap J_E|}(2i)^{[E:\mathbb{Q}]}\chi_0(g_i)\int_{Z_0}\iota_J^*\left(\det(\alpha_z)^{-\widehat{m}-\widehat{k}-1}j(\alpha_z,i)^{\widehat{k}+2}f(g_0g_i\alpha_z)\right)$$

$$\times\chi_L(\det(\alpha_zg_0g_i))^{-1}|\det(\alpha_zg_0g_i)|_{\mathbb{A}_E}^{-[k+2m]}\det(\alpha_z)^{-2m-k}(-1)^{|J\cap J_E|}(-2i)^{[E:\mathbb{Q}]}$$

$$\times\left(\prod_{\sigma\in J_E}y_\sigma^{k_\sigma}\right)dx_\infty dy_\infty d\mu_0(g_0)$$

$$=\sum_{i=1}^{t}\chi_0(g_i)(2i)^{[L:\mathbb{Q}]}(-1)^{[E:\mathbb{Q}]}\int_{Z_0}\iota_J^*\left(\det(\alpha_z)^{-\widehat{m}-\widehat{k}-1}j(\alpha_z,i)^{\widehat{k}+2}f(g_0g_i\alpha_z)\right)$$

$$\times\chi_L(\det(\alpha_zg_i))^{-1}|\det(\alpha_zg_0g_i)|_{\mathbb{A}_E}^{-[k+2m]}\det(\alpha_z)^{-2m-k}\prod_{\sigma\in J_E}y_\sigma^{k_\sigma}dxdyd\mu_0(g_0),$$

where $d\mu_0$ is the Haar measure on $R_{E/F}\mathrm{GL}_2(\mathbb{A}_f)$ giving $K_0(\mathfrak{c}_E)$ volume one. $\square$

9.4 The twisting correspondence

In Section 5.11 we defined the twist of a modular form f by a Hecke character $\theta = \theta^u | \cdot |^w$ (where $w \in \mathbb{Z}$ and where θ^u is of finite order). It was constructed so that the L-function of $f \otimes \theta$ was the twist of the L-function of f. In this section we show that a similar twisting operation may be performed on cycles. For this purpose, we approximately follow [Ram3, MurR] (who approximately followed [Hid1] Section 5, [Hid2] Section 6, who approximately followed [Shim2, Shim3]); see also [Rib]). We construct a *twisting correspondence* $\mathcal{T}$ which, when applied to the differential form $\omega(f)$ corresponding to a modular form f, yields the differential form $\omega(f \otimes \theta)$ that corresponds to the twisted modular form. Since it is a geometric correspondence, $\mathcal{T}$ can also be applied to the cycle $Z = Z_0(\mathfrak{c}_E)$ to obtain a new cycle Z_θ, thus giving a cycle-level analog of the twisting operation. Our construction differs slightly from that of [Ram3, MurR] in order to take into account the local coefficient system.

As in the preceding section, let χ_E be a Hecke character for the real field E and let $\chi = \chi_L = \chi_E \circ N_{L/E}$ be the resulting Hecke character for L. Let $\mathfrak{c}$ be divisible by the conductor of χ. Let θ^u be another Hecke character of L, of finite order, with conductor $\mathfrak{b}$ and let $w \in \mathbb{Z}$. Define $\theta := \theta^u | \cdot |^w_{\mathbb{A}_L}$. We obtain characters $\chi_0, \chi_0^\vee$ on $K_0(\mathfrak{c})$ and $\theta_0 = \theta_0^u, \theta_0^\vee = (\theta^u)_0^\vee$ on $K_0(\mathfrak{b})$. Let $\widehat{\kappa} = (\widehat{k}, \widehat{m}) \in \mathcal{X}(L)$ as defined above in Section 9.1, and define

$$\widehat{\kappa} - w := (\widehat{k}, \widehat{m} - w\mathbf{1}) \in \mathcal{X}(L).$$

The characters $\chi_0^\vee$, $\theta_0^\vee$, $\chi \circ \det$, and $\theta \circ \det$ are all trivial on the following compact open subgroup of $G(\mathbb{A}_f)$,

$$K_{11}(\mathfrak{c}\mathfrak{b}^2) := \left\{ \begin{pmatrix} a & b \\ c & d \end{pmatrix} \in K_0(\mathfrak{c}\mathfrak{b}^2) : a, d \equiv 1 \pmod{\mathfrak{c}\mathfrak{b}^2} \right\}.$$

Let $Y_{11}(\mathfrak{c}\mathfrak{b}^2)$ denote the corresponding Hilbert modular variety, that is,

$$Y_{11}(\mathfrak{c}\mathfrak{b}^2) = G(\mathbb{Q}) \backslash G(\mathbb{A}) / K_\infty K_{11}(\mathfrak{c}\mathfrak{b}^2).$$

The inclusion $K_{11}(\mathfrak{c}\mathfrak{b}^2) \subset K_0(\mathfrak{c}\mathfrak{b}^2) \subset K_0(\mathfrak{c})$ induces natural covering maps $\pi_1 : Y_{11}(\mathfrak{c}\mathfrak{b}^2) \to Y_0(\mathfrak{c}\mathfrak{b}^2)$ and $\pi_2 : Y_{11}(\mathfrak{c}\mathfrak{b}^2) \to Y_0(\mathfrak{c})$.

Recall from Section 5.11 the definitions of $\Upsilon = \mathfrak{b}^{-1}\widehat{\mathcal{O}}_L$, $\widetilde{\Upsilon} = \Upsilon/\widehat{\mathcal{O}}_L$ and $\theta_\mathfrak{b}$. For $t \in \Upsilon$ define $u_t = u(t) \in G(\mathbb{A})$ by $u(t)_v = \begin{pmatrix} 1 & 0 \\ 0 & 1 \end{pmatrix}$ if $v \nmid \mathfrak{b}$ and $u(t)_v = \begin{pmatrix} 1 & t_v \\ 0 & 1 \end{pmatrix}$ if $v | \mathfrak{b}$. We now describe a correspondence

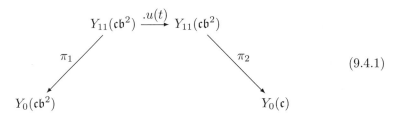

(9.4.1)

9.4. The twisting correspondence

and a lift of this correspondence to the local systems

$$\mathcal{L}(\widehat{\kappa} - w, (\chi\theta^2)_0) = \mathcal{L}(\widehat{\kappa} - w) \otimes \mathcal{E}((\chi\theta^2)_0^\vee) \quad \text{and} \quad \mathcal{L}(\widehat{\kappa}, \chi_0) = \mathcal{L}(\widehat{\kappa}) \otimes \mathcal{E}(\chi_0^\vee)$$

on $Y_0(\mathfrak{c}\mathfrak{b}^2)$ and $Y_0(\mathfrak{c})$ respectively.

Lemma 9.3. *Let* $t \in \Upsilon$. *The mapping (which we denote by* $\cdot u_t$)

$$[g, v] \mapsto [gu_t, \theta(\det g)\theta_\mathfrak{b}(t)^{-1}v] \tag{9.4.2}$$

gives a well-defined, canonical isomorphism

$$\pi_1^* \mathcal{L}(\widehat{\kappa} - w, \chi_0 \theta_0^2) \to \pi_2^* \mathcal{L}(\widehat{\kappa}, \chi_0).$$

Equivalently, the mapping $P_{2,t} := \pi_2 \circ (\cdot u(t))$ *defines a mapping*

$$P_{2,t} : \pi_1^* \mathcal{L}(\widehat{\kappa} - w, \chi_0 \theta_0^2) \to \mathcal{L}(\widehat{\kappa}, \chi_0).$$

Proof. Assume first that θ^u is trivial, so $\mathfrak{b} = \mathcal{O}_L$ and $\theta(b) = |b|_{\mathbb{A}_L}^w$ for all $b \in \mathbb{A}_L^\times$. In this case $\widetilde{\Upsilon} = \{1\}$ and we write P_2 rather than $P_{2,t}$. We need to show that the mapping $[g, v] \mapsto [g, |\det g|_{\mathbb{A}_L}^w v]$ gives a well-defined mapping from the vector bundle

$$\mathcal{L}(\widehat{\kappa} - (0, w), \chi_0) = \operatorname{Sym}^{\widehat{k}}(V^\vee) \otimes \det_\infty^{-\widehat{m}+w\mathbf{1}} \otimes \mathcal{E}(\chi_0^\vee)$$

to the vector bundle

$$\mathcal{L}(\widehat{\kappa}, \chi_0) = \operatorname{Sym}^{\widehat{k}}(V^\vee) \otimes \det_\infty^{-\widehat{m}} \otimes \mathcal{E}(\chi_0^\vee)$$

(see Proposition 6.3). This involves checking the required equivariance with respect to K_∞. Since the mapping P_2 does not involve the factors $\operatorname{Sym}^{\widehat{k}} \otimes \mathcal{E}(\chi_0^\vee)$ it suffices to check the equivariance with respect to the relevant powers of the line bundle $\det_\infty$. A vector in $\det_\infty^{-\widehat{m}+w\mathbf{1}}$ is an equivalence class $[g, v]$ with $g \in G(\mathbb{A})$ and $v \in \mathbb{C}$ where $[gk_\infty, v] \sim [g, \det(k_\infty)^{-\widehat{m}+w\mathbf{1}}v]$ for all $k_\infty \in K_\infty$. Applying P_2 gives

$$P_2([gk_\infty, v]) = [gk_\infty, |\det g|_{\mathbb{A}_L}^w |\det k_\infty|^w v]$$
$$\sim [g, \det k_\infty^{-\widehat{m}} |\det g|_{\mathbb{A}_L}^w |\det k_\infty|^w v]$$
$$= P_2([g, \det(k_\infty)^{-\widehat{m}+w\mathbf{1}}v])$$

which completes the verification that the mapping (9.4.2) is K_∞-equivariant.

Now assume that $w = 0$ and $\theta = \theta^u$ is nontrivial. In this case checking the K_∞ and $G(\mathbb{Q})$-equivariance of P_2 is trivial, so it suffices to describe the lift to the one-dimensional local systems $\mathcal{E}(\chi_0^\vee)$ and $\mathcal{E}((\chi\theta^2)_0^\vee)$. Let $k = \begin{pmatrix} a & b \\ c & d \end{pmatrix} \in K_{11}(\mathfrak{c}\mathfrak{b}^2)$. It suffices to show that $P_{2,t}([gk, v]) = P_{2,t}([g, \chi_0^\vee(k)\theta_0^\vee(k)^2 v])$. As in Section 5.11, the elements ct and ct^2 are integral. Therefore the element

$$u_t^{-1} k u_t = \begin{pmatrix} a - ct & b + (a-d)t - ct^2 \\ c & d + ct \end{pmatrix} \quad \text{is in } K_{11}(\mathfrak{c}\mathfrak{b}^2) \subset K_0(\mathfrak{c}\mathfrak{b}^2).$$

Therefore $\chi_0^\vee(u_t^{-1}ku_t) = 1$ and

$$\begin{aligned}P_{2,t}([gk,v]) &= [gku_t, \theta(\det k)\theta(\det g)\theta_\mathfrak{b}(t)^{-1}v]\\ &= [gu_t, \theta(\det g)\theta_\mathfrak{b}(t)^{-1}v]\\ &= P_{2,t}([g, \chi_0^\vee(k)\theta_0^\vee(k)^2\theta_\mathfrak{b}(t)^{-1}v])\end{aligned}$$

since $\chi_0^\vee(k) = \theta_0^\vee(k) = \theta(\det k) = 1$. □

Let $\widetilde{\Upsilon}$ be a collection of representatives for Υ modulo $\widehat{\mathcal{O}}_L$, as in Section 5.11. Given a Hecke character θ with conductor $\mathfrak{b}$, define the *twisting correspondence* $\mathcal{T}^{\mathrm{coh}}(\mathfrak{c}\mathfrak{b}^2)$ to be the disjoint union over $t \in \widetilde{\Upsilon}$ of the correspondences (9.4.1), which gives the following diagram.

$$\begin{array}{ccc}\mathcal{T}^{\mathrm{coh}}(\mathfrak{c}\mathfrak{b}^2) & \xrightarrow{P_2} & Y_0(\mathfrak{c})\\ \pi_1 \downarrow & & \\ Y_0(\mathfrak{c}\mathfrak{b}^2) & & \end{array}$$

where $P_2 = \sum_{t \in \widetilde{\Upsilon}} P_{2,t}$. Denote the action of $\mathcal{T}^{\mathrm{coh}}(\mathfrak{c}\mathfrak{b}^2)$ on a section s of $\mathcal{L}(\widehat{\kappa}, \chi_0)$ by

$$s|\mathcal{T}^{\mathrm{coh}}(\mathfrak{c}\mathfrak{b}^2) = (\pi_1)_* P_2^*(s) = \sum_{t \in \widetilde{\Upsilon}} (\pi_1)_* u_t^* \pi_2^*(s).$$

The following proposition states that the twist of a modular form in $M_\kappa^{\mathrm{coh}}(K_0(\mathfrak{c}), \chi)$ may be realized, up to a constant, by the action of the twisting correspondence, see equation (5.11.12).

Recall from Section 9.1 that $\kappa = (k,m) \in \mathcal{X}(E)$ and $\widehat{\kappa} = (\widehat{k}, \widehat{m}) \in \mathcal{X}(L)$ are weights. As in Section 9.4 let $\theta = \theta^u |\cdot|^w$ be a Hecke character where θ^u is of finite order and has conductor $\mathfrak{b}$, and where $w \in \mathbb{Z}$.

Proposition 9.4. *Let $f \in M_{\widehat{\kappa}}^{\mathrm{coh}}(K_0(\mathfrak{c}), \chi)$ (where the conductor of χ divides $\mathfrak{c}$). Let $J \subset \Sigma(L)$ and let $\omega_J(f)$ be the differential form associated to f in Proposition 6.5 and equation (6.11.2). Let*

$$f \otimes \theta = (f^{-\iota} \otimes \theta)^{-\iota} \in M_{\widehat{\kappa}-w}^{\mathrm{coh}}(K_0(\mathfrak{c}\mathfrak{b}^2), \chi\theta^2)$$

be the twist of f by θ as described in Definition 5.13 and in Proposition 5.11. Then

$$\omega_J(f)|\mathcal{T}^{\mathrm{coh}}(\mathfrak{c}\mathfrak{b}^2) = (\pi_1)_* P_2^* \omega_J(f) = C_2 \omega_J(f \otimes \theta) \tag{9.4.3}$$

where

$$C_2 = [K_0(\mathfrak{c}\mathfrak{b}^2) : K_{11}(\mathfrak{c}\mathfrak{b}^2)]G((\theta^u)^{-1})|\mathrm{N}_{L/\mathbb{Q}}(\mathcal{D}_{L/\mathbb{Q}})|^{-w}. \tag{9.4.4}$$

Here, $G((\theta^u)^{-1})$ is the Gauss sum defined in equation (5.11.5).

9.4. The twisting correspondence

Proof. We will prove the following slightly more general statement. Let $s(g) = [g, F(g)]$ be a section of $\mathcal{L}(\widehat{\kappa}, \chi_0) = \mathcal{L}(\widehat{\kappa}) \otimes \mathcal{E}(\chi_0^\vee)$ on $Y_0(\mathfrak{c})$. Then $s' := s|\mathcal{T}^{\mathrm{coh}}(\mathfrak{c}\mathfrak{b}^2)$ is a section of $\mathcal{L}(\widehat{\kappa} - w, \chi_0 \theta_0^2)$ on $Y_0(\mathfrak{c}\mathfrak{b}^2)$. Setting $s'(g) = [g, F'(g)]$ we have:

$$F'(g) = \left(F | \mathcal{T}^{\mathrm{coh}}(\mathfrak{c}\mathfrak{b}^2)\right)(g) = \left[K_0(\mathfrak{c}\mathfrak{b}^2) : K_{11}(\mathfrak{c}\mathfrak{b}^2)\right] \theta(\det g)^{-1} \sum_{t \in \widetilde{\Upsilon}} \theta_\mathfrak{b}(t) F(g u_t). \tag{9.4.5}$$

Equation (9.4.3) follows from this statement; the factor $|N_{L/\mathbb{Q}} \mathcal{D}_{L/\mathbb{Q}}|^{-w}$ comes from the definition of the twist of f by $|\cdot|^w$, see equation (5.11.2), and the factor $G((\theta^u)^{-1})$ comes from equation (5.11.4).

To prove equation (9.4.5), first consider the case that θ^u is trivial so that $\theta(b) = |b|_{\mathbb{A}_L}^w$ for all $b \in \mathbb{A}_L$. In this case both sides of equation (9.4.5) are equal to $|\det(g)|_{\mathbb{A}_L}^{-w} F(g)$ so there is nothing to prove.

So we may assume that $w = 0$ and $\theta = \theta^u$. As in the proof of Lemma 9.3, it suffices to prove the formula for the restriction $s|_{G(\mathbb{A}_f)/K_0(\mathfrak{c})}$, so we may simply consider s to be a section of $\mathcal{E}(\chi_0^\vee)$. First we need an auxiliary calculation.

Fix $t \in \Upsilon$ and suppose that $s(g K_0(\mathfrak{c})) = [g, F(g)]$ is a section of $\mathcal{E}(\chi_0^\vee)$ over $G(\mathbb{A}_f)/K_0(\mathfrak{c})$, so that $F : G(\mathbb{A}) \to \mathbb{C}$ satisfies $F(gk) = \chi_0^\vee(k)^{-1} F(g)$ for all $k \in K_0(\mathfrak{c})$. Let $k = \begin{pmatrix} a & b \\ c & d \end{pmatrix} \in K_0(\mathfrak{c})$. Since the elements ct and ct^2 are integral, the element

$$\widehat{k} := \begin{pmatrix} a - \frac{act}{d} & b - \frac{act^2}{d} \\ c & d + ct \end{pmatrix} \text{ is in } K_0(\mathfrak{c}).$$

If $v \nmid \mathfrak{b}$ then $k_v . u(t_v) . \widehat{k}_v^{-1} = \begin{pmatrix} 1 & 0 \\ 0 & 1 \end{pmatrix}$, while

$$k_v . u(t_v) . \widehat{k}_v^{-1} = \begin{pmatrix} 1 & a_v d_v^{-1} t_v \\ 0 & 1 \end{pmatrix} \quad \text{if } v | \mathfrak{b}.$$

Hence $k u_t \widehat{k}^{-1} = u_s$ where $s = a d^{-1} t \in \Upsilon$ and

$$F(g k u_t) = F(g k u_t \widehat{k}^{-1} \widehat{k}) = F(g u_s) \chi_0^\vee(\widehat{k}^{-1}) = F(g u_s) \chi_0^\vee(k)^{-1}. \tag{9.4.6}$$

Now let us compute $(\pi_1)_* P_2^* F$. First, we have:

$$(P_2^* F)(g) = \sum_{t \in \widetilde{\Upsilon}} \theta(\det g)^{-1} \theta_\mathfrak{b}(t) F(g u_t).$$

We will use equation (6.1.2) to push this forward by the mapping π_1.

Let $k_1, \ldots, k_m \in K_0(\mathfrak{c}\mathfrak{b}^2)$ be a collection of coset representatives for the quotient $K_0(\mathfrak{c}\mathfrak{b}^2)/K_{11}(\mathfrak{c}\mathfrak{b}^2)$ and set $k_i = \begin{pmatrix} a_i & b_i \\ c_i & d_i \end{pmatrix}$. Then

$$(\pi_1)_* P_2^* F(x) = \sum_{i=1}^m (\chi \theta^2)_0^\vee(k_i) P_2^* F(x k_i).$$

As in equation (9.4.6),
$$F(xk_iu_t) = \chi_0^\vee(k_i)^{-1}F(xu(a_id_i^{-1}t)) = \chi(a_i)^{-1}F(xu_{s_i})$$
where $s_i = a_id_i^{-1}t$. Therefore
$$(\pi_1)_*P_2^*F(x) = \sum_{i=1}^m \chi(a_i)\theta(a_i)^2 \sum_{t\in\widetilde{\Upsilon}} \theta(\det xk_i)^{-1}\theta_{\mathfrak{b}}(t)F(xk_iu_t)$$
$$= \sum_{i=1}^m \theta(a_i)^2 \sum_{t\in\widetilde{\Upsilon}} \theta(a_id_i)^{-1}\theta(\det x)^{-1}\theta_{\mathfrak{b}}(t)F(xu(a_id_i^{-1}t))$$
$$= \sum_{i=1}^m \sum_{t\in\widetilde{\Upsilon}} \theta(a_id_i^{-1})\theta(\det x)^{-1}\theta_{\mathfrak{b}}(t)F(xu(a_id_i^{-1}t))$$
$$= [K_0(\mathfrak{c}\mathfrak{b}^2) : K_{11}(\mathfrak{c}\mathfrak{b}^2)] \sum_{t\in\widetilde{\Upsilon}} \theta(\det x)^{-1}\theta_{\mathfrak{b}}(a_id_i^{-1}t)F(xu(a_id_i^{-1}t))$$
$$= [K_0(\mathfrak{c}\mathfrak{b}^2) : K_{11}(\mathfrak{c}\mathfrak{b}^2)] \theta(\det x)^{-1} \sum_{s\in\widetilde{\Upsilon}} \theta_{\mathfrak{b}}(s)F(xu_s)$$

because multiplication by $a_id_i^{-1}$ permutes the elements of $\widetilde{\Upsilon}$, which proves (9.4.5). $\square$

Remarks.
(1) In the other direction, if $g \mapsto [g, H(g)]$ is a section of $\mathcal{L}(\widehat{\kappa} - w, \chi_0\theta_0^2)$ on $Y_0(\mathfrak{c}\mathfrak{b}^2)$ then
$$((P_2)_* \pi_1^* H)(g) = [K_0(\mathfrak{c}) : K_{11}(\mathfrak{c}\mathfrak{b}^2)]\, \theta(\det g) \sum_{t\in\widetilde{\Upsilon}} \theta_{\mathfrak{b}}^{-1}(t) H(gu_t). \quad (9.4.7)$$

(2) Exactly the same formula (9.4.2) also gives an isomorphism
$$\pi_1^*\mathcal{E}((\chi_0\theta_0^2)^\vee) \to \pi_2^*\mathcal{E}((\chi_0)^\vee)$$
or equivalently, a mapping $P_2 : \pi_1^*\mathcal{E}((\chi_0\theta_0^2)^\vee) \to \mathcal{E}(\chi_0^\vee)$ because χ_0 and θ_0 are trivial on $K_{11}(\mathfrak{c}\mathfrak{b}^2)$.

(3) For modular forms in $M_{\widehat{\kappa}}(K_0(\mathfrak{c}), \chi)$ (rather than $M_{\widehat{\kappa}}^{\text{coh}}(K_0(\mathfrak{c}), \chi)$), change equation (9.4.2) to
$$[g, v] \mapsto [gu_t, \theta(\det g)^{-1}\theta_{\mathfrak{b}}(t)^{-1}v]$$
which also gives an isomorphism $\pi_1^*\mathcal{E}(\chi_0\theta_0^2) \to \pi_2^*\mathcal{E}(\chi_0)$. Let $\mathcal{T}(\mathfrak{c}\mathfrak{b}^2)$ denote the resulting correspondence. It converts a section $s(g) = [g, F(g)]$ of

$\mathcal{L}^\vee(\widehat{\kappa}(\chi_0^{-1})) = \mathcal{L}(\widehat{\kappa}) \otimes \mathcal{E}(\chi_0^{-1})$ (see Remarks following Proposition 6.4) into the section

$$(F|\mathcal{T}(\mathfrak{cb}^2))(g) = (\pi_1)_* P_2^*(F)(g)$$
$$= \left[K_0(\mathfrak{cb}^2) : K_{11}(\mathfrak{cb}^2)\right] \theta(\det g) \sum_{t \in \widetilde{\Upsilon}} \theta_{\mathfrak{b}}(t) F(gu_t),$$

of $\mathcal{L}(\widehat{\kappa} - w) \otimes \mathcal{E}(\chi_0^{-1}\theta_0^{-2})$.

9.5 Twisting the cycle $Z_0(\mathfrak{c}_E)$

We now describe how to twist the diagonal chain $Z_0(\mathfrak{c}_E)$ by the character $\theta = \theta^u| \cdot |^w$. First, by replacing the Hecke character χ (and its conductor $\mathfrak{c}$) in Section 9.2 by the character $\chi\theta^2$ (with its conductor $\mathfrak{cb}^2$), and by replacing the weight $\widehat{\kappa} = (\widehat{k}, \widehat{m}) \in \mathcal{X}(L)$ with the weight $\widehat{\kappa} - w = (\widehat{k}, \widehat{m} - w\mathbf{1})$ we obtain a canonical flat section s of the local system

$$\mathcal{L}^\vee(\widehat{\kappa} - w, (\chi\theta^2)_0)$$

over the diagonal cycle $Z_0(\mathfrak{c}_E\mathfrak{b}_E^2) \subset Y_0(\mathfrak{cb}^2)$, where $\mathfrak{b}_E := \mathfrak{b} \cap \mathcal{O}_E$. Apply the resulting twisting correspondence $\mathcal{T}^{\mathrm{coh}}(\mathfrak{cb}^2)$ (going from left to right in the diagram (9.4.1)) to this cycle to obtain

$$Z_\theta := (P_2)_* \pi_1^* \left(Z_0(\mathfrak{c}_E\mathfrak{b}_E^2)\right) \subset Y_0(\mathfrak{c}).$$

The (flat) section s becomes a (flat) section

$$s_\theta := (P_2)_* \pi_1^*(s) \text{ of } \mathcal{L}^\vee(\widehat{\kappa}, \chi_0)$$

over the cycle Z_θ.

Let $f \in S_{\widehat{\kappa}}^{\mathrm{coh}}(K_0(\mathfrak{c}), \chi)$, let $J \subset \Sigma(L)$ and let $\omega_J(f)$ be the corresponding differential form on $Y_0(\mathfrak{c})$ with values in $\mathcal{L}(\widehat{\kappa}, \chi_0)$. As in Section 9.3, the pairing from Proposition 7.10 may be used to multiply the section s with the differential form $\omega_J(f)$, giving a differential form with complex coefficients. Hence the integral $\int_{Z_\theta} \omega_J(f) \in \mathbb{C}$ is defined. We now prove the twisted version of Proposition 9.2.

Proposition 9.5. *The cycle* (Z_θ, s_θ) *has a canonical lift to intersection homology,*

$$[Z_\theta] = (P_2)_* \pi_1^*([Z_0(\mathfrak{c}_E\mathfrak{b}_E^2)]) \in IH_n(X_0(\mathfrak{c}_E), \mathcal{L}^\vee(\widehat{\kappa}, \chi_0)),$$

and for any $f \in S_{\widehat{\kappa}}^{\mathrm{coh}}(K_0(\mathfrak{c}), \chi)$ *and for any* $J \subset \Sigma(L)$ *we have:*

$$\langle [\omega_J(f)], [Z_\theta] \rangle_K = \int_{Z_\theta} \omega_J(f) = C_2 \int_{Z_0} \omega_J(f \otimes \theta)$$

where C_2 *is given in equation* (9.4.4). *If J is not a type for E then this integral vanishes. Assume J is a type for E and take $w = m_\sigma + k_\sigma/2$ (which is independent* of $\sigma \in \Sigma(E)$). Then $f \otimes \theta \in S^{\text{coh}}_{(\widehat{k},-\widehat{k}/2)}(K_0(\mathfrak{c}\mathfrak{b}^2), \chi\theta^2)$ and

$$\langle [\omega_J(f)], [Z_\theta] \rangle_K = C_3 \sum_{i=1}^t (\chi_L \theta^2)_0(g_i)$$

$$\times \int_{Z_0(\mathfrak{c}\mathfrak{b}^2 \cap \mathcal{O}_E)} \iota_J^* \left(\det(\alpha_z)^{-\widehat{k}/2-1} j(\alpha_z, i)^{\widehat{k}+2} (f \otimes \theta)(g_0 g_i \alpha_z) \right)$$

$$\times \chi_L \theta^2 (\det(\alpha_z g_i))^{-1} \prod_{\sigma \in J_E} y_\sigma^{k_\sigma + 2} d\mu_{\mathfrak{c}_E},$$

where $d\mu = d\mu_{\mathfrak{c}\mathfrak{b}^2 \cap \mathcal{O}_E}$ is the canonical measure on $Z_0(\mathfrak{c}_E \mathfrak{b}_E^2) = Y_0(\mathfrak{c}_E \mathfrak{b}_E^2)$ and where

$$C_3 = (2i)^{[L:\mathbb{Q}]} (-1)^{[E:\mathbb{Q}]} [K_0(\mathfrak{c}\mathfrak{b}^2) : K_{11}(\mathfrak{c}\mathfrak{b}^2)] G((\theta^u)^{-1}) |\mathrm{N}_{L/\mathbb{Q}} \mathcal{D}_{L/\mathbb{Q}}|^{-(k_\sigma + 2m_\sigma)/2}.$$

If J is not a type for E then this integral vanishes.

Proof. By Proposition 9.2 the cycle $Z_0(\mathfrak{c}_E \mathfrak{b}_E^2)$ has a canonical lift to intersection homology. Since the mappings π_1, π_2 extend to (finite) mappings on the Baily-Borel compactifications, the same holds for the cycle Z_θ.

To compute $\int_{Z_\theta} \omega_J(f)$, start with $Z_0(\mathfrak{c}_E \mathfrak{b}_E^2)$ on the left side of diagram 9.4.1. Set $Z_{11}(\mathfrak{c}_E \mathfrak{b}_E^2) = \pi_1^*(Z_0(\mathfrak{c}_E \mathfrak{b}_E^2))$, then let $Z_{11}(t)$ be the push-forward by $u(t)$. Using equation (6.1.4) we obtain

$$\int_{(\pi_2)_* Z_{11}(t)} \omega_J(f) = \int_{Z_{11}(t)} \pi_2^* \omega_J(f) = \int_{Z_{11}(\mathfrak{c}_E \mathfrak{b}_E^2)} \omega_{11}(t).$$

where $\omega_{11}(t) = (\cdot u_t)^* \pi_2^* \omega_J(f)$. Set $C_4 := [K_0(\mathfrak{c}\mathfrak{b}^2) : K_{11}(\mathfrak{c}\mathfrak{b}^2)]$. Then

$$C_4 \int_{Z_0(\mathfrak{c}_E \mathfrak{b}_E^2)} (\pi_1)_* \omega_{11}(t) = \int_{(\pi_1)_* \pi_1^* Z_0(\mathfrak{c}_E \mathfrak{b}_E^2)} (\pi_1)_* \omega_{11}(t)$$

$$= \int_{Z_{11}(\mathfrak{c}_E \mathfrak{b}_E^2)} \pi_1^*(\pi_1)_* \omega_{11}(t) = C_4 \int_{Z_{11}(\mathfrak{c}_E \mathfrak{b}_E^2)} \omega_{11}(t).$$

Putting these together, and summing over $t \in \widetilde{\Upsilon}$, and using Proposition 9.4 gives

$$\int_{Z_\theta} \omega_J(f) = \int_{Z_0(\mathfrak{c}_E \mathfrak{b}_E^2)} (\pi_1)_* P_2^* \omega_J(f) = C_2 \int_{Z_0(\mathfrak{c}_E \mathfrak{b}_E^2)} \omega_J(f \otimes \theta)$$

Finally, if $w = m_\sigma + k_\sigma/2$ (which is independent of σ) then upon replacing $Z_0(\mathfrak{c}_E)$ by $Z_0(\mathfrak{c}_E \mathfrak{b}_E^2)$ and upon replacing f by $f \otimes \theta$, Proposition 9.2 gives the value of this integral. $\square$

Chapter 10
The Full Version of Theorem 1.3

10.1 Statement of results

We use the same setup as in the previous chapter: L/E is a quadratic extension of totally real fields, $\mathrm{Gal}(L/E) = \langle 1, \varsigma \rangle$, $\mathrm{Gal}(L/E)^\wedge = \langle 1, \eta \rangle$, $\mathfrak{c} \subset \mathcal{O}_L$ is an ideal, and
$$n := [L:\mathbb{Q}], \quad d := d_{L/E}, \quad \mathcal{D} := \mathcal{D}_{L/E}, \quad \mathfrak{c}_E := \mathfrak{c} \cap \mathcal{O}_E.$$

We write $L = E(\sqrt{\Delta})$ for a totally positive $\Delta \in E$. There is a distinguished type $J_E = \left\{\sigma \in \Sigma(L) : \sigma(\sqrt{\Delta}) < 0\right\}$ for L/E. We assume that the set $\Sigma(L)$ is ordered so that for $z \in \mathfrak{h}^{\Sigma(L)}$ we have
$$dz = dz_{\sigma_1} \wedge dz_{\sigma'_1} \wedge \cdots \wedge dz_{\sigma_{[E:\mathbb{Q}]}} \wedge dz_{\sigma'_{[E:\mathbb{Q}]}}$$
where $J'_E := J_E^\varsigma =: \{\sigma'_1, \ldots, \sigma'_{[E:\mathbb{Q}]}\}$.

Let $\chi_E : E^\times \backslash \mathbb{A}_E^\times \to \mathbb{C}^\times$ be a Hecke character. Let $\chi = \chi_E \circ \mathrm{N}_{L/E}$ be the resulting character of $L^\times \backslash \mathbb{A}_L^\times$ and let χ_0 be the resulting character of $K_0(\mathfrak{c})$. Assume that the conductor of χ divides $\mathfrak{c}$ and that $\chi_E(b_\infty) = b_\infty^{-k-2m}$ for all $b \in \mathbb{A}_E^\times$. Let $\kappa = (k, m)$ be a weight for E and let $\widehat{\kappa} = (\widehat{k}, \widehat{m})$ be the corresponding weight for L. We obtain a "diagonal" cycle $Z_0 := Z_0(\mathfrak{c}_E)$ with a canonical flat section S. Let θ^u be a Hecke character of L, of finite order, with conductor $\mathfrak{b} = \mathfrak{f}(\theta^u)$ and let $w = m_\sigma + k_\sigma/2$ (which is independent of the choice of σ). Define $\theta = \theta^u| \cdot |_{\mathbb{A}_L}^w$. We assume throughout this chapter that
$$\theta|_{\mathbb{A}_E^\times} = \chi_E \eta.$$

For a proof that such a character θ exists, see [Hid8, Lemma 2.1]. Set
$$\mathfrak{c}' = \mathfrak{c}\mathfrak{b}^2 \quad \text{and} \quad \mathfrak{c}'_E := \mathfrak{c}' \cap \mathcal{O}_E.$$

The character θ may be used to twist the cycle $Z_0(\mathfrak{c}'_E)$ and its section S to obtain a cycle Z_θ and flat section S_θ of the local system $\mathcal{L}^\vee(\widehat{\kappa}, \chi_0)$, which gives rise to an intersection homology class $[Z_\theta] \in I^m H_n(X_0(\mathfrak{c}), \mathcal{L}^\vee(\widehat{\kappa}, \chi_0))$.

Let $f \in S_{\widehat{\kappa}}(K_0(\mathfrak{c}), \chi)$ be a simultaneous eigenform for all Hecke operators. Let $J \subset \Sigma(L)$. Then the modular form $f^{-\iota} \in S_{\widehat{\kappa}}^{\mathrm{coh}}(K_0(\mathfrak{c}), \chi)$ determines a differential form $\omega_J(f^{-\iota})$ which gives a "middle" intersection cohomology class $[\omega_J(f^{-\iota})] \in I^m H^n(X_0(\mathfrak{c}), \mathcal{L}(\kappa, \chi_0))$. Write

$$\mathfrak{b}' := \prod \{\mathfrak{p} : \mathfrak{p} \mid d_{L/E} \text{ and } \mathfrak{p} \nmid \mathfrak{f}(\theta)\}$$

where $\mathfrak{f}(\theta)$ denotes the conductor of θ. In Section 10.2 we will prove the following theorem:

Theorem 10.1. *Assume as above that $\theta|_{\mathbb{A}_E^\times} = \chi_E \eta$ and that $J \subseteq \Sigma(L)$ is a type for L/E. If f is a base change of a Hilbert cusp form g on E with nebentypus χ_E, then*

$$\langle [\omega_J(f^{-\iota})], [Z_\theta] \rangle_K$$

is equal to

$$\frac{c_1}{2^{[E:\mathbb{Q}]}} L^{*, d_{L/E} \mathfrak{b} \cap \mathcal{O}_E}(\mathrm{Ad}(g) \otimes \eta, 1) L_{\mathfrak{b}'}(\mathrm{As}(f \otimes \theta^{-1}), 1)$$

where

$$c_1 = \frac{C_3 [K_0(\mathfrak{c}'_E) : K_1(\mathfrak{c}'_E)] \mathrm{N}_{E/\mathbb{Q}}(\mathfrak{c}'_E)^2 d_{E/\mathbb{Q}}^{7/2} \mathrm{N}_{E/\mathbb{Q}}(d_{L/E}) \mathrm{Res}_{s=1} \zeta^{d_{L/E} \mathfrak{c}'_E}(s)}{R_E 2^{\{k+2\}} 2^{-2} |(\mathcal{O}_E / \mathfrak{c}'_E)^\times|}$$

with C_3 defined as in Proposition 9.5 and where R_E is the regulator of E.

If $J \subset \Sigma(L)$ is not a type for L/E or f is not a base change of a Hilbert cusp form g on E with nebentypus χ_E, then

$$\langle [\omega_J(f^{-\iota})], [Z_\theta] \rangle_K = 0.$$

Here $\{k\} := \sum_{\sigma \in \Sigma(E)} k_\sigma$ for $k \in \mathbb{Z}^{\Sigma(E)}$.

We now prepare some notation so that we may state the full version of Theorem 1.3. Let

$$P_{\mathrm{new}} : I^m H_n(X_0(\mathfrak{c}), \mathcal{L}(\widehat{\kappa}, \chi_0)) \longrightarrow I^m H_n^{\mathrm{new}}(X_0(\mathfrak{c}), \mathcal{L}(\widehat{\kappa}, \chi_0))$$

be the canonical projection, where $I^m H_n^{\mathrm{new}}(X_0(\mathfrak{c}), \mathcal{L}(\widehat{\kappa}, \chi_0))$ is the subspace spanned by classes that are f-isotypical under the action of the Hecke algebra for some newform $f \in S_{\widehat{\kappa}}^{\mathrm{new}}(K_0(\mathfrak{c}), \chi)$. Setting $\gamma = P_{\mathrm{new}} W_{\mathfrak{c}}^{*-1}[Z_\theta]$ in our notation from Section 8.3, we have

$$\Phi_{P_{\mathrm{new}} W_{\mathfrak{c}}^{*-1}[Z_\theta], \chi_E} \in I^m H_n^{\chi_E}(X_0(\mathfrak{c}), \mathcal{L}(\widehat{\kappa}, \chi_0)) \otimes S_\kappa^+(\mathcal{N}(\mathfrak{c}), \chi_E).$$

by Theorem 8.4. Finally for $g \in S_{\widehat{\kappa}}^{\mathrm{new}}(K_0(\mathfrak{c}), \chi)$ let

$$A(g, J) := \frac{[\omega_J(g^{-\iota}), \omega_{\Sigma(L)-J}(g^{-\iota})]_{IH^*}}{L^*(\mathrm{Ad}(g), 1)} = \frac{T(g, J)(g, g)_P}{L^*(\mathrm{Ad}(g), 1)} \quad (10.1.1)$$

10.1. Statement of results

This nonzero constant depends only on L, the weight, the nebentypus, the level and $W(g)$ (see (7.4.9)) by Theorem 5.16 and Theorem 7.11. Assuming Theorem 10.1 for the moment, we prove the "full" version of Theorem 1.3:

Theorem 10.2. *If* $\mathfrak{m} + \mathrm{N}_{L/E}(\mathfrak{c})d_{L/E} = \mathfrak{n} + \mathrm{N}_{L/E}(\mathfrak{c})d_{L/E} = \mathcal{O}_E$ *and* $\mathfrak{m}, \mathfrak{n}$ *are both norms from* $\mathcal{O}_L$, *then the* $\mathfrak{m}$*th Fourier coefficient of*

$$[W_{\mathfrak{c}}^{*-1}[Z_\theta](\mathfrak{n}), \Phi_{P_{\mathrm{new}}W_{\mathfrak{c}}^{*-1}[Z_\theta], \chi_E}]_{IH_*}$$

is

$$\frac{c_1^2}{4}(-1)^{[E:\mathbb{Q}]} \sum_f \frac{\left(L^{*, d_{L/E}\mathfrak{f}(\theta) \cap \mathcal{O}_E}(\mathrm{Ad}(f) \otimes \eta, 1) L_{\mathfrak{b}'}(\mathrm{As}(\widehat{f} \otimes \theta^{-1}), 1)\right)^2}{A(\widehat{f}, \emptyset) L^*(\mathrm{Ad}(\widehat{f}), 1)} \lambda_f(\mathfrak{n}) \overline{\lambda_f(\mathfrak{m})},$$

where the sum is over the normalized newforms f *on* E *of nebentypus* χ_E *whose base change* $\widehat{f}$ *to* L *is an element of* $S_{\widehat{k}}^{\mathrm{new}}(K_0(\mathfrak{c}), \chi)$. *If either*

$$\mathfrak{m} + \mathrm{N}_{L/E}(\mathfrak{c})d_{L/E} \neq \mathcal{O}_E \quad \text{or} \quad \mathfrak{n} + \mathrm{N}_{L/E}(\mathfrak{c})d_{L/E} \neq \mathcal{O}_E,$$

then the $\mathfrak{m}$*th Fourier coefficient is zero.*

Here

$$\mathcal{N}(\mathfrak{c}) := \mathfrak{m}_2 \mathfrak{d}_{L/E}(\mathfrak{c}_E) \prod_{\mathfrak{p} | \mathfrak{c}_E} \mathfrak{p}$$

where $\mathfrak{m}_2 \subset \mathcal{O}_E$ is an ideal divisible only by dyadic primes, which we may take to be $\mathcal{O}_E$ if $\mathfrak{c} + 2\mathcal{O}_L = \mathcal{O}_L$ and $\mathfrak{d}_{L/E}$ is an ideal divisible only by primes ramifying in L/E.

Proof. By Theorem 8.5, the $\mathfrak{m}$th Fourier coefficient of $[W_{\mathfrak{c}}^{*-1}[Z_\theta](\mathfrak{n}), \Phi_{W_{\mathfrak{c}}^{*-1}[Z_\theta]}]_{IH_*}$ is zero if $\mathfrak{m} + \mathrm{N}_{L/E}(\mathfrak{c}')d_{L/E} \neq \mathcal{O}_E$ or $\mathfrak{n} + \mathrm{N}_{L/E}(\mathfrak{c}')d_{L/E} \neq \mathcal{O}_E$. Again by Theorem 8.5, if $\mathfrak{m}$ and $\mathfrak{n}$ are norms and $\mathfrak{m} + \mathrm{N}_{L/E}(\mathfrak{c}')d_{L/E} = \mathfrak{n} + \mathrm{N}_{L/E}(\mathfrak{c}')d_{L/E} = \mathcal{O}_E$ then this coefficient is equal to

$$\frac{1}{4} \sum_{J \subset \Sigma(E)} \sum_f \frac{\int_{Z_\theta} \omega_J(\widehat{f}^{-\iota}) \int_{Z_\theta} \omega_{\Sigma(L) - J}(\widehat{f}^{-\iota})}{T(\Sigma(L) - J, \widehat{f})(\widehat{f}, \widehat{f})_P} \lambda_f(\mathfrak{n}) \overline{\lambda_f(\mathfrak{m})}.$$

The theorem now follows immediately from Theorem 10.1 and Theorem 5.16 together with the observation that $T(\Sigma(L) - J, \widehat{f}) = (-1)^{|J|} T(\emptyset, \widehat{f})$. We only remark that one uses the fact that the number of types for L/E is $2^{[E:\mathbb{Q}]}$. □

10.2 Rankin-Selberg integrals

In the last chapter, we defined certain cycles Z_θ on $Y_0(\mathfrak{c})$ and gave an integral expression for $\langle [\omega_J(f^{-\iota})], [Z_\theta] \rangle_K$. In this section we apply the Rankin-Selberg method to this integral expression to prove Theorem 10.1. Our argument follows [Hid8, Section 6].

Let $J \subseteq \Sigma(L)$ be a type for L/E (see Section 9.1), and assume as in Theorem 10.2 that $\theta|_{\mathbb{A}_E^\times} = \chi_E \eta$. For each $f \in S_{\widehat{\kappa}}(K_0(\mathfrak{c}), \chi)$ and $s \in \mathbb{C}$ let

$$I(f, s) := \int_{E_+^\times \backslash \mathbb{A}_{E,+}^\times} \int_{E \backslash \mathbb{A}_E} (f \otimes \theta^{-1})(\iota_J \begin{pmatrix} y & x \\ 0 & 1 \end{pmatrix}) |y|_{\mathbb{A}_E}^s dx d^\times y$$

when this integral is defined (e.g., for $\text{Re}(s)$ sufficiently large). Here $\mathbb{A}_{E,+}^\times$ (resp. $E_+^\times$) is the set of $(b_v) \in \mathbb{A}_E^\times$ (resp. $\alpha \in E^\times$) such that $b_\sigma > 0$ (resp. $\sigma(\alpha) > 0$) for all $\sigma \in \Sigma(E)$. Moreover dx and dy are the Haar measures of Section C.4. We have the following lemma:

Lemma 10.3. *If $f \in S_{\widehat{\kappa}}(K_0(\mathfrak{c}), \chi)$ is a simultaneous eigenform for all Hecke operators, then*

$$I(f, s) = \frac{|d_{E/\mathbb{Q}} \mathrm{N}_{E/\mathbb{Q}}(d_{L/E})|^{\frac{1}{2}|s+2|} |d_{E/\mathbb{Q}}|^{1/2} \Gamma_E(k + (s+2)\mathbf{1})}{(4\pi)^{\{k\} + [E:\mathbb{Q}](s+2)}}$$

$$\times \sum_{\mathfrak{m} \subset \mathcal{O}_E} a(\mathfrak{m}\mathcal{O}_L, f \otimes \theta^{-1}) \mathrm{N}_{E/\mathbb{Q}}(\mathfrak{m})^{-s-2}.$$

Here $\Gamma_E(k) := \prod_{\sigma \in \Sigma(L)} \Gamma(k_\sigma)$, and, as above, $\{k\} := \sum_{\sigma \in \Sigma(L)} k_\sigma$ for $k \in \mathbb{C}^{\Sigma(L)}$.

Proof. To ease notation, write $g := f \otimes \theta^{-1}$. Using a slight modification of the proof of Theorem 5.8 one sees that the Fourier expansion of $g \circ \iota_J$ is

$$|y|_{\mathbb{A}_L} \sideset{}{'}\sum_{\xi \in L^\times} a(\xi y \mathcal{D}_{L/\mathbb{Q}}, g) q_{(\widehat{k}, -\widehat{k}/2)}(\xi y, \xi x)$$

where the $'$ indicates that the sum is over those $\xi \in L^\times$ such that $\sigma(\xi) < 0$ for $\sigma \in J$ and $\sigma(\xi) > 0$ for $\sigma \in J' := J^c$. Thus

$$\int_{E_+^\times \backslash \mathbb{A}_{E,+}^\times} \int_{E \backslash \mathbb{A}_E} g(\iota_J \begin{pmatrix} y & x \\ 0 & 1 \end{pmatrix}) |y|_{\mathbb{A}_E}^s dx d^\times y$$

$$= \int_{E_+^\times \backslash \mathbb{A}_{E,+}^\times} \int_{E \backslash \mathbb{A}_E} \sideset{}{'}\sum_{\xi \in L^\times} a(\xi y \mathcal{D}_{L/\mathbb{Q}}, g) |(\xi y)^{\widehat{k}/2}|$$

$$\times \exp(-2\pi \Sigma_{\sigma \in \Sigma(L)} |\sigma(\xi) y_\sigma|) e_L(\xi x) |y|_{\mathbb{A}_E}^{s+2} dx d^\times y.$$

10.2. Rankin-Selberg integrals

The character $e_L(\xi x)$ restricted to $\mathbb{A}_E$ is trivial if and only if $\xi = -\varsigma(\xi)$. In view of this, the well-known computation of the volume of $E\backslash \mathbb{A}_E$ yields

$$\int_{E\backslash \mathbb{A}_E} e_L(\xi x)dx = \begin{cases} \sqrt{d_{E/\mathbb{Q}}} & \text{if } \xi = -\varsigma(\xi) \\ 0 & \text{otherwise.} \end{cases}$$

Thus

$$\int_{E_+^\times\backslash \mathbb{A}_{E,+}^\times} \int_{E\backslash \mathbb{A}_E} {\sum_{\xi\in L^\times}}' a(\xi y \mathcal{D}_{L/\mathbb{Q}}, g)|(\xi y)^{\widehat{k}/2}|$$

$$\times \exp(-2\pi\Sigma_{\sigma\in\Sigma(L)}|\sigma(\xi)y_\sigma|)e_L(\xi x)|y|_{\mathbb{A}_E}^{s+2}dxd^\times y$$

$$= \sqrt{d_{E/\mathbb{Q}}}\int_{E_+^\times\backslash \mathbb{A}_{E,+}^\times} {\sum_{\substack{\xi\in L^\times \\ \xi=-\varsigma(\xi)}}}' a(\xi y\mathcal{D}_{L/\mathbb{Q}}, g)|(\xi y_\infty)^{\widehat{k}/2}|$$

$$\times \exp(-2\pi\Sigma_{\sigma\in\Sigma(L)}|\sigma(\xi)y_\sigma|)|y|_{\mathbb{A}_E}^{s+2}d^\times y$$

Choose $u_J \in \mathcal{O}_E - 0$ such that $\sigma(u_J\Delta^{1/2}) > 0$ if and only if $\sigma \in J$. Note that the requirement that ξ satisfies $\xi = -\varsigma(\xi)$ and $\sigma(\xi) < 0$ if and only if $\sigma \in J$ is equivalent to the requirement that $\xi = \xi' u_J^{-1}\Delta^{-1/2}$ for some $\xi' \in E_+^\times$. Translating the positivity conditions on ξ to positivity conditions on ξ', we obtain

$$\int_{E_+^\times\backslash \mathbb{A}_{E,+}^\times} {\sum_{\substack{\xi\in L^\times \\ \xi=-\varsigma(\xi)}}}' a(\xi y\mathcal{D}_{L/\mathbb{Q}}, g)|(\xi y_\infty)^{\widehat{k}/2}|\exp(-2\pi\Sigma_{\sigma\in\Sigma(L)}|\sigma(\xi)y_\sigma|)|y|_{\mathbb{A}_E}^{s+2}dxd^\times y$$

$$= \int_{E_+^\times\backslash \mathbb{A}_{E,+}^\times} \sum_{\xi'\in E_+^\times} a(\xi' u_J^{-1}\Delta^{-1/2}y\mathcal{D}_{L/\mathbb{Q}}, g)|(\xi' u_J^{-1}\Delta^{-1/2}y_\infty)^{\widehat{k}/2}|$$

$$\times \exp(-2\pi\Sigma_{\sigma\in\Sigma(L)}|\sigma(\xi' u_J^{-1}\Delta^{-1/2})y_\sigma|)|y|_{\mathbb{A}_E}^{s+2}d^\times y$$

$$= \int_{\mathbb{A}_{E,+}^\times} a(u_J^{-1}\Delta^{-1/2}y\mathcal{D}_{L/\mathbb{Q}}, g)|(u_J^{-1}\Delta^{-1/2}y_\infty)^{\widehat{k}/2}|$$

$$\times \exp(-2\pi\Sigma_{\sigma\in\Sigma(L)}|\sigma(u_J^{-1}\Delta^{-1/2})y_\sigma|)|y|_{\mathbb{A}_E}^{s+2}d^\times y$$

By writing this global integral as an infinite product of local integrals, we see that the above is equal to

$$\int_{\mathbb{A}_{E,+}^\times} a(u_J^{-1}\Delta^{-1/2}y\mathcal{D}_{L/\mathbb{Q}}, g)|(u_J^{-1}\Delta^{-1/2}y_\infty)^{\widehat{k}/2}| \tag{10.2.1}$$

$$\times \exp(-2\pi\Sigma_{\sigma\in\Sigma(L)}|\sigma(u_J^{-1}\Delta^{-1/2})y_\sigma|)|y|_{\mathbb{A}_E}^{s+2}d^\times y$$

$$= \left(\prod_{\sigma\in\Sigma(E)}\int_0^\infty |\sigma(u_J^{-1}\Delta^{-1/2})y|^{k_\sigma}\exp(-4\pi|\sigma(u_J^{-1}\Delta^{-1/2})y|)|y|_\infty^{s+2}\frac{dy}{|y|}\right)$$

$$\times \prod_\mathfrak{p}\int_{E_\mathfrak{p}^\times} a(yu_J^{-1}\Delta^{-1/2}\mathcal{D}_{L/\mathbb{Q}}, g)|y|_\mathfrak{p}^{s+2}d^\times y_\mathfrak{p}$$

Here we are using the fact that the coefficients $a(\cdot, g)$ are multiplicative in the sense that $a(\mathfrak{mn}, g) = a(\mathfrak{m}, g)a(\mathfrak{n}, g)$ if $\mathfrak{m} + \mathfrak{n} = \mathcal{O}_L$. We evaluate the factor corresponding to the infinite places and each of the factors corresponding to finite places separately. First, the factor corresponding to the infinite places is

$$\left(\prod_{\sigma \in \Sigma(E)} \int_0^\infty |\sigma(u_J^{-1}\Delta^{-1/2})y|^{k_\sigma} \exp(-4\pi|\sigma(u_J^{-1}\Delta^{-1/2})y|)|y|_\infty^{s+2} \frac{dy}{|y|} \right)$$

$$= \prod_{\sigma \in \Sigma(E)} |\sigma(u_J^{-1}\Delta^{-1/2})|^{-s-2}(4\pi)^{-k_\sigma-s-2}\Gamma(k_\sigma + s + 2).$$

By definition of the different, we may write $u_J^{-1}\Delta^{-1/2}\mathcal{D}_{L/\mathbb{Q}} = \mathcal{D}_0 \mathcal{O}_L$ for a fractional ideal $\mathcal{D}_0 \subset E$. Moreover

$$|N_{E/\mathbb{Q}}(\mathcal{D}_0)| = |d_{E/\mathbb{Q}} N_{E/\mathbb{Q}}(d_{L/E})|^{\frac{1}{2}} \prod_{\sigma \in J} \sigma(u_J^{-1}\Delta^{-1/2})|.$$

With this in mind we see that the factor corresponding to the finite places satisfies

$$\prod_{\mathfrak{p}} \int_{E_{v(\mathfrak{p})}^\times} a(u_J^{-1}\Delta^{-1/2}y\mathcal{D}_{L/\mathbb{Q}}, g)|y|_{v(\mathfrak{p})}^{s+2} d^\times y \quad (10.2.2)$$

$$= \prod_{\mathfrak{p}} \int_{E_{v(\mathfrak{p})}^\times} a(y, g)|(\mathcal{D}_0)_{v(\mathfrak{p})}^{-1} y|_{v(\mathfrak{p})}^{s+2} d^\times y$$

$$= \prod_{\mathfrak{p}} |(\mathcal{D}_0)_{v(\mathfrak{p})}|_{v(\mathfrak{p})}^{-s-2} \sum_{j=1}^\infty a(\mathfrak{p}^j, g) N_{E/\mathbb{Q}}(\mathfrak{p})^{-js-2j}$$

$$= \left(|d_{E/\mathbb{Q}} N_{E/\mathbb{Q}}(d_{L/E})|^{\frac{1}{2}} \prod_{\sigma \in J} \sigma(u_J^{-1}\Delta^{-1/2})| \right)^{s+2} \prod_{\mathfrak{p}} \sum_{j=1}^\infty a(\mathfrak{p}^j, g) N_{E/\mathbb{Q}}(\mathfrak{p})^{-js-2j}.$$

Combining (10.2.1) and (10.2.2) yields the lemma. $\square$

One can use Rankin's method to give another expression for $I(f, s)$. In order to state it precisely, we first set some notation. Let B' be the algebraic $\mathbb{Q}$-group whose points in a $\mathbb{Q}$-algebra A are

$$B'(A) := \left\{ \begin{pmatrix} a & b \\ 0 & 1 \end{pmatrix} : a \in (E \otimes_\mathbb{Q} A)^\times \text{ and } b \in (E \otimes_\mathbb{Q} A) \right\} \subseteq \operatorname{Res}_{E/\mathbb{Q}} \operatorname{GL}_2(A).$$

Let $\operatorname{Res}_{E/\mathbb{Q}} \operatorname{GL}_2(\mathbb{R})^+$ be the identity component of $\operatorname{Res}_{E/\mathbb{Q}} \operatorname{GL}_2(\mathbb{R})$. Set

$$\operatorname{Res}_{E/\mathbb{Q}} \operatorname{GL}_2(\mathbb{A})^+ := \operatorname{Res}_{E/\mathbb{Q}} \operatorname{GL}_2(\mathbb{R})^+ \operatorname{Res}_{E/\mathbb{Q}} \operatorname{GL}_2(\mathbb{A}_f)$$
$$\operatorname{Res}_{E/\mathbb{Q}} \operatorname{GL}_2(\mathbb{Q})^+ := \operatorname{Res}_{E/\mathbb{Q}} \operatorname{GL}_2(\mathbb{Q}) \cap \operatorname{Res}_{E/\mathbb{Q}} \operatorname{GL}_2(\mathbb{A})^+$$
$$B'(\mathbb{Q})^+ := B'(\mathbb{Q}) \cap \operatorname{Res}_{E/\mathbb{Q}} \operatorname{GL}_2(\mathbb{A})^+$$
$$B'(\mathbb{A})^+ := B'(\mathbb{A}) \cap \operatorname{Res}_{E/\mathbb{Q}} \operatorname{GL}_2(\mathbb{A})^+$$

10.2. Rankin-Selberg integrals

Moreover let $\mathcal{O}_{E,+}^{\times} := \mathcal{O}_E^{\times} \cap E_+^{\times}$. Let $\mathcal{N} : \mathrm{Res}_{E/\mathbb{Q}}\mathrm{GL}_2(\mathbb{A}) \to \mathbb{A}_E^{\times}$ be defined by

$$\mathcal{N}(\alpha) := \begin{cases} y & \text{if } \alpha = \begin{pmatrix} y & * \\ 0 & 1 \end{pmatrix} bu \text{ for } y, b \in \mathbb{A}_E^{\times}, u \in K_{E,\infty} K_0(\mathfrak{c}_E') \\ 0 & \text{otherwise.} \end{cases}$$

Now let $\mathcal{E}$ be the Eisenstein series defined by

$$\mathcal{E}(\alpha, s) := \sum \left\{ |\mathcal{N}(\gamma\alpha)|_{\mathbb{A}_E}^s : \gamma \in (\mathcal{O}_{E,+}^{\times} B'(\mathbb{Q})^+) \backslash \mathrm{Res}_{E/\mathbb{Q}} \mathrm{GL}_2(\mathbb{Q})^+ \right\}$$
$$= \sum \left\{ |\mathcal{N}(\gamma\alpha)|_{\mathbb{A}_E}^s : \gamma \in (\mathcal{O}_E^{\times} B'(\mathbb{Q})) \backslash \mathrm{Res}_{E/\mathbb{Q}} \mathrm{GL}_2(\mathbb{Q}) \right\}.$$

This Eisenstein series is absolutely convergent if $\mathrm{Re}(s) > 1$, as can be verified using (10.2.4) below.

As above, let $Z_0(\mathfrak{c}_E')$ be the image of $\mathrm{Res}_{E/\mathbb{Q}}\mathrm{GL}_2(\mathbb{A}) \hookrightarrow \mathrm{Res}_{L/\mathbb{Q}}\mathrm{GL}_2(\mathbb{A}) \to Y_0(\mathfrak{c}')$, where the first arrow is the diagonal embedding and the second is the canonical projection. Then we have the following proposition:

Proposition 10.4. *For $\mathrm{Re}(s) > 0$ we have*

$$I(f,s) = \int_{Z_0(\mathfrak{c}_E')} \det(\iota_J(\alpha_z))^{-\widehat{k}/2-1} g^{-\iota}(\iota_J(\alpha)) j(\iota_J(\alpha_z), \mathbf{i})^{\widehat{k}+2} \mathcal{N}(\iota_J(\alpha_z))^{k+2}$$
$$\times \mathcal{E}(\alpha, s+1) d\mu_{\mathfrak{c}_E'}.$$

The proof is essentially the same as the computation in [Hid5, Section 4] (see also [Hid8]).

Proof. As above, write $g = f \otimes \theta^{-1}$ to ease notation. We have

$$I(f,s) = \int_{B'(\mathbb{Q})^+ \backslash B'(\mathbb{A})^+} g(\iota_J(\alpha)) |\det(\alpha)|_{\mathbb{A}_E}^{s+1} d\mu_{B'}(\alpha)$$

where $d\mu_{B'}$ is the measure defined in Section 5.7 and $\mathrm{Re}(s) > 0$. Now

$$g(\iota_J(\alpha)) = g(\det(\iota_J(\alpha))^{-1} \iota_J(\alpha)) \chi \theta^{-2}(\det(\iota_J(\alpha))).$$

When restricted to $\alpha \in \mathrm{Res}_{E/\mathbb{Q}}\mathrm{GL}_2(\mathbb{A})$ this is simply $g^{-\iota}(\iota_J(\alpha))$. Thus

$$I(f,s) = \int_{B'(\mathbb{Q})^+ \backslash B'(\mathbb{A})^+} g^{-\iota}(\iota_J(\alpha)) |\det(\alpha)|_{\mathbb{A}_E}^{s+1} d\mu_{B'}(\alpha)$$
$$= \int_{B'(\mathbb{Q})^+ \backslash B'(\mathbb{A})^+} g^{-\iota}(\iota_J(\alpha)) |\mathcal{N}(\alpha)|_{\mathbb{A}_E}^{s+1} d\mu_{B'}(\alpha).$$

We wish to use the decomposition

$$\mathrm{Res}_{E/\mathbb{Q}}\mathrm{GL}_2(\mathbb{A})^+ = \mathrm{Res}_{E/\mathbb{Q}}\mathrm{GL}_2(\mathbb{Q})^+ B'(\mathbb{A})^+ K_0(\mathfrak{c}_E') K_{E,\infty} \qquad (10.2.3)$$

(see [Hid5, Section 4]) to rewrite $I(f,s)$ as an integral over $Z_0(\mathfrak{c}'_E)$. For this purpose, note that the invariance properties of $g^{-\iota}$ together with the fact that $\chi\theta^{-2}$ is trivial when restricted to $\mathbb{A}_E^\times$ implies that for $\alpha \in B(\mathbb{A})^+ K_0(\mathfrak{c}'_E) K_{E,\infty}$

$$\det(\iota_J(\alpha_\infty))^{-\widehat{k}/2-1} g^{-\iota}(\iota_J(\alpha)) j(\iota_J(\alpha_\infty), \mathbf{i})^{\widehat{k}+2} \mathcal{N}(\iota_J(\alpha_\infty))^{k+2}$$

is invariant under left multiplication by $B'(\mathbb{Q})^+$ and right multiplication by $K_{E,\infty} K_0(\mathfrak{c}'_E)$. Moreover $|\mathcal{N}(\alpha)|_{\mathbb{A}_E}^{s+1}$ is invariant under left multiplication by $B'(\mathbb{Q})^+$ and right multiplication by $K_{E,\infty} K_0(\mathfrak{c}'_E)$. Therefore applying our previous expression for $I(f,s)$ we have that $I(f,s)$ is equal to

$$\int_{B'(\mathbb{Q})^+ \backslash B'(\mathbb{A})^+} g^{-\iota}(\iota_J(\alpha)) |\mathcal{N}(\alpha)|_{\mathbb{A}_E}^{s+1} d\mu_{B'}(\alpha)$$

$$= \int_{B'(\mathbb{Q})^+ \backslash B'(\mathbb{A})^+} \det(\iota_J(\alpha_\infty))^{-\widehat{k}/2-1} g^{-\iota}(\iota_J(\alpha)) j(\iota_J(\alpha_\infty), \mathbf{i})^{\widehat{k}+2}$$
$$\times \mathcal{N}(\iota_J(\alpha_\infty))^{k+2} |\mathcal{N}(\alpha)|_{\mathbb{A}_E}^{s+1} d\mu_{B'}(\alpha)$$

$$= \int_{B'(\mathbb{Q})^+ \backslash D_E^+ \times K_0(\mathfrak{c}'_E)/K_0(\mathfrak{c}'_E)} \det(\iota_J(\alpha_z))^{-\widehat{k}/2-1} g^{-\iota}(\iota_J(\alpha)) j(\iota_J(\alpha_z), \mathbf{i})^{\widehat{k}+2}$$
$$\times \mathcal{N}(\iota_J(\alpha_z))^{k+2} |\mathcal{N}(\alpha)|_{\mathbb{A}_E}^{s+1} d\mu_{\mathfrak{c}'_E}$$

where $D_E := B'(\mathbb{R})/K_{E,\infty}$, D_E^+ is the connected component of D_E containing $\mathbf{i}$, where $\alpha_z \in \mathrm{Res}_{E/\mathbb{Q}} \mathrm{GL}_2(\mathbb{R})^+$ is chosen so that $\alpha_z \mathbf{i} = z$, and where $\alpha = \alpha_z \alpha_0$ for some $\alpha_0 \in \mathrm{Res}_{E/\mathbb{Q}} \mathrm{GL}_2(\mathbb{A}_f)$. In the last equality we have used the comments on the relationship between $d\mu_{B'}$ and $d\mu_{\mathfrak{c}'_E}$ contained in Section 5.7. Using the fact that $\mathcal{O}_{E,+}^\times$ is naturally a subgroup of the group of diagonal matrices in $K_{E,\infty} K_0(\mathfrak{c}'_E)$ together with (10.2.3), we have that the above is equal to

$$\int_{Z_0(\mathfrak{c}'_E)} \sum_\gamma \det(\iota_J(\gamma\alpha_z))^{-\widehat{k}/2-1} g^{-\iota}(\iota_J(\gamma\alpha)) j(\iota_J(\gamma\alpha_z), \mathbf{i})^{\widehat{k}+2}$$
$$\times \mathcal{N}(\iota_J(\gamma\alpha_z))^{k+2} |\mathcal{N}(\gamma\alpha)|_{\mathbb{A}_E}^{s+1} d\mu_{\mathfrak{c}'_E}.$$

where the sum is over

$$\gamma \in \mathcal{O}_{E,+}^\times B'(\mathbb{Q})^+ \backslash \mathrm{Res}_{E/\mathbb{Q}} \mathrm{GL}_2(\mathbb{Q})^+.$$

This infinite sum converges absolutely for $\mathrm{Re}(s) > 0$, a fact that follows from the absolute convergence of the Eisenstein series $\mathcal{E}(x,s)$ for $\mathrm{Re}(s) > 1$. Now for $\gamma_\infty \in \mathrm{Res}_{E/\mathbb{Q}} \mathrm{GL}_2(\mathbb{R})$ one has the cocycle relation $j(\gamma_\infty \alpha_z, \mathbf{i}) = j(\gamma_\infty, z) j(\alpha_z, \mathbf{i})$ and one verifies that

$$\mathcal{N}(\gamma_\infty \alpha_z) = \frac{\det(\gamma_\infty)}{|j(\gamma, z)|^2} \mathcal{N}(\alpha_z). \qquad (10.2.4)$$

10.2. Rankin-Selberg integrals

Using these facts, with the same sum over γ, the above integral becomes

$$\int_{Z_0(\mathfrak{c}'_E)} \det(\iota_J(\alpha_z))^{-\widehat{k}/2-\mathbf{1}} g^{-\iota}(\iota_J(\alpha)) j(\iota_J(\alpha_z), \mathbf{i})^{\widehat{k}+2}$$
$$\times \mathcal{N}(\iota_J(\alpha_z))^{\widehat{k}+2} \sum_\gamma |\mathcal{N}(\gamma\alpha)|_{\mathbb{A}_E}^{s+1} d\mu_{\mathfrak{c}'_E}$$
$$= \int_{Z_0(\mathfrak{c}'_E)} \det(\iota_J(\alpha_z))^{-\widehat{k}/2-\mathbf{1}} g^{-\iota}(\iota_J(\alpha)) j(\iota_J(\alpha_z), \mathbf{i})^{\widehat{k}+2}$$
$$\times \mathcal{N}(\iota_J(\alpha_z))^{k+2} \mathcal{E}(\alpha, s+1) d\mu_{\mathfrak{c}'_E}.$$

As noted above, the sum defining the Eisenstein series $\mathcal{E}(\alpha, s)$ converges absolutely for $\mathrm{Re}(s) > 1$ so these formal manipulations are justified. $\square$

We can now prove Theorem 10.1:

Proof of Theorem 10.1. By Proposition 9.5, we can assume that $J \subset \Sigma(L)$ is a type for L/E. By Lemma 10.3 and the discussion in Section 5.12.4, for s with $\mathrm{Re}(s) > 0$ we have

$$\zeta^{\mathfrak{c}' \cap \mathcal{O}_E}(2(s+1)) I(f, s) = c_2 L(\mathrm{As}(f \otimes \theta^{-1}), s+1). \qquad (10.2.5)$$

where

$$c_2 = \frac{(d_{E/\mathbb{Q}} N_{E/\mathbb{Q}}(d_{L/E})^{1/2})^{s+2} d_{E/\mathbb{Q}}^{1/2} \Gamma_E(k + (s+2)\mathbf{1})}{(4\pi)^{\{k\} + [E:\mathbb{Q}](s+2)}}.$$

By Proposition 10.4, we therefore have

$$c_2 L(\mathrm{As}(f \otimes \theta^{-1}), s+1) \qquad (10.2.6)$$
$$= \zeta^{\mathfrak{c}'_E}(2(s+1)) \int_{Z_0(\mathfrak{c}'_E)} \det(\iota_J(\alpha_z))^{-\widehat{k}/2-\mathbf{1}} g^{-\iota}(\iota_J(\alpha)) j(\iota_J(\alpha_z), \mathbf{i})^{\widehat{k}+2}$$
$$\times \mathcal{N}(\iota_J(\alpha_z))^{k+2} \mathcal{E}(\alpha, s+1) d\mu_{\mathfrak{c}'_E}$$

It follows from the analytic properties of the Asai L-function (see Section 5.12.4) and the absolute convergence of $\mathcal{E}(\alpha, s)$ for $s > 1$ that both sides of (10.2.6) are holomorphic functions of s in the half-plane $\mathrm{Re}(s) > 1$. Moreover, the fact that $\mathcal{E}(\alpha, s)$ can be meromorphically continued to whole complex plane with a simple pole at $s = 1$ (see [Hid5, Section 6] or [Hid6, Section 9.1 Theorem 1] or [Hid8]) implies the same is true of the right-hand side (we could have also applied the results of [Ram2] to come to this conclusion).

By (RES2) and the equation directly above it in the appendix of [Hid8] one has

$$\mathrm{Res}_{s=1} \zeta^{\mathfrak{c}'_E}(2s) \mathcal{E}(\alpha, s) = \mathrm{N}(\mathfrak{c}'_E)^{-2} |(\mathcal{O}_E/\mathfrak{c}'_E)^\times| \frac{2^{[E:\mathbb{Q}]-2} \pi^{[E:\mathbb{Q}]} R_E}{d_{E/\mathbb{Q}}}$$

where R_E is the regulator of E. Taking residues at $s = 0$ on both sides of (10.2.6) we obtain

$$c_4 \Gamma_E(k+2) \mathrm{Res}_{s=1} L(\mathrm{As}(f \otimes \theta^{-1}), s)$$
$$= \int_{Z_0(\mathfrak{c}'_E)} \det(\iota_J(\alpha_z))^{-\widehat{k}/2-1} g^{-\iota}(\iota_J(\alpha)) j(\iota_J(\alpha_z), \mathbf{i})^{\widehat{k}+2} \mathcal{N}(\iota_J(\alpha_z))^{k+2} d\mu_{\mathfrak{c}'_E}$$

where
$$c_4 = \frac{N_{E/\mathbb{Q}}(\mathfrak{c}'_E)^2 d_{E/\mathbb{Q}}^{7/2} N_{E/\mathbb{Q}}(d_{L/E})}{R_E(4\pi)^{\{k+2\}} 2^{[E:\mathbb{Q}]-2} \pi^{[E:\mathbb{Q}]} |(\mathcal{O}_E/\mathfrak{c}'_E)^\times|}.$$

We now wish to apply Proposition 9.5 to relate this last integral to the pairing $\langle [\omega_J(f)], [Z_\theta] \rangle_K$. For this purpose we remark that in the case at hand the character $\chi \theta^{-2}$ is trivial when restricted to $\mathbb{A}_E^\times$. This implies that the additional summation over g_i in Proposition 9.5 just multiplies the integral there by a factor. With this in mind we see that

$$c_4 \Gamma_E(k+2) \mathrm{Res}_{s=1} L(\mathrm{As}(f \otimes \theta^{-1}), s) = C_3^{-1} [K_0(\mathfrak{c}'_E) : K_{11}(\mathfrak{c}'_E)]^{-1} \langle [\omega_J(f)], [Z_\theta] \rangle_K \tag{10.2.7}$$

where C_3 is defined as in Proposition 9.5.

We claim that if $\langle [\omega_J(f)], [Z_\theta] \rangle_K \neq 0$ then f is a base change of a modular form on E. In order to see this, note first that $\langle [\omega_J(f)], [Z_\theta] \rangle_K \neq 0$ implies that $\mathrm{Res}_{s=1} L(\mathrm{As}((f \otimes \theta^{-1})), s) \neq 0$. Write $g = f \otimes \theta^{-1}$ to ease notation. As recalled in Section 5.12.4,

$$L^{\mathcal{D}_{L/E}\mathfrak{c}'}(g \times g^\varsigma, s) = L^{d_{L/E}(\mathfrak{c}' \cap \mathcal{O}_E)}(\mathrm{As}(g) \otimes \eta, s) L^{d_{L/E}(\mathfrak{c}' \cap \mathcal{O}_E)}(\mathrm{As}(g), s). \tag{10.2.8}$$

Let $\pi(g), \pi(f)$, etc. be the (unitary) cuspidal automorphic representations attached to g, f, etc. With our normalization, $L^{\mathcal{D}_{L/E}\mathfrak{c}'}(g \times (g^\varsigma), s)$ has a pole at $s = 1$ if and only if the contragredient $\pi(g)^\vee$ satisfies $\pi(g)^\vee \cong \pi(g)^\varsigma$ [JaS2, Proposition 3.6]. This implies that

$$\pi(f^\varsigma \otimes (\theta^\varsigma)^{-1}) \cong \pi(f \otimes \theta^{-1})^\varsigma$$
$$\cong \pi(f \otimes \theta^{-1})^\vee$$
$$\cong \pi(f) \otimes \chi^{-1}\theta.$$

Thus $\pi(f^\varsigma) \otimes (\theta^\varsigma \theta)^{-1} \chi \cong \pi(f)$. It is easy to check that $(\theta^\varsigma \theta)^{-1} \chi = \chi_{\mathrm{triv}}$, so we conclude that $\pi(f^\varsigma) \cong \pi(f)$. By the theory of base change [Lan], this implies that f is a base change of a Hilbert modular form h on E. Thus our claim is proven if we show that $L(\mathrm{As}(g), s)$ has a pole at $s = 1$ if and only if $L^{d_{L/E}(\mathfrak{c}' \cap \mathcal{O}_E)}(\mathrm{As}(g), s)$ does. To prove this, we show that any pole s_0 of $L_\mathfrak{p}(\mathrm{As}(g), s)$ satisfies $\mathrm{Re}(s_0) < 1$. Suppose that $\mathfrak{p} \nmid \mathfrak{f}(\pi(g)) \cap \mathcal{O}_E$. Then for any prime $\mathfrak{P}$ above $\mathfrak{p}$ we have the well-known estimate

$$N_{L/\mathbb{Q}}(\mathfrak{P})^{-1/5} < |a_{i,\mathfrak{P}}(g)| < N_{L/\mathbb{Q}}(\mathfrak{P})^{1/5}$$

10.2. Rankin-Selberg integrals

(see [Sha, (4.1.3)]). We note that one actually knows $|a_{i,\mathfrak{P}}(g)| = 1$ by the Ramanujan conjecture (see [Bla]) but we won't need this. The absolute convergence then follows from an elementary bounding argument. If $\mathfrak{p} \mid \mathfrak{f}(\pi(g)) \cap \mathcal{O}_E$, then $\pi(g)$ is ramified, and in this case we have strong bounds on the $\mathfrak{P}$-power Fourier coefficients of g which easily imply the desired holomorphicity (see [She, Theorem 3.3]).

We now show that, under the assumption that $\langle [\omega_J(f^{-\iota})], Z_\theta \rangle_K \neq 0$, the Hilbert modular form h on E whose base change to L is f has nebentypus χ_E. Its nebentypus is either χ_E or $\chi_E \eta$. Suppose that the nebentypus of h is $\chi_E \eta$. By Proposition 5.17, we obtain

$$L^{d_{L/E}(\mathfrak{c} \cap \mathcal{O}_E)}(\mathrm{As}(f \otimes \theta^{-1}), s) = L^{d_{L/E}(\mathfrak{c} \cap \mathcal{O}_E)}(\mathrm{Ad}(h), s) L^{d_{L/E}(\mathfrak{c} \cap \mathcal{O}_E)}(\eta, s). \quad (10.2.9)$$

As proven above, the left-hand side has a pole at $s = 1$. On the other hand,

$$L^{d_{L/E}(\mathfrak{c} \cap \mathcal{O}_E)}(\mathrm{Ad}(h), s) \zeta^{d_{L/E}(\mathfrak{c} \cap \mathcal{O}_E)}(s) = L^{d_{L/E}(\mathfrak{c} \cap \mathcal{O}_E)}(h \times (h \otimes (\chi_E \eta)^{-1}), s)$$

for $\mathrm{Re}(s) > 1$. Since the pole of the right-hand side at $s = 1$ is simple [JaS2, Proposition 3.6], it follows that $L^{d_{L/E}(\mathfrak{c} \cap \mathcal{O}_E)}(\mathrm{Ad}(h), s)$ has no pole at $s = 1$, contradicting (10.2.9).

Now assume that $f = \widehat{h}$ is the base change of a Hilbert modular form h on E with nebentypus χ_E. In this case (10.2.7) and Proposition 5.17 imply that

$$c_4 \Gamma_E(k+2) \mathrm{Res}_{s=1} \zeta^{d_{L/E} \mathfrak{c}'_E}(s) L^{d_{L/E} \mathfrak{b} \cap \mathcal{O}_E}(\mathrm{Ad}(h) \otimes \eta, 1) L_{\mathfrak{b}'}(\mathrm{As}(f \otimes \theta^{-1}), 1)$$
$$= C_3^{-1} [K_0(\mathfrak{c}'_E) : K_{11}(\mathfrak{c}'_E)]^{-1} \langle [\omega_J(f)], [Z_\theta] \rangle_K$$

where $\mathfrak{b}' = \prod_{\substack{\mathfrak{p} \mid d_{L/E} \\ \mathfrak{p} \nmid \mathfrak{f}(\theta) \cap \mathcal{O}_E}} \mathfrak{p}$. Using the definition of the adjoint Gamma factors from Section 5.12.3 we see that this implies

$$c_1 L^{*, d_{L/E} \mathfrak{b} \cap \mathcal{O}_E}(\mathrm{Ad}(h) \otimes \eta, 1) L_{\mathfrak{b}'}(\mathrm{As}(f \otimes \theta^{-1}), 1) = \langle [\omega_J(f)], [Z_\theta] \rangle_K. \quad \square$$

Chapter 11

Eisenstein Series with Coefficients in Intersection Homology

Thus far we have ignored classes in $I^m H_{\text{inv}}^{[L:\mathbb{Q}]}(X_0(\mathfrak{c}), \mathcal{L}(\widehat{\kappa}, \chi_0))$ and their Poincaré duals in intersection homology. We now take up the study of these classes. We will assume throughout this chapter that the following conditions hold:

- We have that $\widehat{\kappa} = (0, \widehat{m}) \in \mathcal{X}(L)$ for some $m = [m]\mathbf{1} \in \frac{1}{2}\mathbb{Z}\mathbf{1}$.
- $\chi_0 = \chi_{\text{triv}0}$, where χ_{triv} is the trivial character.

As stated in Section 7.2, the group $I^m H_{\text{inv}}^{[L:\mathbb{Q}]}(X_0(\mathfrak{c}), \mathcal{L}(\widehat{\kappa}, \chi_0))$ is zero unless these conditions hold, so these assumptions are loss of generality. We also continue to assume the notational conventions of Section 8.1.

11.1 Eisenstein series

It turns out that the formal Fourier series associated to certain elements of

$$I^m H_{\text{inv}}^{[L:\mathbb{Q}]}(X_0(\mathfrak{c}), \mathcal{L}(\widehat{\kappa}, \chi_0))$$

are Eisenstein series with coefficients in intersection homology. In order to make this precise, we first set notation for certain Eisenstein series. For each $\mathfrak{c}_E \subset \mathcal{O}_E$ and character $\vartheta : E^\times \backslash \mathbb{A}_E^\times \to \mathbb{C}^\times$ that is trivial at the infinite places and whose conductor divides $\mathfrak{c}_E$, let

$$\mathcal{E}'_{\mathfrak{c}_E, \vartheta}\left(\begin{pmatrix} y & x \\ 0 & 1 \end{pmatrix}\right)$$

be the Fourier series given by

$$|y|_{\mathbb{A}_E} \sum_{\substack{\xi \in E^\times \\ \xi \gg 0}} \sigma'_{\mathfrak{c}_E, \vartheta}(\xi y \mathcal{D}_{E/\mathbb{Q}}) q_\kappa(\xi x, \xi y).$$

Here for each ideal $\mathfrak{c}_E \subset \mathcal{O}_E$, the function $\sigma'_{\mathfrak{c}_E,\vartheta}$ is defined on the set of fractional ideals of E by setting

$$\sigma'_{\mathfrak{c}_E,\vartheta}(\mathfrak{a}) := \sigma'_{\mathfrak{c}_E,\vartheta,[m]}(\mathfrak{a}) := \begin{cases} \sum_{\substack{\mathfrak{b}\subseteq\mathfrak{a} \\ \mathfrak{a}/\mathfrak{b}+\mathfrak{c}_E=\mathcal{O}_E}} \vartheta(\mathfrak{a}/\mathfrak{b}) \mathrm{N}_{E/\mathbb{Q}}(\mathfrak{b})^{1-[m]} & \text{if } \mathfrak{a} \text{ is integral} \\ 0 & \text{otherwise.} \end{cases}$$

In the definition of $\sigma'_{\mathfrak{c}_E,\vartheta}$, the sum is over integral ideals. We have also committed a standard abuse of notation, in that whenever we wrote $\vartheta(\mathfrak{b})$ for some ideal $\mathfrak{b} \subset \mathcal{O}_E$ we should have written $\vartheta(b)$ for some idèle trivial at the infinite places and the places dividing $\mathfrak{c}'_E$ whose associated ideal is $\mathfrak{b}$. This is well defined by the assumption that the conductor of ϑ divides $\mathfrak{c}_E$.

The following proposition is well known:

Proposition 11.1 (Shimura). *Let $\mathfrak{c}_E \subset \mathcal{O}_E$. If $\mathfrak{c}_E \neq \mathcal{O}_E$, then*

$$\mathcal{E}'_{\mathfrak{c}_E,\vartheta} \in M_{(0,m)}(K_0(\mathfrak{c}_E),\vartheta).$$

One can obtain this proposition, for example, from the results contained in [Hid5, Section 6]) after twisting by an appropriate power of $|\cdot|_{\mathbb{A}_E}$ (see also [Shim4, Proposition 3.4] and [Hid6, Theorem 1, Section 9.1]).

11.2 Invariant classes revisited

Let $\mathfrak{c} \subset \mathcal{O}_L$ be an ideal. In (7.2.4) we defined

$$I^m H_i^{\mathrm{inv}}(X_0(\mathfrak{c}), \mathcal{L}((0,\widehat{m}),\chi_0))$$

to be

$$\mathcal{Z}_\varsigma\left(\bigoplus_\phi H^i(\mathfrak{g}, K_\infty; \mathbb{C}\phi \otimes L((0,\widehat{m}),\chi_{\mathrm{triv0}}))\right),$$

the sum is over $\phi = \chi \circ \det : G(\mathbb{Q})\backslash G(\mathbb{A}) \to \mathbb{C}^\times$ such that the quasicharacter χ is unramified at all finite places and satisfies $\chi_\infty(b_\infty) = b_\infty^{-\widehat{m}}$. We now make this space more explicit. For each subset $J \subset \Sigma(L)$, consider the differential form on $\mathfrak{h}^{\Sigma(L)}$ given by

$$v_J := \bigwedge_{\sigma \in J} y_\sigma^{-2} dx_\sigma \wedge dy_\sigma.$$

Then, by [Har, Section 3.2], we have

$$H^{2i+1}(\mathfrak{g}, K_\infty; \mathbb{C}\phi \otimes L((0,\widehat{m}),\chi_{\mathrm{triv0}})) = 0$$

and

$$H^{2i}(\mathfrak{g}, K_\infty; \mathbb{C}\phi \otimes L((0,\widehat{m}),\chi_{\mathrm{triv0}})) = \Big\langle v_J \otimes \phi : |J| = i \Big\rangle.$$

11.3 Definition of the $V_{\chi_E}(\mathfrak{m})$

Let $\chi_E : E^\times \backslash \mathbb{A}_E^\times \to \mathbb{C}^\times$ be a quasi-character satisfying $\chi_E(b_\infty) = b_\infty^{-m}$. We assume for simplicity that

$$\chi_E = \chi_E'^2 \text{ for some quasi-character } \chi_E' : E^\times \backslash \mathbb{A}_E^\times \to \mathbb{C}^\times$$

We assume moreover that $\chi_E \circ N_{L/E}$ is unramified at all finite places. Fix an ideal $\mathfrak{c} \subset \mathcal{O}_L$. For any $J \subset \Sigma(L)$ of order $[L:\mathbb{Q}]/2$ set

$$V_J = \mathcal{Z}_\mathfrak{c} \left(v^J \otimes (\chi_E \circ N_{L/E} \circ \det) \right) \in I^m H_{[L:\mathbb{Q}]}(X_0(\mathfrak{c}), \mathcal{L}(\widehat{\kappa}, \chi_{\text{triv}0})). \quad (11.3.1)$$

We wish to define classes $V_{J,\chi_E}(\mathfrak{m})$ in analogy with the definition in (8.3.3).

Let $\mathfrak{m} \subset \mathcal{O}_E$. If $\mathfrak{m} = N_{L/E}(\mathfrak{m}')$ for some $\mathfrak{m}' \subset \mathcal{O}_L$, $\mathfrak{m} + d_{L/E}(\mathfrak{c} \cap \mathcal{O}_E) = \mathcal{O}_E$, define $V_{J,\chi_E}(\mathfrak{m}) := \widehat{T}_{\mathfrak{c},\chi_E}(\mathfrak{m})_* V_J$, otherwise, define $V_{J,\chi_E}(\mathfrak{m}) = 0$.

Here $\widehat{T}_{\mathfrak{c}_E,\chi_E}(\mathfrak{m})$ is the Hecke operator introduced in Section 8.2.

11.4 Statement and proof of Theorem 11.2

Assume as above that $\chi_E : E^\times \backslash \mathbb{A}_E^\times \to \mathbb{C}^\times$ is a quasi-character such that $\chi_E = \chi_E'^2$ for a quasicharacter $\chi_E' : E^\times \backslash \mathbb{A}_E^\times \to \mathbb{C}^\times$. We assume moreover that $\chi_E \circ N_{L/F}$ is unramified at every finite place and that $\chi_E(b_\infty) = b_\infty^{-m}$. Suppose $\kappa = (0, m) \in \mathcal{X}(L)$ and $\mathfrak{c}_E \subset \mathcal{O}_L$. Choose an auxilliary ideal $\mathfrak{b}' \subset \mathcal{O}_E$ such that $\mathfrak{b}' \neq \mathcal{O}_E$ if $(\mathfrak{c} \cap \mathcal{O}_E)d_{L/E} = \mathcal{O}_E$, and set $\mathfrak{c}_E := \mathfrak{b}'(\mathfrak{c} \cap \mathcal{O}_E)d_{L/E}$. Finally assume that $J \subset \Sigma(L)$ has order $[L:\mathbb{Q}]/2$ and define

$$\Phi_{V_J,\chi_E,\mathfrak{b}} := V_J \otimes \frac{1}{2}\left(\mathcal{E}'_{\mathfrak{c}_E,1} \otimes \chi_E' \eta + \mathcal{E}'_{\mathfrak{c}_E,1} \otimes \chi_E' \right)$$

$$\in I^m H_{[L:\mathbb{Q}]}^{\text{inv}}(X_0(\mathfrak{c}), \mathcal{L}(\widehat{\kappa}, \chi_0)) \otimes M_\kappa(K_0(\mathcal{N}'(\mathfrak{c})), \chi_E)$$

Here $\mathcal{N}'(\mathfrak{c}) := d_{L/E}(\mathfrak{c} \cap \mathcal{O}_E)^3 \mathfrak{b}'$, and the twisting operator $\otimes \chi_E'$ was defined in Proposition 5.11.

The main result of this chapter is the following:

Theorem 11.2. *If $\mathfrak{m} + \mathfrak{c}_E = \mathcal{O}_E$ and $\mathfrak{m}$ is either a norm from $\mathcal{O}_L$ or $\eta(\mathfrak{m}) = -1$ then the $\mathfrak{m}$th Fourier coefficient of $\Phi_{V_J,\chi_E,\mathfrak{b}}$ is $V_{J,\chi_E}(\mathfrak{m})$.*

Remarks.

(1) In analogy with Theorem 8.3, Theorem 11.2 admits a generalization where

$$I^m H_{[L:\mathbb{Q}]}(X_0(\mathfrak{c}), \mathcal{L}(\widehat{\kappa}, \chi))$$

is replaced by an arbitrary Hecke module. We omit the (straightforward) details.

(2) The auxilliary ideal $\mathfrak{b}'$ was introduced for simplicity so that we always work with an Eisenstein series that does not have a constant term at ∞.

The notion of the degree of a Hecke operator will be useful in the proof of Theorem 11.2. For any $\mathbb{Z}$-algebra $A \subseteq \mathbb{C}$ we recall the homomorphism

$$\deg : \mathbb{T}_{\mathfrak{c}} \otimes_{\mathbb{Z}} A \longrightarrow A$$

defined by letting $\deg(K_0(\mathfrak{c})\gamma K_0(\mathfrak{c}))$ be the number of summands in a decomposition

$$K_0(\mathfrak{c})\gamma K_0(\mathfrak{c}) = \sum_{i=1}^{\deg(K_0(\mathfrak{c})\gamma K_0(\mathfrak{c}))} \gamma_i K_0(\mathfrak{c})$$

for some $\gamma_i \in R(\mathfrak{c})$ and extending A-linearly. See [Shim1] for a proof that deg is well defined and is an algebra homomorphism. One can check[1] that

$$\deg(T_{\mathfrak{c}}(\mathfrak{m})) = \sigma'_{\mathfrak{c},1}(\mathfrak{m}) := \sigma'_{\mathfrak{c},1,[m]}(\mathfrak{m}), \tag{11.4.1}$$

where by the subscript 1 we mean the trivial character and we define the function σ for ideals of $\mathcal{O}_L$ as we did for ideals of $\mathcal{O}_E$ in Section 11.1 above.

Proof of Theorem 11.2. We must show that if $\mathfrak{n} + \mathfrak{c}_E = \mathcal{O}_E$ and $\eta(\mathfrak{n}) = -1$ or $\mathfrak{n}$ is a norm from $\mathcal{O}_L$ then

$$V_{J,\chi_E}(\mathfrak{n}) = \frac{\chi'_E(\mathfrak{n})}{2} \left(\eta(\mathfrak{n})\sigma'_{\mathfrak{c}_E,1}(\mathfrak{n}) + \sigma'_{\mathfrak{c}_E,1}(\mathfrak{n})\right) V_J \tag{11.4.2}$$

If $\eta(\mathfrak{n}) = -1$ this is obvious, so we assume that $\mathfrak{n}$ is a norm from $\mathcal{O}_L$ for the remainder of the proof. We defined $\widehat{T}_{\mathfrak{c},\chi_E}$ multiplicatively, so it suffices to prove (11.4.2) for prime powers.

Suppose that $\mathfrak{p} \subset \mathcal{O}_E$ is a prime coprime to $d_{L/E}(\mathfrak{c} \cap \mathcal{O}_E)\mathfrak{b}$ splitting as $\mathfrak{p} = \mathfrak{P}\overline{\mathfrak{P}}$ in $\mathcal{O}_L$. Then we have

$$V_{J,\chi_E}(\mathfrak{p}^r) := \widehat{T}_{\mathfrak{c}_E}(\mathfrak{p}^r)_* V_J = \chi'_E(N_{L/E}(\mathfrak{P}^r)) \deg(T_{\mathfrak{c}}(\mathfrak{P}^r)))V_J$$
$$= \chi'_E(\mathfrak{p}^r)\sigma'_{\mathfrak{c},1}(\mathfrak{P}^r)V_J = \chi'_E(\mathfrak{p}^r)\sigma'_{\mathfrak{c}_E,1}(\mathfrak{p}^r)V_J$$

It is easy to see that (11.4.2) in the case $\mathfrak{n} = \mathfrak{p}^r$ follows from this.

Now assume that $\mathfrak{p} \subset \mathcal{O}_E$ is a prime coprime to $d_{L/E}(\mathfrak{c} \cap \mathcal{O}_E)\mathfrak{b}$ that is inert in L/E (we denote by $\mathfrak{P}$ the integral closure of $\mathfrak{p}$ in $\mathcal{O}_L$). We have

$$V_{J,\chi_E}(\mathfrak{p}^{2r}) := \widehat{T}_{\mathfrak{c}_E}(\mathfrak{p}^{2r})_* V_J$$
$$= T_{\mathfrak{c}}(\mathfrak{P}^r)_* V_J + \chi_E(\mathfrak{p})N_{E/\mathbb{Q}}(\mathfrak{p})T_{\mathfrak{c}}(\mathfrak{P}^{r-1})_* V_J$$
$$= \chi'_E(N_{L/E}(\mathfrak{P}^r)) \deg(T_{\mathfrak{c}}(\mathfrak{P}^r))V_J$$
$$\quad + \chi_E(\mathfrak{p})N_{E/\mathbb{Q}}(\mathfrak{p})\chi'_E(N_{L/E}(\mathfrak{P}^{r-1})) \deg(T_{\mathfrak{c}}(\mathfrak{P}^{r-1}))V_J$$
$$= \left(\chi'_E(\mathfrak{p}^{2r})\sigma'_{\mathfrak{c},1}(\mathfrak{P}^r) + (\chi'^2_E)(\mathfrak{p})N_{E/\mathbb{Q}}(\mathfrak{p})\chi'_E(\mathfrak{p}^{2r-2})\sigma'_{\mathfrak{c},1}(\mathfrak{P}^{r-1})\right)V_J$$
$$= \chi'_E(\mathfrak{p}^{2r})\sigma'_{\mathfrak{c}_E,1}(\mathfrak{p}^{2r})V_J.$$

This implies (11.4.2) in the case $\mathfrak{n} = \mathfrak{p}^{2r}$ for inert $\mathfrak{p}$. $\square$

---
[1] See [Bu, (6.4) p. 494] for the case of $T_{\mathfrak{c}}(\mathfrak{p})$ when $\mathfrak{p} \nmid \mathfrak{c}$.

Appendix A

Proof of Proposition 2.4

A.1 Cellular cosheaves

Let K be a finite regular cell complex. A *cellular cosheaf*[1] is a gadget $\mathbf{E}$ that assigns to each cell $\sigma \in K$ an R module $\mathbf{E}_\sigma$ and to each face $\tau < \sigma$ a homomorphism $\Phi_{\sigma\tau} : \mathbf{E}_\sigma \to \mathbf{E}_\tau$ such that whenever $\tau < \omega < \sigma$ are faces we have $\Phi_{\sigma\tau} = \Phi_{\omega\tau} \circ \Phi_{\sigma\omega}$. Thus, $\mathbf{E}$ is a contravariant functor from the category (which we also denote by K) whose objects are the cells of K and whose morphisms $\tau \to \sigma$ are inclusions of faces $\tau < \sigma$. A *simplicial local system* is a cellular cosheaf such that all the homomorphisms $\Phi_{\sigma\tau}$ are isomorphisms.

An elementary r chain with coefficients in a cellular cosheaf $\mathbf{E}$ is an equivalence class of formal products $a\sigma$ where $\sigma \in K$ is an r-dimensional cell and $a \in \mathbf{E}_\sigma$, modulo the identification $a\sigma \sim (-a)\sigma'$ where σ' is the same cell but with the opposite orientation. The chain module $C_r(K, \mathbf{E})$ is the collection of all finite formal linear combinations of elementary cellular r chains. The boundary of an elementary r chain $a\sigma$ is

$$\partial(a\sigma) = \sum_\tau \Phi_{\sigma\tau}(a)\tau$$

where the sum is taken over those $\tau < \sigma$ of dimension $r - 1$.

Let $H_r(K, \mathbf{E})$ be the homology of the complex $C_*(K, \mathbf{E})$. It is called the homology of K with coefficients in the cosheaf $\mathbf{E}$. If K' is a (finite) refinement of K then the cosheaf $\mathbf{E}$ on K determines a cosheaf $\mathbf{E}'$ on K' by declaring $\mathbf{E}'(\sigma) = \mathbf{E}(\tau)$ where $\sigma \in K'$, and $\tau \in K$ is the unique cell containing the interior σ^o. Then refinement determines a natural injection $C_r(K, \mathbf{E}) \to C_r(K', \mathbf{E}')$.

If σ is a cell in a convex linear cell complex K and if $\widehat{\sigma} \in \sigma^o$ is a point in its interior, define the *stellar subdivision* of K with respect to $\widehat{\sigma}$ to be the convex

---

[1] Recall that every simplex σ has a canonical open neighborhood, namely the open star $St^o(\sigma)$ of σ. The reason for the terminology "cellular cosheaf" is that if $\tau < \sigma$ then the open stars satisfy the reverse containment: $St^o(\sigma) \subset St^o(\tau)$.

linear cell complex in which the cell σ has been divided into the collection of cones $\widehat{\sigma}*\tau$ (where τ varies over the cells in $\partial\sigma$), together with the zero-dimensional cell $\widehat{\sigma}$.

This definition differs slightly from that in some other texts such as [Hud]. Even if K is a simplicial complex, the resulting subdivision K' is only a cell complex. The first barycentric subdivision of K may be obtained by starring (i.e., taking the stellar subdivision) with respect to all the simplices of K in order of increasing dimension.

Proposition A.1. *Let $\mathbf{E}$ be a cellular cosheaf on a convex linear cell complex K and let K' be a refinement of K with corresponding cosheaf $\mathbf{E}'$. Then the refinement mapping induces a canonical isomorphism*

$$H_r(K, \mathbf{E}) \cong H_r(K', \mathbf{E}')$$

for all r.

Proof. (Most of the classical proofs of the invariance of homology under subdivision are either very complicated, or else they do not work in this setting.) Let us consider the case that K' is a stellar subdivision of K with respect to a single barycenter $\widehat{\sigma} \in \sigma^o$ of a cell of dimension r. In this case, we need to show that the homology of the quotient complex $D_* = C_*(K', \mathbf{E}')/C_*(K, \mathbf{E})$ vanishes. Let L denote the subcomplex of K corresponding to $\partial\sigma$. For $j \neq 0, r$ the quotient $D_j = C_j(K', \mathbf{E}')/C_j(K, \mathbf{E})$ may be identified with the group $C_{j-1}(L, \mathbf{E}_\sigma)$ (with constant coefficients), while $D_0 = \mathbf{E}_\sigma$ and $D_r = C_{r-1}(L, \mathbf{E}_\sigma)/\mathbf{E}_\sigma$, this last being the quotient under the diagonal embedding $\mathbf{E}_\sigma \to C_{r-1}(L, \mathbf{E}_\sigma)$. In other words, the complex D_* is the chain complex (with constant coefficients) for the $r-1$-dimensional sphere, augmented in degrees 0 and r, so $H_r(D_*) = 0$ for all r.

The stellar subdivisions of K form a cofinal system in the directed set of all finite refinements of K, so we conclude that the homology $H_*(K, \mathbf{E})$ is invariant under refinement. $\square$

We remark for completeness that a simplicial *sheaf* is a *covariant* functor from the category K to the category of R-modules. The cohomology of a simplicial sheaf is defined in a way that is dual to the homology of a cosheaf.

A.2 Proof of Lemma 2.3

Let K be a regular cell complex, L a closed subcomplex, set $X = |K|$ and $Y = |K| - |L|$ with $i : Y \to X$ the inclusion mapping. Let $\mathbf{E}$ be a local coefficient system on Y. Then we obtain a cellular cosheaf $i_!\mathbf{E}$ on X which assigns to any cell $\sigma \in K$ the group

$$i_!\mathbf{E}(\sigma) = \begin{cases} \mathbf{E}_\sigma & \text{if } \sigma^o \subset Y \\ 0 & \text{otherwise.} \end{cases}$$

It follows immediately from the definitions that the chain groups $C_r(K, i_!\mathbf{E}) = \widehat{C}_r^K(Y, \mathbf{E})$ are identical, hence $H_r(K, i_!\mathbf{E}) \cong \widehat{H}_r^K(Y, \mathbf{E})$. Proposition A.1 then implies that the cellular Borel-Moore homology groups $H_r^K(Y, \mathbf{E})$ are invariant under (finite) refinements, which proves Lemma 2.3. □

A.3 Proof of Proposition 2.4

In this section, K is a finite convex linear cell complex, L is a closed subcomplex, $X = |K|$, $Y = |K| - |L|$, and $\mathbf{E}$ is a local coefficient system (of R modules) on Y. Set $A = |L| \subset X$. We have a triangulation T of Y (with infinitely many simplices) that refines K and we wish to consider the map on homology that is induced from the injection $\widehat{C}_r^K(Y, \mathbf{E}) \to C_r^{BM,T}(Y, \mathbf{E}) \to C_r^{BM}(Y, \mathbf{E})$.

First let us consider the case that E extends as a local system over all of X. Then the chain group $\widehat{C}_i^K(Y, \mathbf{E})$ is canonically isomorphic to the quotient group $C_i(K, \mathbf{E})/C_i(L, \mathbf{E})$, since both complexes have bases that are indexed by i-dimensional simplices $\sigma \in K$ such that $\sigma \notin L$, in other words, such that $\sigma^o \subset Y$. It is also easy to see that the boundary homomorphisms are compatible with this isomorphism, hence $\widehat{H}_r^K(Y, \mathbf{E}) \cong H_r(X, A; \mathbf{E})$. On the other hand, the mapping $C_r(K, \mathbf{E})/C_r(L, \mathbf{E}) \to C_r^{BM}(Y, \mathbf{E})$ also induces an isomorphism on homology, which may be checked by building up X from A, one simplex at a time.

Now consider the general case, when E does not necessarily extend as a local system over X. By taking the barycentric subdivision if necessary, we may assume that K is a *simplicial* complex and that L is a *full* subcomplex, meaning that if σ is a simplex of K and all its vertices are in L then $\sigma \in L$. Let M be the subcomplex of K consisting of simplices that are disjoint from $A = |L|$. Then M is also a full subcomplex of K (for if σ is a simplex of K whose vertices are in M, then the simplex σ is disjoint from A so it lies in M). Let us now consider the simplices of K that are neither in M or L. The vertices of such a simplex σ fall into two classes: those in L and those in M. Thus we have isolated two faces $\tau_L \in L$ and $\tau_M \in M$ of σ such that $\sigma = \tau_L * \tau_M$ is the join of these faces. In other words, every point $x \in \sigma$ lies on a unique line segment I_x with one endpoint in τ_L and the other endpoint in τ_M. In this way, the mapping $|L| \to \{0\}$ and $|M| \to \{1\}$ extends uniquely, linearly, and continuously to a simplicial mapping $f : X \to [0, 1]$. (See, for example, [Mun] Lemma 70.1, p. 414.) Let us denote by $X_S = f^{-1}(S)$ for any subset $S \subset [0, 1]$. Then $Y = X_{[0,1)}$. The simplicial mapping f is necessarily a product mapping over the open interval $(0, 1)$. We obtain in this way a "collaring" $X_{(0,1)} \cong X_{\{\frac{1}{2}\}} \times (0, 1)$ of a neighborhood of infinity.

It follows that there is a piecewise linear homeomorphism $Y \to X_{[0,\frac{1}{2})}$, which then induces canonical isomorphisms

$$H_r^{BM}(Y, \mathbf{E}) \to H_r^{BM}(X_{[0,\frac{1}{2})}, \mathbf{E}) \cong H_r(X_{[0,\frac{1}{2}]}, X_{\{\frac{1}{2}\}}; \mathbf{E}) \qquad (A.3.1)$$

since $\mathbf{E}$ extends across $X_{\{\frac{1}{2}\}}$.

We now wish to make a similar identification of $\widehat{H}_r^K(Y, \mathbf{E})$.

Let us refine the interval $[0,1]$ by adding a vertex $\{\frac{1}{2}\}$, and then refine K into a regular cell complex such that $f^{-1}(\frac{1}{2})$ is a subcomplex: each simplex $\sigma = \tau_L * \tau_M$ is decomposed into two cells, $\sigma \cap f^{-1}([0, \frac{1}{2}])$ and $\sigma \cap f^{-1}([\frac{1}{2}, 1])$ whose intersection is the cell $\sigma \cap f^{-1}(\frac{1}{2})$. Let K' denote this (cellular) refinement of K and let $\widehat{C}_r^{K'}(Y, \mathbf{E})$

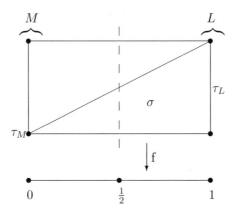

Figure A.3: Decomposing simplices of K

be the cellular Borel-Moore chains on Y with respect to this pseudo-cellulation K' of Y. The chain group decomposes into three subgroups $\widehat{C}_r^{K'}(Y, \mathbf{E}) = A_r \oplus B_r \oplus C_r$, each of which consists of formal linear combinations of r-dimensional cells $\sigma \in K'$ with $\sigma^o \subset X_{[0, \frac{1}{2})}$ in case A_r, with $\sigma^o \subset X_{\{\frac{1}{2}\}}$ in case B_r, and with $\sigma^o \subset X_{(\frac{1}{2}, 1)}$ in case C_r. The boundary homomorphism decomposes as follows.

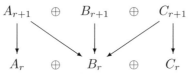

Each cell in C_r is the product of a cell in B_{r-1} with the open interval $(\frac{1}{2}, 1)$ so the subcomplex $B \oplus C$ is acyclic. Hence the left arrow in the following diagram induces an isomorphism on homology groups. The right arrow in this diagram is an isomorphism of chain complexes:

$$A \oplus B \oplus C \longrightarrow (A \oplus B \oplus C)/(B \oplus C) \longleftarrow (A \oplus B)/B$$

Altogether, we obtain a canonical isomorphism

$$\widehat{H}_r^{K'}(Y, \mathbf{E}) \cong H_r(A \oplus B/B) = H_r(X_{[0, \frac{1}{2}]}, X_{\frac{1}{2}}; \mathbf{E}).$$

Combining this with the isomorphisms of Lemma 2.3 and equation (A.3.1) gives the desired isomorphism

$$\widehat{H}_r^K(Y, \mathbf{E}) \cong \widehat{H}_r^{K'}(Y, \mathbf{E}) \cong H_r^{BM}(Y, \mathbf{E}). \qquad \square$$

Appendix B

Recollections on Orbifolds

If a compact group acts with finite isotropy on a smooth manifold, then the quotient space is an orbifold. The singularities in such a space are "mild" in that it is possible to develop a theory of differential forms on such a space which supports the theorems of Stokes and de Rham. In this section we review the definition and basic properties of orbifolds, or V-manifolds, as described in [Sat], [ChR], and elegantly reformulated in [MP]. A good general reference is [Adem]. See also [Kaw], [Bai], and [Mo2]. Although none of this material is new, we have filled in some technical details, particularly with respect to refinement of an orbifold atlas. As a consequence, several hypotheses appearing in [Sat], [Kaw], and [ChR] may be omitted from the definition of an orbifold. There are slight differences among the various definitions and terminologies. The orbifolds considered here are *reduced* in the sense of [ChR].

B.1 Effective actions

Let G be a finite group of diffeomorphisms of an n-dimensional connected manifold M. Assume the action is effective: every element $g \in G$ acts non-trivially except for the identity element $1 \in G$. Write $g \cdot m$ for the action of $g \in G$ on $m \in M$ and let $dg(m) : T_m M \to T_{g \cdot m} M$ be the differential of this mapping. For each $m \in M$ let $G_m \subset G$ denote the isotropy group of G, and for each $g \in G$ let M^g denote the set of points in M that are fixed by g. A choice of G-invariant Riemannian metric on M determines a system of geodesics on M such that the exponential mapping is G-equivariant in the sense that the following diagram commutes over some neighborhood W_m of the origin in $T_m M$:

$$\begin{array}{ccc} T_m M & \xrightarrow{dg(m)} & T_{g \cdot m} M \\ \exp \downarrow & & \downarrow \exp \\ M & \xrightarrow{g} & M \end{array} \qquad (\text{B.1.1})$$

See [Koba, Section VI Proposition 1.1 and Section IV Proposition 2.5]. This has several immediate consequences.

(A) For any $g \in G$ the fixed point set M^g is a union of smoothly embedded closed submanifolds of M.

For if $g \cdot m = m$ then there exists a neighborhood U_m of m such that $M^g \cap U_m = \exp(V)$ where $V = \{v \in T_mM \mid dg(m)(v) = v\} \cap W_m$ is an open set in a vector space.

(B) Let $g \in G$. If there exists a point $m \in M^g$ such that $dg(m) : T_xM \to T_xM$ is the identity mapping, then $g = 1$.

For, if $dg(m) = I$ then g fixes a whole neighborhood of m. Therefore the set of points $y \in M$ such that $dg(y) = I$ is both open and closed, so g acts trivially on M. Since the action is effective, $g = 1$.

(C) The set M^0 of points on which G acts freely is open and dense in M.

In fact M^0 is the complement of the finitely many closed submanifolds M^g (for $g \neq 1$), each of which has codimension ≥ 1.

Lemma B.1 ([MP]). *Let M' be another connected n-dimensional manifold with an effective action by a finite group G'. Let $i : M/G \to M'/G'$ be an embedding. Suppose $f : M \to M'$ is a smooth embedding which covers the mapping i. Then*

(1) *There exists a unique mapping $\lambda : G \to G'$ such that f is equivariant with respect to λ (meaning that $\lambda(g) \cdot f(m) = f(g \cdot m)$).*

(2) *The mapping λ is an injective group homomorphism. Its image is*

$$Im(\lambda) = \{g' \in G' \mid g' \cdot f(M) = f(M)\}.$$

For each $m \in M$ the mapping λ induces an isomorphism of isotropy groups $G_m \cong G'_{f(m)}$.

(3) *If there exists $g' \in G'$ such that $f(M) \cap g' \cdot f(M) \neq \phi$ then $f(M) = g' \cdot f(M)$ and g' is in the image of λ.*

(4) *If $h : M \to M'$ is another smooth embedding that covers the same embedding i then there exists a unique $g' \in G'$ such that $h(m) = g' \cdot f(m)$ for all m.*

Let $\mathcal{M}_S$ be the category of n-manifolds with finite symmetry. An object in $\mathcal{M}_S$ is a pair (M, G) where M is a smooth connected n-manifold and G is a finite group acting effectively on M. A morphism $(M, G) \to (M', G')$ is a smooth equivariant open embedding $f : M \to M'$ which induces an open embedding $i : M/G \to M'/G'$. In this case we say that the embedding f *covers* the embedding i.

Proof of Lemma B.1. We will prove (4) first. Assume we are given mappings $f, h : M \to M'$ which cover the mapping i. If M^0 and $(M')^0$ denote the subsets on which G and G' act freely then the set $W = M^0 \cap f^{-1}((M')^0) \cap h^{-1}((M')^0)$ is open and

B.1. Effective actions

dense in M; the projection $\pi : W \to W/G$ is a smooth unramified covering; and the restriction $i : W/G \to (M')^0/G'$ is a smooth embedding of smooth manifolds. Fix a point $m_0 \in W$. Then $f(m_0)$ and $h(m_0)$ lie in the same fiber $(\pi')^{-1}(i\pi(m_0))$ where $\pi' : M' \to M'/G'$ is the projection. Therefore there is a unique $g'_0 \in G'$ such that $h(m_0) = g'_0 \cdot f(m_0)$. We will show that $h(m) = g'_0 \cdot f(m)$ for all $m \in M$.

From covering space theory we know that $h(m) = g'_0 \cdot f(m)$ for all m in the connected component of W that contains the point m_0, and this equality also holds in the closure of this connected component. The same remarks apply to every connected component of W. Let W_0 denote the union of those connected components of W such that $h(m) = g'_0 f(m)$ for all m in the closure $\overline{W}_0$ of W_0.

Recall that W is the complement of a finite collection of closed submanifolds of M having dimension $\leq n-1$. Choose a point $m_1 \in M$ on one of these submanifolds that separates the region W_0 from some other region, say W_1, for which the corresponding group element $g'_1 \neq g'_0$.

Choose a G'-invariant Riemannian metric $b(\cdot, \cdot)$ on M' and consider its pullback $f^*(b)$ to M. This is a smooth metric on M and its restriction to W is the pullback $f^*(b) = \pi^* i^*(\bar{b})$ of a smooth metric $\bar{b}$ on M^0/G'. Therefore $f^*(b)$ is G-invariant on W so by continuity it is G invariant on all of M. Similarly, the metric $h^*(b)$ coincides with $f^*(b)$ on W so they coincide everywhere, and h is also an isometry. By [Koba] Section VI Proposition 1.1 we again have a diagram which commutes over some neighborhood of the origin in $T_{m_1}M$:

$$\begin{array}{ccc} T_{m_1}M & \xrightarrow{d(g'_0 f)(m)} & T_{f(m_1)}M' \\ \exp \downarrow & & \downarrow \exp \\ M & \xrightarrow{g'_0 \cdot f} & M' \\ \pi \downarrow & & \downarrow \pi' \\ M/G & \xrightarrow{i} & M'/G' \end{array} \quad (B.1.2)$$

and there is a similar diagram for h. On the one hand $h(m) = g'_0 \cdot f(m)$ for m in the region $\overline{W}_0$ (hence $dh(m_1) = dg'_0 \circ df(m_1)$) but on the other hand $h(m) = g'_1 \cdot f(m)$ for m in the region $\overline{W}_1$ (hence $dh(m_1) = dg'_1 \circ df(m_1)$). Therefore the group element $(g'_1)^{-1} g'_0$ fixes the point $f(m_1)$ and acts on $T_{m_1}M'$ by the identity mapping, so by (B) above, $g'_1 = g'_0$. In summary, the set of points m where $h(m) = g'_0 \cdot f(m)$ is both open and closed, hence $h = g'_0 \cdot f$. This completes the proof of part (4) of the lemma.

To prove part (1), let us return to the original basepoint $m_0 \in M$. Since G acts freely on m_0 and since G' acts freely on $f(m_0)$, for any $g \in G$ there is a unique element, call it $\lambda(g) \in G'$ such that $f(g \cdot m_0) = \lambda(g) \cdot f(m_0)$. The mappings $m \mapsto f(g \cdot m)$ and $m \mapsto \lambda(g) \cdot f(m)$ both cover the mapping i and they agree at the point m_0 so by the preceding paragraph, they coincide everywhere. Moreover this implies that $f(g_1 g_2 \cdot m) = \lambda(g_1) \cdot f(g_2 \cdot m) = \lambda(g_1) \cdot \lambda(g_2) \cdot f(m)$ which proves that λ

is a group homomorphism. It is clearly injective: if $\lambda(g) = 1$ then $f(gx_0) = f(x_0)$. But f is an embedding so $gx_0 = x_0$ hence $g = 1$. This proves part (2).

For part (3), suppose $f(M) \cap g' \cdot f(M) \neq \phi$. Then the set $f^{-1}(g' \cdot f(M))$ is open and non-empty. Since M^0 is dense in M there exist $m_1, m_2 \in M^0$ such that $f(m_1) = g' \cdot f(m_2)$. It follows that $\pi(m_1) = \pi(m_2)$ so there exists a unique $g \in G$ such that $m_2 = g \cdot m_1$. The embeddings $m \mapsto f(g \cdot m)$ and $m \mapsto g' \cdot f(m)$ agree at the point m_1 so they coincide. Therefore $g' = \lambda(g)$ and hence $g' \cdot f(M) = f(g \cdot M) = f(M)$. $\square$

This lemma allows us to remove several hypotheses in [Sat] and [Kaw] concerning the definition of an orbifold.

B.2 Definitions

Throughout the remainder of this appendix, fix a locally compact Hausdorff space X and a regular, commutative Noetherian ring R (with unit) of finite cohomological dimension (e.g., a principal ideal domain). An *R-orbifold chart* (also called a *local uniformization*) on X is a collection $\mathcal{C} = (U, M, G, \phi)$ where $U \subset X$ is a connected open subset, (M, G) is an object in $\mathcal{M}_S$ such that every rational integer dividing $|G|$ is invertible in R, and $\phi : M \to X$ is a continuous G-invariant mapping which induces a homeomorphism $\bar{\phi} : M/G \to U \subset X$ onto an open subset U of X. Suppose $\mathcal{C} = (U, M, G, \phi)$ and $\mathcal{C}' = (U', M', G', \phi')$ are charts such that $U \subset U'$. We say these charts are *compatible*, and we write $\mathcal{C} \to \mathcal{C}'$, if there exists a morphism $f : (M, G) \to (M', G')$ in $\mathcal{M}_S$ that covers the inclusion $i : U \to U'$. In this case we also write $f : (U, M, G, \phi) \to (U', M', G', \phi')$ and we refer to f as an *embedding of charts* or as a *morphism of charts*. By Lemma B.1 such a morphism $f : M \to M'$, if one exists, is uniquely determined up to the action by elements of G'.

Let us say that an open covering $\mathcal{U}$ of X is *good* if each $U \in \mathcal{U}$ is connected and if $\mathcal{U}$ is closed under pairwise intersections: if $U_1, U_2 \in \mathcal{U}$ and $U_1 \cap U_2 \neq \phi$ then $U_1 \cap U_2 \in \mathcal{U}$.

Definition B.2.1. An *R-orbifold atlas* $\mathfrak{U}$ on X consists of a good open cover $\mathcal{U}$ and an assignment, for each $U \in \mathcal{U}$ of an R-orbifold chart (U, M, G, ϕ) over U, such that: if $U \subset U'$ are elements of $\mathcal{U}$ then the charts (U, M, G, ϕ) and (U', M', G', ϕ') are compatible. We say that $\mathfrak{U}$ is an orbifold atlas *over* the cover $\mathcal{U}$.

We use this somewhat restrictive notion of a "good" open cover, which requires all pairwise intersections $U \cap U'$ to be connected (or empty), in order to facilitate the proof of Proposition B.3 below. However this condition may be weakened to the more standard, but equivalent condition

- If $U, U' \in \mathcal{U}$ and if $x \in U \cap U'$ then there exists $V \in \mathcal{U}$ such that $x \in V \subset U \cap U'$.

In fact, any open cover of an orbifold X admits a "good" refinement [Mo1].

B.2. Definitions

Remark. If X admits a R-orbifold atlas then X is a R-homology manifold.

One annoying aspect of these definitions is the fact that, given an R-orbifold atlas and an arrangement of open sets sets in $\mathcal{U}$,

$$\begin{array}{ccc} U_1 & \longrightarrow & U_2 \\ \downarrow & & \downarrow \\ U_3 & \longrightarrow & U_4 \end{array}$$

it may be impossible to choose the morphisms so that the corresponding diagram commutes:

$$\begin{array}{ccc} (M_1, G_1) & \longrightarrow & (M_2, G_2) \\ \downarrow & & \downarrow \\ (M_3, G_3) & \longrightarrow & (M_4, G_4) \end{array}$$

Consequently some constructions become quite difficult when using the definition of orbifold as given in [Kaw].

On the other hand, suppose $U_1 \subset U_2 \subset U_3$ are open sets in X and suppose R-orbifold charts $\mathcal{C}_i = (U_i, M_i, G_i, \phi_i)$ are given over each of these (with $1 \leq i \leq 3$). It is easy to see that if $\mathcal{C}_1 \to \mathcal{C}_2$ (meaning that these charts are compatible) and if $\mathcal{C}_2 \to \mathcal{C}_3$ then $\mathcal{C}_1 \to \mathcal{C}_3$. Moreover we have the following:

Lemma B.2. *Let $\mathcal{C}_1, \mathcal{C}_2, \mathcal{C}_3$ be charts with $U_1 \subset U_2 \subset U_3$ as above. Suppose $\mathcal{C}_1 \to \mathcal{C}_3$ and $\mathcal{C}_2 \to \mathcal{C}_3$. Then $\mathcal{C}_1 \to \mathcal{C}_2$.*

Proof. Let $f_1 : (M_1, G_1) \to (M_3, G_3)$ and $f_2 : (M_2, G_2) \to (M_3, G_3)$ be morphisms that cover the inclusions $U_1 \subset U_3$ and $U_2 \subset U_3$ respectively. Let M_1^0 be the set of points m such that G_3 acts freely on $f_1(m)$, cf. Lemma B.1 part (2). Then there exists $g_3 \in G_3$ such that $g_3 \cdot f_1(m) \in f_2(M_2)$. We claim that $g_3 \cdot f_1(M_1) \subset f_2(M_2)$, from which it will follow that $f_2^{-1} \circ (g_3 \cdot f_1) : M_1 \to M_2$ is a morphism covering the inclusion $U_1 \subset U_2$.

From the theory of covering spaces we know that $f_2(M_2)$ contains the image $g_3 \cdot f_1(M_1)$ of the connected component M_1^m of M_1^0 that contains the point m. So, just as in the proof of Lemma B.1 we must consider the behavior of the mapping f_3 at a point $m_1 \in M_1$ that separates several regions of M_1^0. A choice of G_3-invariant Riemannian metric on M_3 determines G_j-invariant Riemannian metrics on M_j (for $j = 1, 2$) and we have a diagram which commutes in some neighborhood of the origins,

$$\begin{array}{ccccc} T_{m_1}M_1 & \xrightarrow{d(g_3 f_1)(m_1)} & T_{m_3}M_3 & \xleftarrow{df_2(m_2)} & T_{m_2}M_2 \\ {\scriptstyle \exp}\downarrow & & {\scriptstyle \exp}\downarrow & & \downarrow{\scriptstyle \exp} \\ M_1 & \xrightarrow[g_3 \cdot f_1]{} & M_3 & \xleftarrow[f_2]{} & M_2 \end{array}$$

where $m_3 = g_3 \cdot f_1(m_1)$ and $m_2 = f_2^{-1}(m_3)$. Since df_2 is an isomorphism, $g_3 \cdot f_1$ takes a whole neighborhood of x_1 into the image, $f_2(M_2)$. Therefore the set of points in M_1 that are taken by $g_3 \cdot f_1$ into $f_2(M_2)$ is both open and closed in M_1. This proves the claim, and hence completes the proof of the lemma. □

B.3 Refinement

A good open covering $\mathcal{U}$ of X is said to refine a good open covering $\mathcal{U}'$ if every $U \in \mathcal{U}$ is contained in some $U' \in \mathcal{U}'$. An R-orbifold atlas $\mathfrak{U}$ over a good open cover $\mathcal{U}$ is said to refine an R-orbifold atlas $\mathfrak{U}'$ over a good open cover $\mathcal{U}'$ if $\mathcal{U}$ refines $\mathcal{U}'$ and if the chart (U, M, G, ϕ) is compatible with the chart (U', M', G', ϕ') whenever $U \in \mathcal{U}$ is contained in $U' \in \mathcal{U}'$.

Proposition B.3. *Let $\mathcal{U}, \mathcal{U}'$ be good open covers such that $\mathcal{U}$ refines $\mathcal{U}'$. Let $\mathfrak{U}'$ be an R- orbifold atlas over $\mathcal{U}'$. Then there exists an R-orbifold atlas over $\mathcal{U}$ that refines $\mathfrak{U}'$.*

Proof. For each $U \in \mathcal{U}$ choose a chart (U', M', G', ϕ') in $\mathfrak{U}'$ such that $U \subset U'$. Choose a connected component M of the fiber product $FP = U \times_{U'} M'$:

$$\begin{array}{ccc} FP & \longrightarrow & M' \\ \phi \downarrow & & \downarrow \phi' \\ U & \longrightarrow & U' \end{array}$$

The group G' acts on FP so we may define $G = \{g' \in G'|\ g'(M) = M\}$. Since this is a subgroup of G', we have that every rational integer dividing $|G|$ is invertible in R. We claim that these choices $\{(U, M, G, \phi)|\ U \in \mathcal{U}\}$ form an atlas $\mathfrak{U}$ that refines the atlas $\mathfrak{U}'$.

First we must show that the charts

$$\mathcal{C}_1 = (U_1, M_1, G_1, \phi_1) \quad \text{and} \quad \mathcal{C}_2 = (U_2, M_2, G_2, \phi_2)$$

are compatible whenever $U_1 \subset U_2$ are elements of $\mathcal{U}$. Let $\mathcal{C}'_1, \mathcal{C}'_2$ be the charts in $\mathfrak{U}'$ that were associated to U_1 and U_2 respectively, and let $\mathcal{C}'_{12}$ be the chart corresponding to $U'_1 \cap U'_2 \in \mathcal{U}'$. Consider the diagram of inclusions and chart compatibilities,

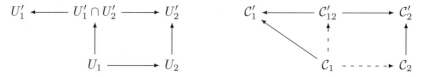

Applying Lemma B.2 to $U_1 \subset U'_1 \cap U'_2 \subset U'_1$ gives $\mathcal{C}_1 \to \mathcal{C}'_{12}$. Applying the same lemma to $U_1 \subset U_2 \subset U'_2$ gives $\mathcal{C}_1 \to \mathcal{C}_2$ as needed. Therefore $\mathfrak{U}$ is an atlas. To show that it is a refinement of $\mathfrak{U}'$ we need to prove that the chart $\mathcal{C} = (U, M, G, \phi)$ in $\mathfrak{U}$ is

B.4. Stratification

compatible with chart $\mathcal{C}'_1 = (U'_1, M'_1, G'_1, \phi'_1)$ in $\mathfrak{U}'$ whenever $U \subset U'_1$. For this purpose let $\mathcal{C}'_2 = (U'_2, M'_2, G'_2, \phi'_2)$ be the chart in $\mathfrak{U}'$ that was associated to U during the construction of the atlas $\mathfrak{U}$. Then there is a chart $\mathcal{C}'_{12} = (U'_1 \cap U'_2, M'_{12}, G'_{12}, \phi'_{12})$ in $\mathfrak{U}'$ that lies over $U'_1 \cap U'_2$ so we have a diagram of inclusions and compatibilities,

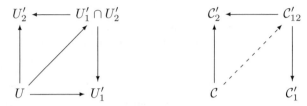

Lemma B.2 applied to $U \subset U'_1 \cap U'_2 \subset U'_2$ implies that $\mathcal{C} \to \mathcal{C}'_{12} \to \mathcal{C}'_1$ as claimed. $\square$

Definition B.3.1. An R-orbifold structure on X is an equivalence class of R-orbifold atlases, two being equivalent if they have a common refinement. The orbifold is *orientable* (resp. *complex*) if, for each chart (U, M, G, ϕ) the manifold M is orientable (resp. complex) and the action of G on M is orientation preserving (resp. holomorphic). The orbifold is *subanalytic* if X is a subanalytic set and for each chart (U, M, G, ϕ) the manifold M is subanalytic, the group G acts subanalytically on M, and the mapping $\phi : M/G \to X$ is subanalytic.

The word "orbifold" replaces "R-orbifold" if the ring R is understood. Notice that any R-orbifold is automatically a $\mathbb{Q}$-orbifold, though not conversely.

B.4 Stratification

Let (U, M, G, ϕ) be a chart on X. Decompose M into *strata* according to the isomorphism type of the isotropy groups, that is, if H is a finite group, set

$$M_H = \{m \in M |\ G_m \cong H\}.$$

It follows from (A) in Section B.1 that M_H is a disjoint union of smoothly embedded submanifolds of M. (Its connected components may have varying dimensions.) The group G preserves M_H so if $m \in M_H$ then $\phi^{-1}(\phi(m)) \subset M_H$. The projection $M_H \to M_H/G$ is a local diffeomorphism so the quotient M_H/G is also a union of smooth manifolds. The open dense subset on which G acts freely is $M^0 = M_{\{1\}}$. If $f : (U, M, G, \phi) \to (U', M', G', \phi')$ is an embedding of charts then $G'_{f(m)} = \lambda(G_m)$ for all $m \in M$ (where $\lambda : G \to G'$ is the injection from Lemma B.1). Hence $f(M_H) \subset M'_H$ and $\phi(M_H) \subset \phi'(M'_H)$ so we may define

$$X_H = \bigcup_{\mathcal{C}} \phi(M_H)$$

where the union is taken over all charts $\mathcal{C}$ in an atlas. Then each connected component of X_H is a smooth manifold, topologically embedded in X; the sets X_H and $X_{H'}$ are disjoint if $H \neq H'$, and the subset $X^0 = \bigcup_{\mathcal{C}} \phi(M^0)$ is open and dense in X.

Proposition B.4. *There exists an embedding of X into Euclidean space so that the decomposition of X into connected components of the various X_H forms a (locally finite) Whitney stratification of X. The topological space X can be triangulated so that the closure of each stratum becomes a closed subcomplex.*

Proof. The quotient of a finite-dimensional real vector space V under the action of a finite group can be embedded as a semi-analytic subset of Euclidean space so that the decomposition into strata V_H satisfies the Whitney conditions. Every Whitney stratified subset of a manifold can be triangulated so that the closure of each stratum is a subcomplex. These two facts can be used to prove the Proposition. Details for the embedding results may be found in [Bie], [Sja, Section 6], [Schwa] [Pro], [Ma2], [Cu]. Details for the triangulation results may be found in [Mo1, Section 1.2], [Ya], [Gre1]. □

B.5 Sheaves and cohomology

In the next few paragraphs we recall the definition of the cohomology of an R-orbifold, and [Sat] the complex of differential forms that may be used to compute it provided $R = \mathbb{R}$ or $R = \mathbb{C}$. In general, the cohomology (in the orbifold sense) of an orbifold X differs from the (singular) cohomology of the underlying topological space X. But if the coefficient ring is the rational or real numbers, then these coincide.

Suppose a finite group G acts on a smooth manifold M with orbit space $\pi : M \to M/G$. A *G-equivariant sheaf*, or G-sheaf (of R-modules) $\mathbf{F}$ on M is a sheaf together with an isomorphism $\phi_g : g^*(\mathbf{F}) \to \mathbf{F}$ for each $g \in G$, such that $\phi_g \circ g^*(\phi_h) = \phi_{hg}$ for all $g, h \in G$. The category of G-sheaves is abelian and it has enough injectives. An *equivariant section* $s : M \to \mathbf{F}$ is a section such that $\phi_g(s(g \cdot m)) = s(m)$ for all $m \in M$ and $g \in G$. Let $\Gamma_G(M, \mathbf{F})$ denote the abelian group of equivariant sections. The *equivariant cohomology* is the right derived functor, $H_G^i = R^i \Gamma_G$. Thus, if $\mathbf{F}$ is a G-sheaf on M then $H_G^i(M, \mathbf{F})$ is obtained as the cohomology of the complex of global invariant sections of any resolution $\mathbf{F} \to \mathbf{I}^0 \to \mathbf{I}^1 \to \cdots$ by injective G-sheaves.

Let EG be a universal space for G with corresponding classifying space $BG = EG/G$. An equivariant sheaf $\mathbf{F}$ on M pulls up to an equivariant sheaf on $EG \times M$ and it passes to a sheaf, which we denote by $\widehat{\mathbf{F}}$, on the Borel construction $EG \times_G M$. There is a natural isomorphism

$$H_G^i(M, \mathbf{F}) \cong H^i(EG \times_G M, \widehat{\mathbf{F}}).$$

B.5. Sheaves and cohomology

Suppose A is a module over a ring in which $|G|$ is invertible and suppose G acts linearly on A. This action determines a locally constant sheaf (or *local system*) $\mathbf{A} = EG \times_G A$ on BG. The resulting cohomology is the *group cohomology* ([Weib] Prop. 6.1.10),

$$H^i(G, A) = H^i_G(\{\text{point}\}, A) \cong H^i(BG, \mathbf{A}) = \begin{cases} A^G & \text{for } i = 0 \\ 0 & \text{otherwise} \end{cases} \quad (B.5.1)$$

where A^G denotes the submodule of invariants in A.

For any G-sheaf $\mathbf{F}$ on M the group G acts on the push-forward $\pi_*(\mathbf{F})$ and we let

$$\overline{\mathbf{F}} = \pi_*(\mathbf{F})^G$$

be the sheaf of invariants. It is the sheafification of the presheaf whose sections over an open set $U \subset M/G$ are $\Gamma_G(\pi^{-1}(U), \mathbf{F})$. If $y \in M$ and if G_y denotes the isotropy group at y then there is a natural identification, $\mathbf{F}_y^{G_y} \cong \overline{\mathbf{F}}_{\pi(y)}$ between the G_y-invariants in the stalk $\mathbf{F}_y$ and the stalk of $\overline{\mathbf{F}}$ at $\pi(y) \in M/G$.

If G acts freely on M then $\mathbf{F} \cong \pi^*(\overline{\mathbf{F}})$ and $H^i_G(X, \mathbf{F}) \cong H^i(X/G, \overline{\mathbf{F}})$. But if G acts with nontrivial isotropy on X then $H^i_G(X, \mathbf{F})$ is usually non-zero (but torsion) for infinitely many values of i. However if $|G|$ is invertible in the coefficient ring then we regain an isomorphism between the two cohomology groups:

Proposition B.5. *Suppose a finite group G acts on a smooth manifold M. Let $\mathbf{F}$ be a sheaf on M of modules over a ring R in which $|G|$ is invertible. Then there is a natural isomorphism*

$$H^i(M/G, \overline{\mathbf{F}}) \cong H^i_G(M, \mathbf{F}).$$

Proof. Let $y \in M$. The mapping $q : EG \times_G M \to M/G$ has for its fiber over the point $\pi(y)$ the classifying space $EG/G_y = BG_y$ of the stabilizer group G_y. The restriction of the sheaf $\widehat{\mathbf{F}}$ to this fiber is the locally trivial sheaf corresponding to the representation of G_y on the stalk $\mathbf{F}_y$. The stalk of $R^b q_*(\widehat{\mathbf{F}})$ may therefore be identified with $H^b(G_y, \mathbf{F}_y)$. So the Leray-Serre spectral sequence for the map q has, as its E_2 page,

$$E_2^{a,b} = H^a(M/G, R^b q_*(\widehat{\mathbf{F}})) \implies H^{a+b}_G(M, \mathbf{F}).$$

By equation (B.5.1) this cohomology sheaf vanishes for $b \ne 0$, while $H^0(G_y, \mathbf{F}_y) \cong \overline{\mathbf{F}}_y$ is the vector space of invariants, $\mathbf{F}_\mathbf{y}^{G_y}$. So the natural morphism $\overline{\mathbf{F}} \to R^0 q_*(\mathbf{F})$ is an isomorphism of sheaves. $\square$

Let X be a locally compact topological space with an R-orbifold atlas $\mathfrak{U}$. Recall from [MP] for example, that a *sheaf* $\mathbf{F}$ on the R-orbifold X is a choice, for each chart (U, M, G, ϕ), of a G-sheaf $\mathbf{F}_\mathbf{U}$ of R-modules on M, and an isomorphism $\psi_f : f^*(\mathbf{F}_{\mathbf{U}'}) \to \mathbf{F}_\mathbf{U}$ of G-sheaves of R-modules whenever $f : (U, M, G, \phi) \to (U', M', G', \phi')$ is an embedding of charts. The morphisms ψ_f are required to be compatible: $\psi_{f'f} = \psi_{f'}\psi_f$ if f, f' are composable morphisms.

A *section s* of the sheaf $\mathbf{F}$ is a choice of *invariant* section $s_U \in \Gamma_G(\phi^{-1}(U), \mathbf{F_U})$ in each chart (U, M, G, ϕ) which are compatible: $\psi_f \circ f^*(s') = s$ for each morphism $f : (U, M, G, \phi) \to (U', M', G', \phi')$ and section $s' \in \Gamma_{G'}(\phi^{-1}(U'), \mathbf{F_{U'}})$, $s \in \Gamma_G(\phi^{-1}(U), \mathbf{F_U})$. In other words, the sections $\Gamma(X, \mathbf{F})$ (in this R-orbifold sense) are precisely the sections $\Gamma(X, \overline{\mathbf{F}})$ (in the topological sense) of the sheaf $\overline{\mathbf{F}}$ which is obtained from the presheaf of invariant sections in each chart.

The category of sheaves (of abelian groups) on the R-orbifold X is abelian and it has enough injectives. The cohomology $H^i(X, \mathbf{F})$ is defined to be the right derived functor $R^i\Gamma(X, \mathbf{F})$. It may be nonzero for infinitely many values of i.

If we start with a sheaf $\mathbf{F_0}$ of R-modules on X (in the topological sense) then it pulls up to a G-equivariant sheaf in each chart (U, M, G, ϕ) so it gives a sheaf $\mathbf{F}$ in the R-orbifold sense, on which each stabilizer group G_y acts trivially. Therefore the resulting sheaf $\overline{\mathbf{F}}$ on X coincides with the original sheaf $\mathbf{F_0}$. Therefore we obtain,

Proposition B.6. *Let $\mathbf{F_0}$ be a sheaf of R-modules on the topological space X. Let $\mathbf{F}$ denote the resulting sheaf in the R-orbifold sense. Then the isomorphism $\overline{\mathbf{F}} \cong \mathbf{F_0}$ induces an isomorphism $H^*(X, \mathbf{F_0}) \cong H^*(X, \mathbf{F})$ between the (singular) cohomology of $(X, \mathbf{F_0})$ and the cohomology in the sense of R-orbifolds of $(X, \mathbf{F})$.* □

B.6 Differential forms

For the remainder of this chapter, we take $R = \mathbb{Q}$ and we drop it from our notation. A *local system* $\mathbf{E}$ of real vector spaces on an orbifold X is a sheaf of real vector spaces such that each $\mathbf{E_U}$ is a local system. Let $\mathbf{E}$ be a local system of real vector spaces on the orbifold X. The sheaf of differential p-forms on the orbifold X is the sheaf which in each chart (U, M, G, ϕ) is given by the sheaf $\Omega_G^p(M, \mathbf{E})$. It is easy to check that this collection satisfies the required compatibility conditions for a sheaf on an orbifold. Thus, a differential p-form with coefficients in $\mathbf{E}$ on X is a choice, for each chart (U, M, G, ϕ), of a smooth differential form $\omega_U \in \Omega^p(M, \mathbf{E_U})$ which is G-invariant, such that $f^*(\omega_{U'}) = \omega_U$ whenever $f : (U, M, G, \phi) \to (U', M', G', \phi')$ is a morphism of charts. If ω, η are G-invariant differential forms then so is $d\omega$ and $\omega \wedge \eta$ so we obtain exterior differentials and products on this complex of differential forms.

The differential p forms $\Omega^p(X, \mathbf{E})$ in fact form a fine sheaf (in the orbifold sense), $\Omega^p(X, \mathbf{E})$ on X. It follows that the de Rham theorem holds:

Proposition B.7. *Let $\mathbf{E}$ be a local system of real vector spaces on the orbifold X. Then the cohomology of the complex of smooth differential forms $\Omega^\bullet(X, \mathbf{E})$ is canonically isomorphic to the cohomology $H^i(X, \mathbf{E})$.* □

According to the remarks in Section B.5 the differential p forms on the orbifold X are precisely the global sections (in the usual sense) of the (topological)

sheaf
$$\overline{\Omega}^p(X, \overline{\mathbf{E}}) \tag{B.6.1}$$
of invariant sections of $\Omega^p(X, \mathbf{E})$. Let us examine this sheaf in more detail.

Let X^0 denote the part of X over which the isotropy groups are trivial. Then the restriction $\overline{\Omega}^p(X, \overline{\mathbf{E}})|X^0$ is canonically isomorphic to the usual sheaf of differential forms $\Omega^p(X^0, \mathbf{E})$. Thus, a differential form ω on the orbifold X is simply a smooth differential form on X^0 which, near the singular points of X, satisfies an equivariance condition:

(EQ) if (U, M, G, ϕ) is a chart, then the pullback of ω to M^0 extends (uniquely) to a smooth, G-invariant differential form on M.

Since the sheaf (B.6.1) is fine, we conclude, in analogy with Proposition B.5 that the cohomology of the complex of differential forms is naturally isomorphic to $H^*(X, \overline{\mathbf{E}})$. In summary, we have:

Proposition B.8 ([Sat])**.** *Let $\mathbf{E}$ be a local system (in the topological sense) of real vector spaces on a locally compact Hausdorff space X. Assume that X is endowed with an orbifold structure. Let $\Omega^\bullet(X, \mathbf{E})$ be the complex of smooth differential forms on X^0 (with coefficients in $\mathbf{E}$) which satisfy the above equivariance condition (EQ) near the singularities of X. Denote by $H^*_{dR}(X, \mathbf{E})$ the cohomology of this complex. Then the inclusion of sheaves $\mathbf{E} \to \overline{\Omega}^\bullet(X, \mathbf{E})$ is a fine resolution of $\mathbf{E}$ and it induces an isomorphism $H^*_{dR}(X, \mathbf{E}) \cong H^*(X, \mathbf{E})$ between the de Rham cohomology and the singular cohomology, and an isomorphism $H^*_{dR,c}(X, \mathbf{E}) \cong H^*_c(X, \mathbf{E})$ between the compactly supported de Rham cohomology and the singular cohomology with compact supports.* $\square$

B.7 Groupoids

For completeness we mention that the theory of orbifolds can be developed in a more natural and global manner within the context of topological groupoids, see [MP], [Mo2], [Mo1], [Adem]. Recall that a groupoid is a category in which every morphism is invertible, and a *topological groupoid* is a groupoid object in the category of Hausdorff topological spaces. In practice this means the following: a topological groupoid is a collection of topological spaces and continuous mappings,

$$\begin{array}{c} X_1 \\ s \Big\uparrow u \Big\downarrow \Big\downarrow t \\ X_0 \longrightarrow X_{-1} = X \end{array}$$

with various properties. First, the pair (X_1, X_0) form a (topological) category such that every morphism is invertible, where the objects are the points of X_0, and the morphisms are the points in X_1. The maps s, t assign to each morphism its source and target respectively, so each element $f \in X_1$ may be thought of as an arrow $f : x \to y$ where $x, y \in X_0$. The mapping $u : X_0 \to X_1$ in the above diagram associates to each $x \in X_0$ the (two-sided) identity (or unit) morphism (so $su(x) = tu(x) = x$ for all $x \in X_0$). The composition of morphisms is a continuous mapping $m : X_1 \times_{X_0} X_1 \to X_1$ with its obvious relations to s, t, u and we usually write ff' rather than $m(f, f')$ for $f, f' \in X_1$ with $s(f) = t(f')$. Associativity of composition says that a certain diagram commutes. The (two-sided) inverse $\iota : X_1 \to X_1$ swaps s and t and acts as an inverse, that is, if $f : x \to y$ then $\iota(f)f = u(x)$ and $f\iota(f) = u(y)$. If $x \in X_0$ then the set

$$G_x = \{f \in X_1 |\ s(f) = t(f) = x\}$$

forms a group, the *isotropy group* of x. The image of the mapping $(s, t) : X_1 \to X_0 \times X_0$ is an equivalence relation and the quotient space of X_0 with respect to this equivalence relation is the "orbit space" X in the above diagram.

Orbifold structures on X correspond to *effective proper étale Lie* groupoids. The "proper étale" part means that s and t are proper local homeomorphisms. The "Lie" part means that the spaces X_1, X_0 are smooth manifolds, the mappings s, t, u, ι, m are smooth, and s and t are submersions. It follows that each isotropy group G_x is compact and discrete, hence finite.

Given a proper étale Lie groupoid X_1, X_0 with quotient space X we obtain an orbifold atlas on X as follows. It can be shown that each point $x \in X_0$ has a neighborhood U such that the isotropy group G_x acts on U and such that the restriction

$$U_1 = \{f \in X_1 |\ f : a \to b \text{ with } a, b \in U\} \rightrightarrows U$$

is isomorphic to the groupoid $G_x \times U \rightrightarrows U$. The groupoid is *effective* if each of these actions $G_x \times U \rightrightarrows U$ is effective. In this case, composing with the projection $U \to U/G_x$ gives a chart on X. The resulting covering admits a "good" refinement. We omit the tedious verification of the claim that restricting these charts to this refinement defines an atlas on X.

If a compact group G acts smoothly on a manifold M such that each stabilizer subgroup G_x is finite then we obtain a proper étale Lie groupoid with $X_1 = G \times M$ and $X_0 = M$ where $s(g, x) = x$ and $t(g, x) = g \cdot x$. The quotient space is $X = M/G$. If such an action exists locally (that is, if X has an orbifold atlas) then these local actions can be patched together to give a proper étale Lie groupoid over X in several different possible ways, [Hae], [MP], [Mo1].

Many operations with orbifolds (morphisms, sheaves, cohomology, derived categories) are most naturally formulated in the language of groupoids.

Appendix C
Basic Adèlic Facts

C.1 Adèles and idèles

Let L be an algebraic number field. A *place* of L is an equivalence class v of valuations on L, two being equivalent if they induce the same topology on the completion L_v. Normalizations for these absolute values are specified in Section C.3. Write $v|\infty$ or $v < \infty$ if v is an archimedean or non-archimedean valuation, respectively. Each non-archimedean place v corresponds to a prime ideal $\mathfrak{p}_v$ in $\mathcal{O}_L$. We also write $\Sigma(L)$ for the set of infinite places. For $v < \infty$ the *valuation ring* is $\mathcal{O}_v = \{x_v \in L_v : |x_v|_v \leq 1\}$.

The adèle ring $\mathbb{A}_L$ is the product $\mathbb{A}_L := \mathbb{A}_{Lf} \times \mathbb{A}_{L\infty}$ where $\mathbb{A}_{L\infty} = \prod_{v|\infty} L_v$ and $\mathbb{A}_{Lf}$ is the restricted direct product

$$\mathbb{A}_{Lf} := {\prod_{v<\infty}}' L_v = \Big\{ (x_v) \in \prod_{v<\infty} L_v : x_v \in \mathcal{O}_v \text{ for almost all } v \Big\}.$$

There is a unique topology on $\mathbb{A}_L$ such that for any finite set $S \supset \Sigma(L)$ of places, the induced topology on

$$\prod_{v \in S} L_v \times \prod_{v \notin S} \mathcal{O}_v$$

agrees with the product topology. We usually identify the field L with its image under the diagonal embedding $L \to \mathbb{A}_L$. Its image is a discrete subgroup of $\mathbb{A}_L$ and the quotient $L \backslash \mathbb{A}_L$ is compact. For $x \in \mathbb{A}_L = \mathbb{A}_{Lf} \times \mathbb{A}_{L\infty}$ we write $x = x_0 x_\infty$ or $x = x_f x_\infty$. If $v < \infty$ corresponds to the prime ideal $\mathfrak{p}$ we sometimes write $x_\mathfrak{p}$ rather than x_v for the v-component of an adèle x. We also write $\widehat{\mathcal{O}}_L = \prod_{v<\infty} \mathcal{O}_v$.

The idèle group $\mathbb{A}_L^\times$ is the group of invertible adèles. It consists of elements (x_v) such that $x_v \neq 0$ for all places v, and such that $x_v \in \mathcal{O}_v^\times$ for almost all places $v < \infty$. There is a unique topology on $\mathbb{A}_L^\times$ such that for any finite set $S \supset \Sigma(L)$ the induced topology on

$$\prod_{v \in S} L_v^\times \times \prod_{v \notin S} \mathcal{O}_v^\times$$

is the product topology. We write $\widehat{\mathcal{O}}_L^\times = \prod_{v<\infty} \mathcal{O}_v^\times$.

As in Section 1.2, any finite idèle $x_0 \in \mathbb{A}_{Lf}^\times$ gives rise to a fractional ideal,

$$[x_0] = \prod_{\mathfrak{p}} \mathfrak{p}^{\mathrm{ord}_{\mathfrak{p}}(x_{\mathfrak{p}})}$$

where the product is over all prime ideals. Then $[x_0 y_0] = [x_0]$ for any $y_0 \in \widehat{\mathcal{O}}_L^\times = \prod_{v<\infty} \mathcal{O}_v^\times$. The ideal $[x_0]$ is integral iff $x_0 \in \widehat{\mathcal{O}}_L = \prod_{v<\infty} \mathcal{O}_v$.

The *inverse different* and the *local inverse different* are the fractional ideals

$$\mathcal{D}_{L/\mathbb{Q}}^{-1} = \{x \in L : \mathrm{Tr}_{L/\mathbb{Q}}(x\mathcal{O}_L) \subset \mathbb{Z}\}$$
$$\mathcal{D}_v^{-1} = \{x_v \in L_v : \mathrm{Tr}_{L_v/\mathbb{Q}_p}(x_v \mathcal{O}_v) \subset \mathbb{Z}_p\}$$

where $v|p$ is a finite place of L. Then $\mathcal{D}_{L/\mathbb{Q}}^{-1} = L \cap \prod_{v<\infty} \mathcal{D}_v^{-1}$. The norm of the different $\mathcal{D}_{L/\mathbb{Q}}$ is the discriminant $d_{L/\mathbb{Q}}$.

Let $\widetilde{\mathcal{D}}_{L/\mathbb{Q}} \in \mathbb{A}_{Lf}^\times$ be any finite idèle such that $[\widetilde{\mathcal{D}}_{L/\mathbb{Q}}^{-1}] = \mathcal{D}_{L/\mathbb{Q}}^{-1}$. It follows immediately that (see the definition of e_{Lf} below):

Lemma C.1. *For any finite idèle $y \in \mathbb{A}_L^\times$, the ideal $[y\widetilde{\mathcal{D}}_{L/\mathbb{Q}}]$ is integral iff $e_{Lf}(yz) = 1$ for all $z \in \widehat{\mathcal{O}}_L$.* □

C.2 Characters of $L\backslash\mathbb{A}_L$

Let L be a number field with the discrete topology. Its Pontrjagin dual $L^\vee$ is the compact abelian group $L\backslash\mathbb{A}_L$, as we will now explain. If v is a place of L lying over a finite prime $p \in \mathbb{Q}$, define the *standard additive character* $e_v : L_v \to \mathbb{C}^\times$ to be the composition

$$L_v \xrightarrow{Tr} \mathbb{Q}_p \longrightarrow \mathbb{Q}_p/\mathbb{Z}_p \longrightarrow \mathbb{Q}/\mathbb{Z} \xrightarrow{e} \mathbb{C}^\times$$

where $e(x) = \exp(-2\pi i x)$, so that

$$e_v(z_v) = \exp(-2\pi i Tr(z_v))$$

for any $z_v \in L_v$. If v is an archimedian place of L set $e_v(z_v) = \exp(2\pi i z_v) \in \mathbb{C}^\times$ for any $z_v \in L_v$. Together, these functions define a continuous additive character

$$e_L : \mathbb{A}_L \to \mathbb{C}^\times \quad \text{by} \quad e_L(z) = \prod_v e_v(z_v)$$

which clearly decomposes as the product $e_{Lf}(z_0)e_{L\infty}(z_\infty)$. The character e_L is trivial on $L \subset \mathbb{A}_L$, and every other continuous additive character of $L\backslash\mathbb{A}_L$ is of the form $z \mapsto e_L(az)$ for some $a \in L$. Thus we obtain a mapping

$$L \times (L\backslash\mathbb{A}_L) \to U(1) \subset \mathbb{C}^\times$$

given by $(a,z) \mapsto e_L(az)$, which identifies $L^\vee \cong L\backslash\mathbb{A}_L$ and $(L\backslash\mathbb{A}_L)^\vee \cong L$.

C.3. Characters of $\mathrm{GL}_1(L)\backslash\mathrm{GL}_1(\mathbb{A}_L)$

Let us normalize the Haar measure (see Section C.4) on $L\backslash\mathbb{A}_L$ so as to have total volume 1. The Fourier transform of a continuous function $h : L\backslash\mathbb{A}_L \to \mathbb{C}^\times$ is the function $\widehat{h} : (L\backslash\mathbb{A}_L)^\vee \to \mathbb{C}$ defined by

$$\widehat{h}(\tau) = \int_{L\backslash\mathbb{A}_L} h(z)\bar\tau(z)dz$$

which can therefore be interpreted as the function $\widehat{h} : L \to \mathbb{C}$ given by

$$\widehat{h}(w) = \int_{L\backslash\mathbb{A}_L} h(z)\bar e_L(wz)dz.$$

Consequently the Fourier inversion formula gives

$$h(z) = \sum_{w \in L} \widehat{h}(w)e_L(wz). \qquad (\text{C.2.1})$$

C.3 Characters of $\mathrm{GL}_1(L)\backslash\mathrm{GL}_1(\mathbb{A}_L)$

The *normalized absolute value* $|\cdot|_\mathbb{A} = |\cdot|_{\mathbb{A}_L} : \mathbb{A}_L \to \mathbb{R}_{\geq 0}$ is the product $|x|_\mathbb{A} = \prod_v |x_v|_v$ where

$$|x|_v = \begin{cases} |x| & \text{if } L_v = \mathbb{R} \\ x\bar x & \text{if } L_v = \mathbb{C} \end{cases}$$

if v is an infinite place, and where $|\pi|_v = 1/q$ if v is a finite place with residue field having q elements, and π is a uniformizing parameter for L_v. Then $|xy|_\mathbb{A} = |x|_\mathbb{A}|y|_\mathbb{A}$. Artin's product theorem says that $|x|_\mathbb{A} = 1$ for all $x \in L^\times$. If $x \in \mathbb{A}_L$ then multiplication by x induces a mapping $\cdot x : \mathbb{A}_L \to \mathbb{A}_L$ which takes the Haar measure μ to the Haar measure $|x|_\mathbb{A}\mu$, cf. Section C.4.

The norm map $\mathrm{N}_{L/\mathbb{Q}} : L \to \mathbb{Q}$ extends to a map $\mathrm{N}_{L/\mathbb{Q}} : \mathbb{A}_L \to \mathbb{A}_\mathbb{Q}$ by

$$(x_v)_v \mapsto \left(\prod_{v|u} \mathrm{N}_u(x_v)\right)_u$$

Then $|x|_{\mathbb{A}_L} = |\mathrm{N}_{L/\mathbb{Q}}(x)|_{\mathbb{A}_\mathbb{Q}}$ for all $x \in \mathbb{A}_L$.

The *ideal class group* of L is the (finite) group of equivalence classes of fractional ideals, under the equivalence relation $\mathfrak{m} \sim \lambda\mathfrak{m}$ for any $\lambda \in L^\times$. The *narrow class group*, which is defined when L is totally real, consists of the equivalence classes of fractional ideals under the same relation, but with the restriction that $\lambda \gg 0$. The *idèle class group*, which is not compact, is the quotient of the idèles $\mathbb{A}_L^\times = \mathrm{GL}_1(\mathbb{A}_L)$ by the principal idèles $L^\times = \mathrm{GL}_1(L)$. The mapping which assigns to any finite idèle $x \in \mathbb{A}_{L,f}^\times$ the corresponding fractional ideal $[x]$ is trivial on $\widehat{\mathcal{O}}_L^\times = \prod_{v<\infty} \mathcal{O}_v^\times$. It takes $L^\times$ to the principal ideals and it induces a bijection

$$L^\times\backslash\mathbb{A}_L^\times/\widehat{\mathcal{O}}_L^\times\mathbb{A}_{L,\infty}^\times \longrightarrow \text{ideal class group}$$

where $\mathbb{A}_{L,\infty}^{\times} = \prod_{v \in \Sigma(L)} L_v^{\times} = (\mathbb{R}^{\times})^{\Sigma(L)}$. Similarly we obtain a bijection

$$L^{\times} \backslash \mathbb{A}_L^{\times} / \widehat{\mathcal{O}}_L^{\times} \mathbb{A}_{L,\infty}^{+} \longrightarrow \text{narrow class group} \tag{C.3.1}$$

where $\mathbb{A}_{L,\infty}^{+} = \prod_{v \in \Sigma(L)} L_v^{>0} = (\mathbb{R}^{>0})^{\Sigma(L)}$.

A *Hecke character* or *quasicharacter* is a continuous homomorphism

$$\phi : L^{\times} \backslash \mathbb{A}_L^{\times} = \mathrm{GL}_1(L) \backslash \mathrm{GL}_1(\mathbb{A}_L) \to \mathbb{C}^{\times}.$$

It can be expressed as a product $\phi = \prod_v \phi_v$ of continuous local characters $\phi_v : L_v^{\times} \to \mathbb{C}^{\times}$. For any $v < \infty$ the restriction $\phi_v | \mathcal{O}_v^{\times}$ takes values in the unit circle, and χ_v is *unramified* if $\phi_v | \mathcal{O}_v^{\times}$ is trivial, in which case there exists a complex number $s \in \mathbb{C}$ such that $\phi_v(x_v) = |x_v|_v^s$ for all $x_v \in L_v^{\times}$. If ϕ_v is ramified then its *conductor* is $\mathfrak{f}(\phi_v) = \mathfrak{p}_v^m$ where $\mathfrak{p}_v$ is the prime ideal corresponding to v, and where $m > 0$ is the smallest integer such that $\phi_v | (1 + \mathfrak{p}_v^m)$ is trivial. In this case there exists a unitary character $\psi_v : L_v^{\times} \to \mathbb{C}^{\times}$ and $s \in \mathbb{C}$ so that $\phi_v(x_v) = \psi_v(x_v) |x_v|_v^s$ for all $x_v \in L_v^{\times}$.

Let $A_{\mathrm{GL}_1} \subseteq \mathrm{GL}_1(\mathbb{A}_L)$ be the connected component of the identity in the maximal $\mathbb{Q}$-split subtorus of $R_{L/\mathbb{Q}} \mathrm{GL}_1$; thus $A_{\mathrm{GL}_1} \cong \mathbb{R}^{>0}$, and

$$\mathrm{GL}_1(\mathbb{A}_F) = A_{\mathrm{GL}_1}{}^0 \mathrm{GL}_1(\mathbb{A}_F),$$

where ${}^0 \mathrm{GL}_1(\mathbb{A}_F)$ is the kernel of $|\cdot|_{\mathbb{A}_L}$ (this is essentially the construction of Section 4.1 in a special case). For any Hecke character ϕ there exists a unique $s \in \mathbb{C}$ such that, for all $x \in \mathbb{A}_L$,

$$\phi(x) = \psi(x) |x|_{\mathbb{A}}^s$$

where $\psi = \otimes_v \psi_v$ is a (unitary) character trivial on A_{GL_1}. The continuity of ϕ implies that ϕ_v is unramified, i.e., trivial on $\mathcal{O}_v^{\times}$, for almost all v. If s is real then the complex conjugate $\overline{\phi}$ satisfies

$$\overline{\phi}(x) = \phi^{-1}(x) |x|^{2s} \quad \text{and} \quad |\phi(x)| = |x|_{\mathbb{A}}^s. \tag{C.3.2}$$

For the Hecke characters in this book the value of s is real (in fact it is a rational integer).

If $\psi = \prod_{v<\infty} \psi_v$ is a Hecke character of finite order, its *conductor*

$$\mathfrak{f} = \prod_{v<\infty} \mathfrak{f}(\psi_v) = \prod_{v<\infty} \mathfrak{p}_v^{n_v} \subset \mathcal{O}_L$$

is the largest integral ideal such that

$$\begin{cases} \psi_v | (1 + \varpi_v^{n_v} \mathcal{O}_v) = 1 & \text{if } \mathfrak{p}_v | \mathfrak{f} \\ \psi_v | \mathcal{O}_v^{\times} = 1 & \text{if } \mathfrak{p}_v \nmid \mathfrak{f} \end{cases}$$

where ϖ_v is a uniformizing parameter for $\mathcal{O}_v$. Consequently ψ defines a one-dimensional representation of

$$\prod_{\mathfrak{p}_v | \mathfrak{f}} \mathcal{O}_v^{\times} / (1 + \varpi_v^{n_v} \mathcal{O}_v) = \prod_{\mathfrak{p}_v | \mathfrak{f}} (\mathcal{O}_v / \varpi_v^{n_v} \mathcal{O}_v)^{\times}.$$

Moreover, if $\mathfrak{m}$ is a fractional ideal relatively prime to the conductor $\mathfrak{f}(\psi)$ of ψ then $\psi(m_0) = \psi(m_0')$ whenever $m_0, m_0' \in \mathbb{A}_L^\times$ are finite idèles such that $[m_0] = [m_0'] = \mathfrak{m}$, which is to say that ψ determines a character on the group of fractional ideals relatively prime to $\mathfrak{f}(\psi)$.

Let L^{ab} be the maximal abelian extension of L. According to class field theory, the Artin map

$$[\cdot, L] : \mathrm{GL}_1(L) \backslash \mathrm{GL}_1(\mathbb{A}_L) \to \mathrm{Gal}(L^{ab}/L)$$

is surjective. A Hecke character has finite order if and only if it factors through the Artin map, which induces a one to one correspondence between Hecke characters of finite order and characters of finite order of $\mathrm{Gal}(L^{ab}/L)$. For $L = \mathbb{Q}$ there is a (multiplicative) group isomorphism

$$\mathrm{GL}_1(\mathbb{Q}) \backslash \mathrm{GL}_1(\mathbb{A}_\mathbb{Q}) \cong \mathbb{R}_+ \times \widehat{\mathbb{Z}}^\times$$

and every Hecke character ψ of finite order factors through the projection to $(\mathbb{Z}/N\mathbb{Z})^\times$ for some N. The conductor of ψ is the smallest positive such N.

If $\phi = \prod_v \phi_v$ is a Hecke character and $\mathfrak{c} \subset \mathcal{O}_L$ is an (integral) ideal we define the *restricted* Hecke character $\phi_\mathfrak{c} = \prod_{\mathfrak{p}_v | \mathfrak{c}} \phi_v$. In particular, if $x \in \mathbb{A}_L^\times$ then $|x|_\mathfrak{c} = \prod_{\mathfrak{p}_v | \mathfrak{c}} |x_v|_v$.

C.4 Haar measure on the adèles

References for this section include [KnL, Sections 5, 6, and 7], [RamV, Section 5]. Recall that a locally compact topological group G admits a left invariant (Haar) measure μ_G which is uniquely determined up to a multiplicative constant. The *modulus function* $\Delta : G \to \mathbb{R}$ is the unique function such that $\mu_G(Ug) = \Delta(g)\mu_G(U)$ for any open set $U \subset G$ and any $g \in G$. It is independent of the choice of (left) Haar measure μ_G. A left Haar measure is also right invariant iff $\Delta \equiv 1$, in which case the topological group G is said to be *unimodular*. Abelian groups, compact groups, and reductive groups are unimodular. In particular, if F is a locally compact field then the group $\mathrm{GL}(n, F)$ is unimodular. A left Haar measure for $\mathrm{GL}(n, \mathbb{R})$ is given by

$$\mu_G(U) = \int_U |\det(X)|^{-n} dX = \int_U |\det(X)|^{-n} \prod_{i,j} dX_{ij} \qquad (\text{C.4.1})$$

where dX denotes Lebesgue measure on $\mathbb{R}^{n^2}$, which is identified with the set of $n \times n$ matrices.

Let G be a locally compact topological group with left Haar measure μ_G. Let $K \subset G$ be a closed subgroup with left Haar measure μ_K. Then there exists a G-invariant measure dx on $D = G/K$ if and only if $\Delta_G | K = \Delta_K$ and in this case, the measure dx may be normalized so that

$$\int_G f d\mu_G = \int_D \int_K f(gk) d\mu_K(k) dx(gK).$$

for any compactly supported continuous function $f : G \to \mathbb{R}$. If G is a real Lie group and K is a closed Lie subgroup then $D = G/K$ is a smooth manifold. If it is orientable then a G-invariant measure on D is the same thing as a G-invariant differential form dx of top degree on D.

Now let $\Gamma \subset G$ be a countable discrete subgroup which acts freely on D. Then counting measure on Γ determines a measure dy on the quotient $Y = \Gamma \backslash D$ which coincides with the restriction of the measure dx to a fundamental domain $\mathcal{F}$ of Γ in D. For any compactly supported continuous function $f : D \to \mathbb{R}$ we have:

$$\int_D f(x)dx = \int_\mathcal{F} \sum_{\gamma \in \Gamma} f(\gamma x) dx = \int_Y \sum_{\gamma \in \Gamma} f(\gamma y) dy.$$

Moreover, the same holds under the weaker assumption that the action is *almost free*, that is, for all $x \in D$ the isotropy group Γ_x is finite, and the set of points $x \in D$ for which $\Gamma_x \neq \{1\}$ has measure zero.

Standard normalizations for Haar measures on adèlic groups are defined as follows. If L is an algebraic number field and v is a finite place of L, there is a unique Haar measure dx_v on L_v such that $\int_{\mathcal{O}_v} dx_v = 1$ where $\mathcal{O}_v$ is the valuation ring of L_v. If v is an infinite place then dx_v denotes Lebesgue measure (on $\mathbb{R}$ or $\mathbb{C}$). Let S be a finite set of places of L that includes all the infinite places. Then the product measure $\prod_v dx_v$ is well defined on the open set

$$\mathbb{A}_S := \prod_{v \in S} L_v \times \prod_{v \notin S} \mathcal{O}_v$$

and there exists a unique Haar measure on $\mathbb{A}_L$ whose restriction to each such $\mathbb{A}_S$ is the above product measure. Similarly, if $v < \infty$ there is a unique Haar measure $d^\times x_v$ on the multiplicative group $L_v^\times$ such that $\int_{\mathcal{O}_v^\times} d^\times x_v = 1$ and in fact

$$d^\times x_v = \frac{q}{q-1} |x|_v^{-1} dx_v$$

where q denotes the order of the residue field. If $v|\infty$ then $d^\times x_v := |x|_v^{-1} dx_v$. Then there exists a unique Haar measure $d_\mathbb{A}^\times x$ on $\mathbb{A}_L^\times$ such that, for every finite set S of places which contains all the infinite places, the restriction of $d_\mathbb{A}^\times x$ to the open set

$$\mathbb{A}_S^\times := \prod_{v \in S} L_v^\times \times \prod_{v \notin S} \mathcal{O}_v^\times$$

is the product measure $\prod_v d^\times x_v$.

Similarly, for $v < \infty$ there is a unique Haar measure dx_v on $\mathrm{GL}_2(L_v)$ normalized so that $\mathrm{GL}_2(\mathcal{O}_v)$ has volume 1. For $v|\infty$ a normalized Haar measure is given by (C.4.1). Then there exists a unique Haar measure $dx_\mathbb{A}$ on $\mathrm{GL}_2(\mathbb{A}_L)$ such that for every finite set S of places, containing all the infinite places, the restriction of $dx_\mathbb{A}$ to the open set

$$\prod_{v \in S} \mathrm{GL}_2(L_v) \times \prod_{v \notin S} \mathrm{GL}_2(\mathcal{O}_v)$$

is the product measure $\prod_v dx_v$.

Appendix D

Fourier Expansions of Hilbert Modular Forms

In this appendix we prove Theorem 5.8, which provides explicit Fourier expansions for Hilbert modular forms. Our approach is based on the theory of Whittaker models. General references for this section include [Ga, Appendix A.2], [Gel, Sect. 3], and [RamV, Chapter 5].

D.1 Statement of the theorem

In this paragraph we give a restatement of Theorem 5.8. Let L be a totally real number field. Let $h \in M_\kappa(K_0(\mathfrak{c}), \chi)$ be a modular form (for notation see Section 5.4). Thus, $\kappa = (k, m) \in \mathcal{X}(L)$ is a weight, $\mathfrak{c} \subset \mathcal{O}_L$ is an ideal, and $\chi : L^\times \backslash \mathbb{A}_L^\times \to \mathbb{C}^\times$ is a (continuous) quasicharacter such that $\chi_\infty(b_\infty) = b_\infty^{-k-2m}$ for all $b \in \mathbb{A}_L^\times$. Consequently $\chi | \cdot |^{k+2m}$ is a (unitary) character, i.e., its image is contained in $U(1) \subseteq \mathbb{C}^\times$.

Theorem 5.8 describes the Fourier expansion of $h\left(\left(\begin{smallmatrix} y & x \\ 0 & 1 \end{smallmatrix}\right)\right)$. Since this is a function of two variables one might expect a double Fourier series. Instead, we take a Fourier expansion with respect to the x variable, with Fourier coefficients that are functions of $y = y_f y_\infty$. It turns out that the dependence on y_∞ is completely determined, and that the dependence on y_f is very coarse. As in Section 5.9, corresponding to the weight $\kappa = (k, m)$, for each $\sigma \in \Sigma(L)$, define

$$W_{m_\sigma} : \mathbb{R}^\times \longrightarrow \mathbb{C} \quad \text{by} \quad W_m(y) = |y|^{-m_\sigma} e^{-2\pi|y|} \tag{D.1.1}$$

For $x \in \mathbb{A}_L$ and $y \in \mathbb{A}_L^\times$, set

$$q_\kappa(x, y) = q_\kappa(x, y_\infty) := e_L(x) \prod_{\sigma \in \Sigma(L)} W_{m_\sigma}(y_\sigma). \tag{D.1.2}$$

Theorem 5.8. Let $h \in M_\kappa(K_0(\mathfrak{c}), \chi)$ be a Hilbert modular form. Then h admits a Fourier series,

$$h\left(\begin{pmatrix} y & x \\ 0 & 1 \end{pmatrix}\right) = |y|_{\mathbb{A}_L}\left(c(y) + \sum_{\substack{\xi \in L^\times \\ \xi \gg 0}} b(\xi y_f) q_\kappa(\xi x, \xi y)\right), \quad (D.1.3)$$

valid for all $x \in \mathbb{A}_L$ and all $y \in \mathbb{A}_L^\times$. Moreover, each coefficient $b(\xi y_f) \in \mathbb{C}$ depends only on the fractional ideal $[\xi y_f]\mathcal{D}_{L/\mathbb{Q}} \in \mathcal{I}_L$ (where $\mathcal{D}_{L/\mathbb{Q}}$ is the different), and $b(\xi y_f)$ vanishes unless this ideal is integral.

Addendum. The constant term $c(y)$ vanishes if h is a cusp form or if $k \notin \mathbb{Z}\mathbf{1}$ or if the ideal $[y_f]\mathcal{D}_{L/\mathbb{Q}}$ is not integral. Otherwise it is a sum,

$$c(y) = c_0(y_f)|y_\infty^{-m}| + c_1(y_f)|y_\infty^{-k-1-m}| \quad (D.1.4)$$

of two terms. Here, $c_0(y_f)$ and $c_1(y_f)$ only depend on the fractional ideal $[y_f]\mathcal{D}_{L/\mathbb{Q}}$. If the functions $F_i(z)$ of (5.4.4) on $\mathfrak{h}^{\Sigma(L)}$ (corresponding to h) are holomorphic, then $c_1(\cdot) = 0$.

D.2 Fourier analysis on $\mathrm{GL}_2(L)\backslash\mathrm{GL}_2(\mathbb{A}_L)$

Fix a Hilbert modular form $h \in M_\kappa(K_0(\mathfrak{c}), \chi)$. The goal of Theorem 5.8 is the computation of $h\left(\begin{pmatrix} y & x \\ 0 & 1 \end{pmatrix}\right)$. For this purpose, fix $g \in \mathrm{GL}_2(\mathbb{A}_L)$ and consider the function defined on $L\backslash\mathbb{A}_L$,

$$\phi(x) = \phi_{h,g}(x) = h\left(\begin{pmatrix} 1 & x \\ 0 & 1 \end{pmatrix} g\right).$$

By equation (C.2.1), it has a Fourier expansion,

$$\phi(x) = \sum_{\xi \in L} \widehat{\phi}(\xi) e_L(\xi x) \quad \text{where} \quad \widehat{\phi}(\xi) = \int_{L\backslash\mathbb{A}_L} h\left(\begin{pmatrix} 1 & z \\ 0 & 1 \end{pmatrix} g\right) \bar{e}_L(\xi z) dz.$$

We wish to describe the Fourier coefficient $\widehat{\phi}(\xi)$ when $\xi \neq 0$. Using the fact that h is left $\mathrm{GL}_2(L)$-invariant and setting $w = \xi z$ and $dw = |\xi|_\mathbb{A} dz = dz$ gives, for $\xi \in L$,

$$\widehat{\phi}(\xi) = \int_{L\backslash\mathbb{A}_L} h\left(\begin{pmatrix} \xi & 0 \\ 0 & 1 \end{pmatrix}\begin{pmatrix} 1 & z \\ 0 & 1 \end{pmatrix} g\right) \bar{e}_L(\xi z) dz = \int_{L\backslash\mathbb{A}_L} h\left(\begin{pmatrix} \xi & w \\ 0 & 1 \end{pmatrix} g\right) \bar{e}_L(w) dw$$

$$= \int_{L\backslash\mathbb{A}_L} h\left(\begin{pmatrix} 1 & w \\ 0 & 1 \end{pmatrix}\begin{pmatrix} \xi & 0 \\ 0 & 1 \end{pmatrix} g\right) \bar{e}_L(w) dw = \widehat{\phi}_{h,g'}(1)$$

where $g' = \begin{pmatrix} \xi & 0 \\ 0 & 1 \end{pmatrix} g$. We may rephrase this by writing, for $\xi \neq 0$,

$$\widehat{\phi}(\xi) = W_h\left(\begin{pmatrix} \xi & 0 \\ 0 & 1 \end{pmatrix} g\right) \quad \text{and} \quad \widehat{\phi}(0) = W_{h0}(g)$$

where

$$W_h(g) := \int_{L\backslash\mathbb{A}_L} h\left(\left(\begin{smallmatrix} 1 & w \\ 0 & 1 \end{smallmatrix}\right)g\right)\bar{e}_L(w)dw \qquad \text{(D.2.1)}$$

$$W_{h0}(g) := \int_{L\backslash\mathbb{A}_L} h\left(\left(\begin{smallmatrix} 1 & w \\ 0 & 1 \end{smallmatrix}\right)g\right)dw. \qquad \text{(D.2.2)}$$

Then, as a function of $g \in \mathrm{GL}_2(\mathbb{A}_L)$, W_h is a *Whittaker function* (see below),

$$W_h\left(\left(\begin{smallmatrix} 1 & x \\ 0 & 1 \end{smallmatrix}\right)g\right) = e_L(x)W_h(g) \quad \text{and} \quad W_{h0}\left(\left(\begin{smallmatrix} 1 & x \\ 0 & 1 \end{smallmatrix}\right)g\right) = W_{h0}(g) \qquad \text{(D.2.3)}$$

for all $x \in \mathbb{A}_L$. If h is cuspidal then $W_{h0}(\cdot) = 0$. We also note that for all $b \in \mathbb{A}_L^\times$,

$$W_h\left(\left(\begin{smallmatrix} b & 0 \\ 0 & b \end{smallmatrix}\right)g\right) = \chi(b)W_h(g) \quad \text{and} \quad W_{h0}\left(\left(\begin{smallmatrix} b & 0 \\ 0 & b \end{smallmatrix}\right)g\right) = \chi(b)W_{h0}(g). \qquad \text{(D.2.4)}$$

Now let us specialize to the case that $g = \left(\begin{smallmatrix} y & 0 \\ 0 & 1 \end{smallmatrix}\right)$. The growth conditions on h imply that the Fourier coefficients $\widehat{\phi}(\xi)$ vanish unless $\xi = 0$ or $\xi \gg 0$. We obtain the expansion:

$$h\left(\begin{smallmatrix} y & x \\ 0 & 1 \end{smallmatrix}\right) = \sum_{\xi \in L} \widehat{\phi}(\xi)e_L(\xi x) = W_{h0}\left(\left(\begin{smallmatrix} y & 0 \\ 0 & 1 \end{smallmatrix}\right)\right) + \sum_{\xi \gg 0} W_h\left(\left(\begin{smallmatrix} \xi y & 0 \\ 0 & 1 \end{smallmatrix}\right)\right)e_L(\xi x) \qquad \text{(D.2.5)}$$

To proceed further we need to separate W_h into finite and archimedean factors.

D.3 Whittaker models

A *Whittaker function* is a smooth mapping $W : \mathrm{GL}_2(\mathbb{A}_L) \to \mathbb{C}$ such that

$$W\left(\left(\begin{smallmatrix} 1 & x \\ 0 & 1 \end{smallmatrix}\right)g\right) = e_L(x)W(g)$$

for all $x \in \mathbb{A}_L$ and all $g \in \mathrm{GL}_2(\mathbb{A}_L)$. Let $\mathcal{W}$ denote the space of all Whittaker functions. Let (π, V_π) be an automorphic representation of $\mathrm{GL}_2(\mathbb{A}_L)$ and let $K'_\infty \subseteq \mathrm{GL}_2(\mathbb{A}_{L\infty})$ be a maximal compact subgroup. A *Whittaker model* of (π, V_π) is a $(\mathfrak{g}, K'_\infty) \times \mathrm{GL}_2(\mathbb{A}_{Lf})$-submodule $\mathcal{W}_\pi \subset \mathcal{W}$ together with a $\mathrm{GL}_2(\mathbb{A}_L)$-equivariant isomorphism $V_\pi \to \mathcal{W}_\pi$. A *Whittaker functional* for (π, V_π) is a continuous linear mapping $\Lambda : V_\pi \to \mathbb{C}$ such that

$$\Lambda\left(\left(\begin{smallmatrix} 1 & x \\ 0 & 1 \end{smallmatrix}\right)\phi\right) = e_L(x)\phi$$

for all $\phi \in V_\pi$ and all $x \in \mathbb{A}_L$. Such a functional determines a $\mathrm{GL}_2(\mathbb{A}_L)$-intertwining operator $H : V_\pi \to \mathcal{W}$ by $H(\phi)(g) = \Lambda(\pi(g)(\phi))$ and hence it determines a Whittaker model, $\mathcal{W}_\pi = H(V_\pi)$. Conversely, a Whittaker model $H : V_\pi \to \mathcal{W}_\pi$ determines a Whittaker functional by $\Lambda(\phi) = H(\phi)(e)$ (where e is the identity in GL_2). A fundamental result of Gelfand, Kazhdan and Shalika states:

Theorem D.1. *If (π, V_π) is an automorphic representation of GL_2 then the space of Whittaker functionals on V_π is at most one-dimensional. Consequently, if π has a Whittaker model, then it has exactly one Whittaker model. The analogous statement also holds over local fields.*

D.4 Decomposition of W_h

Now let $h \in M_\kappa(K_0(\mathfrak{c}), \chi)$ be a modular form as in Section D.2 and let (π, V_π) be the $(\mathfrak{g}, K'_\infty) \times \mathrm{GL}_2(\mathbb{A}_{Lf})$-submodule of $\mathcal{A}(\mathrm{GL}_2, \chi)$ that is generated by h. Throughout this section we will assume that h is a simultaneous eigenform of $T_\mathfrak{c}(\mathfrak{p})$ for almost all prime ideals $\mathfrak{p} \subset \mathcal{O}_L$. In this case π is irreducible so it is an automorphic representation (see Section E.1). It has a natural Whittaker model

$$H : V_\pi \to \mathcal{W}_\pi \subset \mathcal{W} \tag{D.4.1}$$

which associates to any $\phi \in V_\pi$ the Whittaker function

$$W_\phi(g) := \int_{L \backslash \mathbb{A}_L} \phi\left(\left(\begin{smallmatrix} 1 & w \\ 0 & 1 \end{smallmatrix}\right) g\right) \bar{e}_L(w) dw.$$

(This is compatible with the previous notation for $W_h(g)$.) The intertwining operator H is in fact $(\mathfrak{g}, K^1_\infty) \times \mathrm{GL}_2(\mathbb{A}_{Lf})$-equivariant. Choose a factorization

$$(\pi, V_\pi) \cong \otimes'_v (\pi_v, V_{\pi,v}) = (\pi_\infty, V_{\pi,\infty}) \otimes (\pi_f, V_{\pi,f})$$

of π into a restricted tensor product of irreducible admissible representations π_v of $\mathrm{GL}_2(L_v)$, where v ranges over the places of L. The modular form h does not necessarily factor into a product of forms $h_\infty h_f$ but the following holds.

Lemma D.2. *If $h \in M_\kappa(K_0(\mathfrak{c}), \chi)$ is a simultaneous eigenform for almost all Hecke operators, then its image $\phi \in V_{\pi,\infty} \otimes V_{\pi,f}$ decomposes,*

$$\phi = \phi_\infty \otimes \phi_f \quad \text{and} \quad W_h = W_{\phi_\infty} W_{\phi_f}$$

where $\phi_\infty \in V_{\pi,\infty}$, where $\phi_f \in V_{\pi,f}$, and where

$$W_{\phi_\infty} : \mathrm{GL}_2(\mathbb{A}_{L\infty}) \longrightarrow \mathbb{C}$$
$$W_{\phi_f} : \mathrm{GL}_2(\mathbb{A}_{Lf}) \longrightarrow \mathbb{C}.$$

are Whittaker functions, meaning that they satisfy the obvious local analog of (D.2.3) *above* [Bu, Theorem 3.5.4], *viz.*

$$W_{\phi_\infty}\left(\left(\begin{smallmatrix} 1 & x \\ 0 & 1 \end{smallmatrix}\right) g\right) = e_\infty(x) W_{\phi_\infty}(g) \tag{D.4.2}$$
$$W_{\phi_f}\left(\left(\begin{smallmatrix} 1 & x \\ 0 & 1 \end{smallmatrix}\right) g\right) = e_f(x) W_{\phi_f}(g). \tag{D.4.3}$$

Proof. The Whittaker model $\mathcal{W}_\pi$ decomposes into a restricted tensor product of local Whittaker models, $\otimes'_v \mathcal{W}_{\pi,v}$. By the global and local uniqueness of Whittaker models, the intertwining operator H preserves these decompositions. Consequently a decomposition $\phi = \phi_\infty \phi_f$ with $\phi_\infty \in V_{\pi,\infty}$ and $\phi_f \in V_{\pi,f}$ will give rise to a decomposition $W_h = W_{\phi_\infty} W_{\phi_f}$ into Whittaker functions.

So it suffices to show that the image $\phi \in V_{\pi,\infty} \otimes V_{\pi,f}$ of h decomposes. Express ϕ as a finite sum,

$$\phi = \sum_i \phi_{i\infty} \otimes \phi_{if} \qquad (D.4.4)$$

where $\phi_{i\infty} \in V_{\pi,\infty}$ and $\phi_{if} \in V_{\pi,f}$. For each i we claim that $\phi_{i\infty} = c_i \phi_{1\infty}$ for some $c_i \in \mathbb{C}^\times$. This suffices to prove the lemma.

Since $K_\infty^1 \cong SO_2(\mathbb{R})^{\Sigma(L)}$ acts on π_∞, we may decompose ϕ into isotypic components under this action. The group $SO_2(\mathbb{R})^{\Sigma(L)}$ is isomorphic to $(\mathbb{R}/\mathbb{Z})^{\Sigma(L)}$, and hence has Pontryagin dual $\mathbb{Z}^{\Sigma(L)}$. By refining the decomposition of ϕ if necessary, we may assume that for each i we have

$$\pi(u_\infty)\phi_{i\infty} = \prod_{\sigma \in \Sigma(L)} e_\sigma(t_{i\sigma}\theta_\sigma)\phi_{i\infty}$$

for some $(t_{i\sigma}) \in \mathbb{Z}^{\Sigma(L)}$, where

$$u_\infty := \left(\begin{pmatrix} \cos(2\pi\theta_\sigma) & \sin(2\pi\theta_\sigma) \\ -\sin(2\pi\theta_\sigma) & \cos(2\pi\theta_\sigma) \end{pmatrix}\right)_{\sigma \in \Sigma(L)} \in SO_2(\mathbb{R})^{\Sigma(L)}.$$

On the other hand, $\pi(u_\infty)\phi = \prod_{\sigma|\infty} e_\sigma(k_\sigma + 2)\phi$ by definition of $M_\kappa(K_0(\mathfrak{c}), \chi)$. By linear independence of characters, we conclude that $t_{i\sigma} = k_\sigma + 2$ for all i and $\sigma \in \Sigma(L)$.

On the other hand, by [Bu, Proposition 2.5.2, Theorem 2.5.4, Theorem 2.5.5], up to a multiplicative constant, there is a unique vector $\psi_\infty \in V_{\pi,\infty}$ such that

$$\pi(u_\infty)\psi_\infty = \prod_{\sigma|\infty} e_\sigma(k_\sigma + 2)\psi_\infty. \qquad (D.4.5)$$

For each i, it follows that $\phi_{i\infty}$ is some nonzero multiple of $\phi_{1\infty}$ (and as a consequence, we see that ϕ_∞ is some multiple of ψ_∞ and satisfies equation (D.4.5)). $\square$

D.5 Computing $W_{\phi\infty}$ and W_{h0}

So far we have shown that a simultaneous eigenform $h \in M_\kappa(K_0(\mathfrak{c}), \chi)$ (for almost all Hecke operators) has a Fourier expansion (D.2.5) and the resulting Fourier coefficients factor: $W_h = W_{\phi\infty}W_{\phi f}$. The "constant term" W_{h0} must be treated separately.

Proposition D.3. *The archimedean part $W_{\phi\infty}$ of the Whittaker function W_h satisfies the following equation, for any $y \in \mathbb{A}_{L\infty}$:*

$$W_{\phi\infty}\left(\begin{pmatrix} y & 0 \\ 0 & 1 \end{pmatrix}\right) = \prod_{\sigma \in \Sigma(L)} |y_\sigma| W_{m_\sigma}(y_\sigma) \qquad (D.5.1)$$

There exists functions $c_0, c_1 : \mathbb{A}_{Lf}^\times \to \mathbb{C}$ such that for all $y \in \mathbb{A}_L$, the constant term is the sum

$$W_{h0}\left(\begin{pmatrix} y & 0 \\ 0 & 1 \end{pmatrix}\right) = c_0(y_f)|y_\infty|^{1-m} + c_1(y_f)|y_\infty|^{-k-m}.$$

Proof. Consider the $(\mathfrak{g}, K_\infty^1)$-module generated by W_{ϕ_∞}; it is a realization of the admissible $(\mathfrak{g}, K_\infty^1)$-module π_∞. Since the map (D.4.1) is an intertwining map, (D.4.5) implies that

$$\pi(u_\infty) W_{\phi_\infty} = \left(\prod_{\sigma|\infty} e_\sigma(k_\sigma + 2)\right) W_{\phi_\infty} \tag{D.5.2}$$

for

$$u_\infty := \left(\begin{pmatrix} \cos(2\pi\theta_\sigma) & \sin(2\pi\theta_\sigma) \\ -\sin(2\pi\theta_\sigma) & \cos(2\pi\theta_\sigma) \end{pmatrix}\right)_{\sigma \in \Sigma(L)} \in \mathrm{SO}_2(\mathbb{R})^{\Sigma(L)}.$$

Since h and W_{ϕ_∞} both satisfy the invariance property (D.5.2), it follows that the constant term W_{h0} satisfies (D.5.2) as well:

$$\pi(u_\infty) W_{h0} = \left(\prod_{\sigma|\infty} e_\sigma(k_\sigma + 2)\right) W_{h0}. \tag{D.5.3}$$

Write $w(a) := W_{\phi_\infty}\left(\begin{pmatrix} a^{1/2} & 0 \\ 0 & a^{-1/2} \end{pmatrix}\right)$ and $w_0(a, g_f) := W_{h0}\left(\begin{pmatrix} a^{1/2} & 0 \\ 0 & a^{-1/2} \end{pmatrix} g_f\right)$, where $a \in \mathbb{R}^{\Sigma(L)}$ and $g_f \in \mathrm{GL}_2(\mathbb{A}_{Lf})$. The Iwasawa decomposition provides coordinates on $\mathrm{GL}_2(\mathbb{A}_{L\infty})$ and we can write

$$g_\infty = \begin{pmatrix} b & 0 \\ 0 & b \end{pmatrix} \begin{pmatrix} a^{1/2} & 0 \\ 0 & a^{-1/2} \end{pmatrix} \begin{pmatrix} 1 & x \\ 0 & 1 \end{pmatrix} \left(\begin{pmatrix} \cos(2\pi\theta_\sigma) & \sin(2\pi\theta_\sigma) \\ -\sin(2\pi\theta_\sigma) & \cos(2\pi\theta_\sigma) \end{pmatrix}\right)_{\sigma \in \Sigma(L)} \tag{D.5.4}$$

with $a, b \in (\mathbb{R}^\times)^{\Sigma(L)}$ and $\theta, x \in \mathbb{R}^{\Sigma(L)}$. Using (D.2.3), (D.2.4), (D.5.2) and (D.5.3), we obtain

$$W_{\phi_\infty}(g_\infty) := w(a) b^{-k-2m} e_\infty(x) \prod_{\sigma \in \Sigma(L)} e_\sigma((k_\sigma + 2)\theta_\sigma) \tag{D.5.5}$$

$$W_{h0}(g_\infty g_f) = w_0(a, g_f) b^{-k-2m} \prod_{\sigma \in \Sigma(L)} e_\sigma((k_\sigma + 2)\theta_\sigma).$$

With respect to these coordinates, the Casimir operator is given by

$$C_\sigma = 2a_\sigma^2 \left(\frac{\partial^2}{\partial x_\sigma^2} + \frac{\partial^2}{\partial a_\sigma^2}\right) - \frac{2a_\sigma}{2\pi} \frac{\partial^2}{\partial x_\sigma \partial \theta_\sigma}$$

(for every $\sigma \in \Sigma(L)$), where we write g_∞ as in (D.5.4). Here we are using [Bu, Chapter 2, (1.29)], keeping in mind that $C_\sigma = -2\Delta$ in the notation of loc. cit. In view of the intertwining morphism (D.4.1) and assumption (1) in the definition

D.5. Computing $W_{\phi\infty}$ and W_{h0}

of $M_\sigma(K_0(\mathfrak{c}), \chi)$, we have that the function w and $w_0(\cdot, g_f)$ satisfy the differential equations

$$w'' + \left(-4\pi^2 + \frac{2\pi(k_\sigma + 2)}{a_\sigma} - \frac{k_\sigma^2/2 + k_\sigma}{2a_\sigma^2}\right)w = 0 \tag{D.5.6}$$

$$w_0(\cdot, g_f)'' - \frac{k_\sigma^2/2 + k_\sigma}{2a_\sigma^2} w_0(\cdot, g_f) = 0 \tag{D.5.7}$$

as functions of the a_σ-variable. There are two linearly independent solutions to (D.5.6) which is essentially a renormalization of Whittaker's equation. One of them is given by $e^{-2\pi a_\sigma} a_\sigma^{(k_\sigma+2)/2}$. By general facts on solutions to Whittaker's equation the second solution has exponential growth as $a \to \infty$ (compare [Andr, Section 4.3, p. 196]). Therefore w cannot be given by this second solution due to the growth conditions we placed on h. In view of this and (D.5.5) we have

$$W_{\phi\infty}\left(\begin{pmatrix} y & 0 \\ 0 & 1 \end{pmatrix}\right) = y^{-(k+2m)/2} y^{(k+21)/2} \prod_{\sigma \in \Sigma(L)} e^{-2\pi|y_\sigma|} \tag{D.5.8}$$

Therefore, absorbing a nonzero constant into W_{ϕ_f} if necessary, we obtain

$$W_{\phi\infty}\left(\begin{pmatrix} y & 0 \\ 0 & 1 \end{pmatrix}\right) = \prod_{\sigma \in \Sigma(L)} |y_\sigma|^{1-m_\sigma} e^{-2\pi|y_\sigma|} = \prod_{\sigma \in \Sigma(L)} |y_\sigma| W_{m_\sigma}(y_\sigma)$$

where W_{m_σ} is defined as above. This proves (D.5.1).

We now consider $w_0(\cdot, g_f)$. Two linearly independent solutions of the differential equation (D.5.7) for $w_0(\cdot, g)$ are given by $a^{\frac{k}{2}+1}$ and $a^{-\frac{k}{2}}$. Using (D.2.4) (with $b = y^{1/2}$ and $\chi_\infty(b_\infty) = b_\infty^{-k-2m}$) together with (D.5.5), we conclude that

$$W_{h0}\left(\begin{pmatrix} y & 0 \\ 0 & 1 \end{pmatrix}\right) = c_0(y_f)|y_\infty|^{1-m} + c_1(y_f)|y_\infty|^{-k-m}$$

for some functions c_0, c_1 on $\mathbb{A}_{Lf}^\times$. Substituting these equations for W_{h0} and $W_{\phi\infty}$ into the Fourier expansion (D.2.5) gives the following expression for $h\left(\begin{pmatrix} y & x \\ 0 & 1 \end{pmatrix}\right)$:

$$c_0(y_f)|y_\infty|^{1-m} + c_1(y_f)|y_\infty|^{-k-m} + |y|_{\mathbb{A}_L} \sum_{\xi \gg 0} b(\xi y_f) q_\kappa(\xi x, \xi y_\infty) \tag{D.5.9}$$

(since $|\xi|_{\mathbb{A}_L} = 1$), where

$$b(\xi y_f) = W_{\phi_f}\left(\begin{pmatrix} \xi y_f & 0 \\ 0 & 1 \end{pmatrix}\right) \prod_{v < \infty} |\xi y_v|_v^{-1}.$$

We now show that $c_1(\cdot) = 0$ in the holomorphic case. Let $t_i(\mathfrak{c})$ be defined as in Section 5.2, and for $z \in \mathfrak{h}^{\Sigma(L)}$ define

$$H_i(z) := \det(\alpha_z)^{m-1} j(\alpha_z, z_0)^{k+21} h(t_i(\mathfrak{c})\alpha_z)$$

where $z_0 = \mathbf{i} = (\sqrt{-1}, \ldots, \sqrt{-1}) \in \mathfrak{h}^{\Sigma(L)}$ and $\alpha_z = \left(\begin{smallmatrix} y & x \\ 0 & 1 \end{smallmatrix}\right) \in GL_2(\mathbb{R})^0$ is chosen so that $\alpha_z z_0 = z = x_\infty + \mathbf{i} y_\infty$, where the implied action is via fractional linear transformations. As recalled in the remark after the definition of $S_\kappa(K_0(\mathfrak{c}), \chi)$ in Section 5.4, the definition of $M_\kappa(K_0(\mathfrak{c}), \chi)$ implies that $H_i(z)$ is independent of the choice of α_z. The Fourier expansion (D.5.9) gives a similar Fourier expansion of $H_i(z)$:

$$H_i(z) = c_0(t_i(\mathfrak{c})) + c_1(t_i(\mathfrak{c}))|y_\infty|^{k-1} + \sum_{\substack{\xi \in L^\times \\ \xi \gg 0}} b(\xi t_i(\mathfrak{c})) \prod_{\sigma \in \Sigma(L)} e^{2\pi i z_\sigma}.$$

If this is holomorphic and if $k \neq \mathbf{1}$ then $c_1 = 0$. If $k = \mathbf{1}$ then the c_1 term can be absorbed into the c_0 term. Hence, possibly at the expense of renormalizing c_0 in the case $k = \mathbf{1}$, we have $c_1 = 0$. $\square$

D.6 Final steps

In this section we complete the proof of Theorem 5.8 by proving that the coefficients $b(y), c_0(y_f), c_1(y_f)$ are well defined at the level of ideals. The invariance property (3) in the definition of $M_\kappa(K_0(\mathfrak{c}), \chi)$ implies that

$$h\left(\left(\begin{smallmatrix} y & x \\ 0 & 1 \end{smallmatrix}\right)\right) = h\left(\left(\begin{smallmatrix} y & x \\ 0 & 1 \end{smallmatrix}\right)\left(\begin{smallmatrix} 1 & z \\ 0 & 1 \end{smallmatrix}\right)\right) = h\left(\left(\begin{smallmatrix} y & yz+x \\ 0 & 1 \end{smallmatrix}\right)\right) \text{ for all } z \in \widehat{\mathcal{O}}_L \subset \mathbb{A}_{Lf} \quad (\text{D.6.1})$$

$$h\left(\left(\begin{smallmatrix} y & x \\ 0 & 1 \end{smallmatrix}\right)\right) = h\left(\left(\begin{smallmatrix} y & x \\ 0 & 1 \end{smallmatrix}\right)\left(\begin{smallmatrix} u & 0 \\ 0 & 1 \end{smallmatrix}\right)\right) = h\left(\left(\begin{smallmatrix} yu & x \\ 0 & 1 \end{smallmatrix}\right)\right) \text{ for all } u \in \widehat{\mathcal{O}}_L^\times \subset \mathbb{A}_{Lf}^\times \quad (\text{D.6.2})$$

In view of the fact that (D.4.1) is an intertwining morphism the relations (D.6.1) and (D.6.2) imply that

$$W_\phi\left(\left(\begin{smallmatrix} \xi y & 0 \\ 0 & 1 \end{smallmatrix}\right)\right) = W_\phi\left(\left(\begin{smallmatrix} \xi y & 0 \\ 0 & 1 \end{smallmatrix}\right)\right) e_L(\xi y z) \quad (\text{D.6.3})$$

$$W_\phi\left(\left(\begin{smallmatrix} \xi u y & 0 \\ 0 & 1 \end{smallmatrix}\right)\right) = W_\phi\left(\left(\begin{smallmatrix} \xi y & 0 \\ 0 & 1 \end{smallmatrix}\right)\right) \quad (\text{D.6.4})$$

for u and z as in (D.6.1) and (D.6.2), respectively. In view of the Fourier expansion (D.2.5), this implies that

$$W_{h0}\left(\left(\begin{smallmatrix} y & 0 \\ 0 & 1 \end{smallmatrix}\right)\right) = W_{h0}\left(\left(\begin{smallmatrix} y & 0 \\ 0 & 1 \end{smallmatrix}\right)\right) e_L(yz) \quad (\text{D.6.5})$$

$$W_{h0}\left(\left(\begin{smallmatrix} uy & 0 \\ 0 & 1 \end{smallmatrix}\right)\right) = W_{h0}\left(\left(\begin{smallmatrix} y & 0 \\ 0 & 1 \end{smallmatrix}\right)\right). \quad (\text{D.6.6})$$

Equation (D.6.4) (resp. equation (D.6.6)) implies that the coefficient $b(\xi y_f)$, (resp. the coefficients $c_0(y_f)$ and $c_1(y_f)$) depends only on the fractional ideal $[\xi y_f]$ (resp. $[y_f]$) or equivalently, on the fractional ideal $[\xi y_f]\mathcal{D}_{L/\mathbb{Q}}$ (resp. $[y_f]\mathcal{D}_{L/\mathbb{Q}}$). Equation (D.6.3) (resp. (D.6.5)) plus Lemma C.1 implies that the coefficient $b(\xi y_f) = 0$ (resp. $c_0(y_f) = c_1(y_f) = 0$) unless these ideals are integral. We repeat that $k \notin \mathbb{Z}\mathbf{1}$ implies that $M_\kappa(K_0(\mathfrak{c}), \chi) = S_\kappa(K_0(\mathfrak{c}), \chi)$ ([Hid7, Theorem 6.7]), in which case h is a cusp form so in this case we also have $c_0(y_f) = c_1(y_f) = 0$.

Thus any simultaneous eigenform of $T_\mathfrak{c}(\mathfrak{p})$ for almost all primes $\mathfrak{p} \subset \mathcal{O}_L$ has a Fourier expansion as claimed in Theorem 5.8. The subalgebra of $\mathbb{T}_\mathfrak{c}$ spanned by $T_\mathfrak{c}(\mathfrak{p})$ for $\mathfrak{p} \nmid \mathfrak{c}$ is commutative, and hence any element of $M_\kappa(K_0(\mathfrak{c}), \chi)$ can be written as a sum of such simultaneous eigenforms. This completes the proof of Theorem 5.8.

Appendix E
Review of Prime Degree Base Change for GL$_2$

The main result of this appendix is Corollary E.12, which is the key ingredient in the proof of Theorem 8.3. This corollary is really just a translation of the main theorem of prime degree base change for GL$_2$ from the language of automorphic representations to the language of automorphic forms combined with some conductor calculations. We therefore begin in Section E.1 by recalling how one attaches an automorphic representation to a Hilbert modular newform (and, in certain instances, a Hilbert modular form to an automorphic representation). This requires the theory of Hecke operators on automorphic representations, so in Section E.2 we pause to review how the Hecke algebra $\mathbb{T}_{\mathfrak{c}}$ recalled in Section 5.6 is related to the Hecke algebra as defined in automorphic representation theory. The dictionary between automorphic forms and automorphic representations would not be complete without Theorem E.4, which gives an equality between the local Euler factors attached to a Hilbert modular form we recalled in (5.9.6) above and the local Euler factors of its associated automorphic representation.

We then enter into the theory of base change in earnest in Section E.4, where we explain how it fits into the (conjectural) framework of Langlands functoriality. We then describe the relevant base changes we use in this work explicitly, starting with prime degree base change for GL$_1$ in Section E.5 for the purpose of proving some lemmas on the conductors of certain characters and their base changes. These lemmas are then used in Section E.6 to prove a proposition on conductors that will enter in to the proof of Corollary E.12. We include Section E.7 on $(\mathfrak{g}, K)$-modules for completeness, and then finish by proving Corollary E.12 in Section E.8.

E.1 Automorphic forms and automorphic representations

In this paragraph we recall a special case of the well-known dictionary between automorphic forms and automorphic representations. The primary reason for this is to make explicit the relationship between various notations and normalizations in the literature that are related to our results. Our presentation borrows heavily from [Bu, Section 3] and [Ku2].

As above, let L be a totally real number field, and let $\Sigma(L)$ be the set of infinite places of L (i.e., embeddings $L \hookrightarrow \mathbb{R}$). Set $G := \operatorname{Res}_{L/\mathbb{Q}}(\operatorname{GL}_2)$, and let $\chi : L^\times \backslash \mathbb{A}_L^\times \to \mathbb{C}^\times$ be a (continuous) quasicharacter. Let $K'_\infty \subseteq G(\mathbb{R})$ be the maximal compact subgroup

$$K'_\infty := \left\{ \begin{pmatrix} a & b \\ -b & a \end{pmatrix} : a^2 + b^2 = \pm 1 \right\} \cong O_2(\mathbb{R})^{\Sigma(L)}$$

It contains the maximal connected compact subgroup $K^1_\infty \cong SO_2(\mathbb{R})^{\Sigma(L)}$ defined in (5.1.3). Let $\mathfrak{g}$ be the Lie algebra of $G(\mathbb{R})$, and let $Z(\mathfrak{g})$ be the center of the universal enveloping algebra of $\mathfrak{g} \otimes \mathbb{C}$. Denote by

$$\mathcal{A}(G, \chi) \tag{E.1.1}$$

the space of *automorphic forms on G with central quasicharacter χ*, that is, functions $\phi : G(\mathbb{A}) \to \mathbb{C}$ satisfying the following conditions:

(1) For $\gamma \in G(\mathbb{Q})$ and $z \in \mathbb{A}_L^\times$, we have

$$\phi\left(\begin{pmatrix} z & 0 \\ 0 & z \end{pmatrix} \gamma \alpha\right) = \chi(z) \phi(\alpha).$$

(2) For each $\alpha_0 \in G(\mathbb{A}_f)$, the function $\alpha_\infty \mapsto \phi(\alpha_\infty \alpha_0)$ is smooth on $G(\mathbb{R})$.

(3) There is a compact open subgroup $K \subseteq G(\mathbb{A}_f)$ such that ϕ is invariant under the action of K by right translation (i.e., $\phi(\alpha) = \phi(\alpha g)$ for $\alpha \in G(\mathbb{A})$, $g \in K$).

(4) The space of functions spanned by right translates of ϕ by K'_∞ is finite-dimensional.

(5) The function ϕ is $Z(\mathfrak{g})$-finite.

(6) The function ϕ is slowly increasing.

We take a moment to explain (5). The Lie algebra $\mathfrak{g}$ acts on the space of functions satisfying (2) and (3) by

$$X \cdot \phi(\alpha) := \frac{d}{dt} \left(\phi(\alpha \exp(tX))\right)|_{t=0}, \tag{E.1.2}$$

see, for example, [Bu, p. 300]. This extends to an action of $\mathfrak{g} \otimes \mathbb{C}$, and moreover to an action of the universal enveloping algebra of $\mathfrak{g} \otimes \mathbb{C}$. We say that ϕ is $Z(\mathfrak{g})$-*finite*

E.1. Automorphic forms and automorphic representations

if the space spanned by the translates of ϕ under $Z(\mathfrak{g})$ is finite-dimensional. For the definition of (6), see [Bu, p. 300]. We also isolate the subspace

$$\mathcal{A}_0(G, \chi) \subseteq \mathcal{A}(G, \chi)$$

consisting of those functions that satisfy the following additional requirement:

(7) The following integral vanishes

$$\int_{U(\mathbb{Q}) \backslash U(\mathbb{A})} f(g\alpha) dg = 0$$

for all $\alpha \in G(\mathbb{A})$, where $U(A) := \{ \begin{pmatrix} 1 & v \\ 0 & 1 \end{pmatrix} : v \in L \otimes_{\mathbb{Q}} A \}$ for $\mathbb{Q}$-algebras A and dg is a Haar measure on $G(\mathbb{A})$.

The space $\mathcal{A}(G, \chi)$ naturally has the structure of a $(\mathfrak{g}, K'_\infty) \times G(\mathbb{A}_f)$-module. In other words, it is a $(\mathfrak{g}, K'_\infty)$-module and a $G(\mathbb{A}_f)$-module, and the two actions commute (see [Bu, Section 2.4] for the precise definition of a $(\mathfrak{g}, K'_\infty)$-module). The action of the pair $(\mathfrak{g}, K'_\infty)$ is given by (E.1.2) and right translation, respectively, and the action of $G(\mathbb{A}_f)$ is given by right translation. The space $\mathcal{A}_0(G, \chi)$ is preserved by this action (see [Ku2, p. 140] and [Bu, Section 2.2 and Theorem 2.9.2]). An *automorphic representation of G with central quasicharacter χ* is an irreducible $(\mathfrak{g}, K'_\infty) \times G(\mathbb{A}_f)$-module that is isomorphic (as a $(\mathfrak{g}, K^1_\infty) \times G(\mathbb{A}_f)$-module) to a subquotient of $\mathcal{A}(G, \chi)$. An automorphic representation is *cuspidal* if it is isomorphic (as a $(\mathfrak{g}, K'_\infty) \times G(\mathbb{A}_f)$-module) to a subquotient of $\mathcal{A}_0(G, \chi)$. Every automorphic representation π admits a factorization as a restricted tensor product

$$\pi \cong \otimes'_v \pi_v,$$

where π_v is an irreducible admissible $(\mathfrak{gl}_2, O_2(\mathbb{R}))$-module for $v|\infty$ and π_v is an irreducible admissible $\mathrm{GL}_2(L_v)$ module for finite places v (see, e.g., [Bu, Section 3.3]). We will recall the classification of these modules in Section E.7 and Section E.6, respectively.

Eigenforms of Hecke operators give rise to automorphic representations. As in Section 5.4, let $\mathfrak{c} \subset \mathcal{O}_L$ be an ideal, let $\kappa = (k, m) \in \mathcal{X}(L)$ be a weight, and let $\chi : L^\times \backslash \mathbb{A}_L^\times \to \mathbb{C}^\times$ be a quasicharacter of conductor dividing $\mathfrak{c}$ satisfying $\chi_\infty(\alpha_\infty) = \alpha_\infty^{-k-2m}$ for $\alpha_\infty \in \mathbb{A}_{L\infty}^\times$. It is easy to check that there is a natural inclusion

$$M_\kappa(K_0(\mathfrak{c}), \chi) \hookrightarrow \mathcal{A}(G, \chi)$$

which restricts to induce an inclusion

$$S_\kappa(K_0(\mathfrak{c}), \chi) \hookrightarrow \mathcal{A}_0(G, \chi).$$

For every simultaneous eigenform $h \in M_\kappa(K_0(\mathfrak{c}), \chi)$ of the Hecke operators $T_\mathfrak{c}(\mathfrak{p})$ for almost all primes $\mathfrak{p}$, let $\pi(h)$ be the $(\mathfrak{g}, K^1_\infty) \times G(\mathbb{A}_f)$-submodule of $\mathcal{A}(G, \chi)$ spanned by the translates of h. As is well known, $\pi(h)$ is then an automorphic representation of G (compare [Bu, Theorems 3.6.1 and 3.7.3]). There is also a partial converse to this statement, and this gives the dictionary between automorphic forms and automorphic representations on G we mentioned above.

Before making this precise, we recall the notion of the *conductor* of an automorphic representation. Suppose that π_v is an irreducible admissible representation of $\mathrm{GL}_2(L_v)$ with central character χ_v. Temporarily let $\mathfrak{p}$ be the prime ideal of $\mathcal{O}_{L,v}$. For any $n \geq a(\chi_v)$ (the exponent of the conductor of χ_v), we may extend χ_v to a character of

$$K_0(\mathfrak{p}^n) := \left\{ \begin{pmatrix} a & b \\ c & d \end{pmatrix} \in \mathrm{GL}_2(\mathcal{O}_{L,v}) : c \in \mathfrak{p}^n \right\}$$

by setting $\chi_v\left(\begin{pmatrix} a & b \\ c & d \end{pmatrix}\right) := \chi_v(d)$. The *conductor* $\mathfrak{f}(\pi)$ of π is then $\mathfrak{p}^{a(\pi)}$, where $a(\pi)$ is the smallest non-negative integer n such that

$$V(\pi_v)_{\chi_v}^{K_0(\mathfrak{p}^n)} := \{f \in V(\pi_v) : \pi_v(g) \cdot f = \chi_v(g)f \text{ for all } g \in K_0(\mathfrak{p}^n)\} \neq 0.$$

Here $V(\pi_v)$ is the space of π_v. The space $V(\pi_v)_{\chi_v}^{K_0(\mathfrak{f}(\pi_v))}$ is one-dimensional by a theorem of Casselman [Cas]. The *conductor* of π is defined to be the product of its local conductors:

$$\mathfrak{f}(\pi) = \prod_{v \nmid \infty} \mathfrak{f}(\pi_v). \tag{E.1.3}$$

Theorem E.1 (Automorphic dictionary). *Let $\kappa = (k, m) \in \mathcal{X}(L)$ be a weight, and let $\chi : L^\times \backslash \mathbb{A}_L^\times \to \mathbb{C}^\times$ be a quasicharacter satisfying $\chi_\infty(a_\infty) = a_\infty^{-k-2m}$. If $h \in S_\kappa(K_0(\mathfrak{c}), \chi)$ is a simultaneous eigenform for $T_\mathfrak{c}(\mathfrak{p})$ for almost all primes $\mathfrak{p} \subset \mathcal{O}_L$, then the automorphic representation $\pi(h)$ satisfies the following:*

(1) *The central character of $\pi(h)$ is χ.*
(2) *The conductor of $\pi(h)$ is equal to the level of $\underline{h}$.*
(3) *For each infinite place $\sigma \in \Sigma(L)$, we have $\pi(h)_\sigma \cong \mathcal{D}_{-k_\sigma - 2m_\sigma}(k_\sigma + 2)$.*

Conversely, if π is a cuspidal automorphic representation of G with central quasicharacter χ satisfying $\pi_\sigma \cong \mathcal{D}_{-k_\sigma - 2m_\sigma}(k_\sigma + 2)$ for each $\sigma \in \Sigma(L)$, then $\pi \cong \pi(h)$ for a unique normalized newform $h \in S_\kappa^{\mathrm{new}}(K_0(\mathfrak{f}(\pi)), \chi)$.

Here $\mathcal{D}_\mu(k)$ is the discrete series representation of Section E.7, and $\underline{h} \in S_\kappa^{\mathrm{new}}(K_0(\mathfrak{c}'), \chi)$ is the normalized (meaning $a(\mathcal{O}_L, h) = 1$, see Section 5.9) newform associated to h, that is, the normalized newform whose Hecke eigenvalues $\lambda_{\underline{h}}(\mathfrak{m})$ satisfy $\lambda_h(\mathfrak{p}) = \lambda_{\underline{h}}(\mathfrak{p})$ for almost all prime ideals $\mathfrak{p} \subset \mathcal{O}_L$ (where $\mathfrak{c}' \supset \mathfrak{c}$ is the level of $\underline{h}$).

Remark. Thus, there is a one to one correspondence between normalized newforms and cuspidal automorphic representations as above. As the proof will show, this theorem depends on the fact that "multiplicity one" holds for G. In other words, every isomorphism class of cuspidal automorphic representation occurs with mul-

E.1. Automorphic forms and automorphic representations

tiplicity one in $\mathcal{A}_0(G,\chi)$. This is not true for automorphic representations on arbitrary reductive groups.

The proof of Theorem E.1 uses some material contained in Section E.2, Section E.6 and Section E.7 below.

Proof of Theorem E.1. We begin by proving the converse statement. Let $\pi = \otimes'_v \pi_v$ be the given cuspidal automorphic representation. There is a $(\mathfrak{g}, K'_\infty) \times G(\mathbb{A}_f)$-intertwining injection $\pi \hookrightarrow \mathcal{A}_0(G,\chi)$; moreover, this injection is unique (see [JaLan, Propositions 10.9 and 11.1.1]). We therefore identify π with its image under this isomorphism. Regard χ as a character of $K_0(\mathfrak{f}(\pi))$ by setting $\chi\left(\left(\begin{smallmatrix} a & b \\ c & d \end{smallmatrix}\right)\right) = \prod_{\mathfrak{p}|\mathfrak{c}} \chi_{v(\mathfrak{p})}(d_{v(\mathfrak{p})})$, where $\chi_{v(\mathfrak{p})}$ is the local character associated to χ at the place $v(\mathfrak{p})$ associated to the prime $\mathfrak{p} \subset \mathcal{O}_L$. Consider the space

$$V' := V(\pi)^{K_0(\mathfrak{f}(\pi))}_{\kappa,\chi} \qquad (\text{E.1.4})$$
$$:= \left\{ h \in V(\pi) : \begin{array}{l} \pi(g)h = \chi(g)h \text{ for all } g \in K_0(\mathfrak{f}(\pi)) \text{ and} \\ \pi(u_\infty)h = h\exp(\sum_{v|\infty} ik_v\theta_v) \text{ for all } u_\infty \in K^1_\infty \end{array} \right\},$$

where $u_\infty = (\ldots, \left(\begin{smallmatrix} \cos\theta_v & \sin\theta_v \\ -\sin\theta_v & \cos\theta_v \end{smallmatrix}\right), \ldots)$ for $\theta \in (\mathbb{R}/2\pi\mathbb{Z})^{\Sigma(L)}$. By the result of Casselman cited above (see above (E.1.3)), together with the fact that the isotypic components of the discrete series under the action of $SO_2(\mathbb{R})$ are one-dimensional (see (E.7.3) below), we conclude that V' is a one-dimensional complex vector space, say $\mathbb{C}h = V'$ for some $h \in \mathcal{A}_0(G_L, \chi)$. We claim that $h \in S_\kappa(K_0(\mathfrak{f}(\pi)), \chi)$. Indeed, all of the conditions for h to be an element of $S_\kappa(K_0(\mathfrak{f}(\pi)), \chi)$ (see Section 5.4) are immediate, except possibly (1), namely that

$$C_\sigma h(\alpha_\infty \alpha_0) = \left(\frac{k_\sigma^2}{2} + k_\sigma\right) h(\alpha_\infty \alpha_0)$$

where C_σ is the Casimir operator acting on the place σ as normalized in (E.7.1) below. This follows from the fact that $\pi_\sigma \cong \mathcal{D}_{-k_\sigma - 2m_\sigma}(k_\sigma + 2)$ and (E.7.2) below.

The last thing we need to check is that h is a newform. We first claim that h is a simultaneous eigenform for all Hecke operators $T_{\mathfrak{f}(\pi)}(\mathfrak{p})^\chi$ and $T_{\mathfrak{f}(\pi)}(\mathfrak{p},\mathfrak{p})^\chi$ for $\mathfrak{p} \nmid \mathfrak{f}(\pi)$. Let S be the set of places of L dividing ∞ and $\mathfrak{f}(\pi)$. Let $\pi(h)^S = \prod_{v \notin S} \pi_v$. Then $\pi(h)^S$ is an irreducible, admissible $GL_2(\mathbb{A}_L^S)$-module that is fixed by $GL_2(\widehat{\mathcal{O}}_L^S)$. Here we are using a superscript S to denote adèles trivial at the places dividing S. Hence $\pi(h)^S$ is a simple $\mathcal{H}(GL_2(\mathbb{A}_L^S)//GL_2(\widehat{\mathcal{O}}_L^S))$-module (see Section E.2). On the other hand, $\mathcal{H}(GL_2(\mathbb{A}_L^S)//GL_2(\widehat{\mathcal{O}}_L^S))$ is commutative [Cart, Proposition 4.1], so we conclude that

$$T_{\mathfrak{f}(\pi)}(\mathfrak{p})^\chi \quad \text{and} \quad T_{\mathfrak{f}(\pi)}(\mathfrak{p},\mathfrak{p})^\chi \in \mathcal{H}(G(\mathbb{A}_f))$$

act as scalars on $\pi(h)^S$ for all primes $\mathfrak{p} \nmid \mathfrak{f}(\pi)$ (see (E.2.4) for notation). This implies that for $\mathfrak{p} \nmid \mathfrak{f}(\pi)$ the operators $T^\chi_{\mathfrak{f}(\pi)}(\mathfrak{p})^\chi$ and $T_{\mathfrak{f}(\pi)}(\mathfrak{p},\mathfrak{p})^\chi$ act as scalars on h itself, and hence, by the discussion above (E.2.4), the Hecke operators $T_{\mathfrak{c}}(\mathfrak{p}), T_{\mathfrak{c}}(\mathfrak{p},\mathfrak{p}) \in \mathbb{T}_{\mathfrak{c}}$

act as scalars on h. Thus h is a Hecke eigenform for all Hecke operators $T_{\mathfrak{c}}(\mathfrak{m})$, $T_{\mathfrak{c}}(\mathfrak{m},\mathfrak{m})$ with $\mathfrak{m} + \mathfrak{f}(\pi) = \mathcal{O}_L$ (compare (5.6.8)). Suppose that h is not a newform. Then let $\underline{h} \in S_\kappa^{\text{new}}(K_0(\mathfrak{c}), \chi)$ for some $\mathfrak{c} \supsetneq \mathfrak{f}(\pi)$ be its associated newform. Then, by newform theory (compare Section 5.8) we may write

$$h(\alpha) = \sum_{g_i} c_i \underline{h}\left(\alpha \begin{pmatrix} d^{-1} & \\ & 1 \end{pmatrix}\right)$$

for some $c_i \in \mathbb{C}$ and $d \in \widehat{\mathcal{O}}_E$ with $\mathfrak{d}\mathfrak{c} | \mathfrak{f}(\pi)$. It follows that $\underline{h} \in V'$, which contradicts the definition of the conductor $\mathfrak{f}(\pi)$; thus h is a newform. This finishes the proof of the converse statement.

Now assume that $h \in S_\kappa(K_0(\mathfrak{c}), \chi)$ is a simultaneous eigenform for $T_{\mathfrak{c}}(\mathfrak{p})$ and $T_{\mathfrak{c}}(\mathfrak{p},\mathfrak{p})$ for almost all primes $\mathfrak{p} \subset \mathcal{O}_L$. Let $\underline{h}$ be its associated newform. Then, as above, $h(\alpha) = \sum_i c_i \underline{h}(\alpha g_i)$ for some $c_i \in \mathbb{C}$ and $g_i \in G(\mathbb{A}_f)$, so $\pi(h) = \pi(\underline{h})$. We may therefore assume, without loss of generality, that h is a newform. It is easy to see that the cuspidal automorphic representation $\pi(h)$ has central character χ. Since $h \in S_\kappa^{\text{new}}(K_0(\mathfrak{c}), \chi)$, we have that

$$C_\sigma h(\alpha_\infty \alpha_0) = \left(\frac{k_\sigma^2}{2} + k_\sigma\right) h(\alpha_\infty \alpha_0).$$

This fact, combined with Proposition E.10 and a simple computation using the explicit description of χ at the infinite places implies that $\pi(h)_\sigma \cong \mathcal{D}_{-k_\sigma - 2m_\sigma}(k_\sigma + 2)$.

Define $V'' := V(\pi(h))_{\kappa,\chi}^{K_0(\mathfrak{f}(\pi))}$ (see (E.1.4)). By what we proved above, $V'' = \mathbb{C}\widetilde{h}$, where $\widetilde{h} \in S_\kappa^{\text{new}}(K_0(\mathfrak{f}(\pi(h))), \chi)$ is the unique newform such that $\pi(\widetilde{h}) = \pi(h)$. We conclude that $h = \widetilde{h}$, and hence $\mathfrak{f}(\pi(h)) = \mathfrak{c}$. □

In Section E.8 below, we will use Theorem E.1 to translate the statement of prime degree base change for GL_2 from the language of automorphic representations to the language of automorphic forms. This is the formulation we used in the proof of Theorem 8.3.

E.2 Hecke operators

Recall from Section 5.6 the algebra $\mathbb{T}_{\mathfrak{c}}$ which acts on $S_\kappa(K_0(\mathfrak{c}), \chi)$ for $\mathfrak{c} \subset \mathcal{O}_L$. This algebra admits a reinterpretation which permits one to generalize its action on $S_\kappa(K_0(\mathfrak{c}), \chi)$ to an action on certain spaces of automorphic forms. We recall this generalization in this section.

As in [Cart], for any locally compact totally-disconnected group G' let

$$\mathcal{H}(G') \tag{E.2.1}$$

be the convolution algebra of complex-valued functions φ on G' satisfying the following conditions:

E.2. Hecke operators

- The function φ is locally constant.
- The function φ is compactly supported.

We will later restrict to the case $G' = G(L_v)$ or $G' = G(\mathbb{A}_f)$. For concreteness, we note that the convolution product is given by

$$(\varphi_1 * \varphi_2)(\alpha) := \int_{G'} \varphi_1(g)\varphi_2(g^{-1}\alpha)dg, \tag{E.2.2}$$

where dg is a suitable Haar measure. Thus, for example, for every $x \in G'$ and every compact open subgroup $K \subseteq G'$ we have elements char_{KxK} of $\mathcal{H}(G')$ defined by

$$\mathrm{char}_{KxK}(\alpha) := \begin{cases} \left(\int_K dg\right)^{-1} & \text{if } \alpha \in KxK \\ 0 & \text{otherwise.} \end{cases}$$

As explained in [Cart], every admissible G'-representation π is in a natural way a $\mathcal{H}(G')$-module. In particular, if G' is unimodular, then the action is given by

$$\varphi \cdot w := \int_{G'} \varphi(g)\pi(g)w\,dg$$

for $w \in V(\pi)$ (see [Bu, p. 316]). Letting $V(\pi)$ denote the space of π, we have the following extremely useful proposition:

Proposition E.2. *An admissible G'-representation π is irreducible if and only if $V(\pi)$ is a simple $\mathcal{H}(G')$-module.*

See [Cart, p. 118] for a proof of Proposition E.2 and also for the following version of Schur's lemma:

Proposition E.3. *If π be a smooth, irreducible representation of G', and the topology of G' has a countable basis, then $\mathrm{End}_{G'}(V(\pi)) \cong \mathbb{C}$.*

We now return to our case of interest, namely the $G(\mathbb{A}_f)$-module $\mathcal{A}(G,\chi)$. As we mentioned above, the fact that $\mathcal{A}(G,\chi)$ is a $G(\mathbb{A}_f)$-module immediately implies that it is a $\mathcal{H}(G(\mathbb{A}_f))$-module.

In particular, we may consider the $\mathcal{H}(G(\mathbb{A}_f))$-submodule

$$\mathcal{A}^{\mathfrak{c}} := \{h \in \mathcal{A}(G,\chi) : \pi(g)h = \chi(g)h \text{ for } g \in K_0(\mathfrak{c})\},$$

where $\chi\left(\left(\begin{smallmatrix} a & b \\ c & d \end{smallmatrix}\right)\right) := \prod_{\mathfrak{p}|\mathfrak{c}} \chi_{v(\mathfrak{p})}(d_{v(\mathfrak{p})})$ for $\left(\begin{smallmatrix} a & b \\ c & d \end{smallmatrix}\right) \in K_0(\mathfrak{c})$ as above. Recall the definition of the Hecke algebra $\mathbb{T}_{\mathfrak{c}}$ from Section 5.6. The elements of $\mathbb{T}_{\mathfrak{c}}$ can be thought of as formal sums of double cosets of the form

$$K_0(\mathfrak{c})xK_0(\mathfrak{c})$$

with $x \in R(\mathfrak{c})$. Here $R(\mathfrak{c})$ is defined to be

$$G(\mathbb{A}_f) \cap \left\{ x \in M_2(\widehat{\mathbb{Z}} \otimes \mathcal{O}_L) : x = \begin{pmatrix} a & b \\ c & d \end{pmatrix} \text{ with } c \in \mathfrak{c} \text{ and } \prod_{\mathfrak{p}|\mathfrak{c}} (d_{v(\mathfrak{p})}) \in \prod_{\mathfrak{p}|\mathfrak{c}} \mathcal{O}_{L,v(\mathfrak{p})}^\times \right\}$$

as in Section 5.6. Any such double coset can be decomposed into a disjoint sum

$$K_0(\mathfrak{c}) x K_0(\mathfrak{c}) = \sum_i x_i K_0(\mathfrak{c}).$$

Thus we have an action of $\mathbb{T}_\mathfrak{c}$ on $\mathcal{A}^\mathfrak{c}$ by defining

$$(h|K_0(\mathfrak{c}) x K_0(\mathfrak{c}))(\alpha) := \sum_i \chi(x_i)^{-1} h(\alpha x_i).$$

Here $\chi(\begin{pmatrix} a & b \\ c & d \end{pmatrix}) = \chi(\prod_{\mathfrak{p}|\mathfrak{c}}(d_{v(\mathfrak{p})}))$. This definition is consistent with our earlier definition of $h|K_0(\mathfrak{c}) x K_0(\mathfrak{c})$ for $h \in M_\kappa(K_0(\mathfrak{c}), \chi) \subseteq \mathcal{A}^\mathfrak{c}$.

We wish to realize these $K_0(\mathfrak{c}) x K_0(\mathfrak{c})$ as elements of the "new" Hecke algebra $\mathcal{H}(G(\mathbb{A}_f))$. Set

$$K_0(\mathfrak{c}) x K_0(\mathfrak{c})^\chi(g) := \chi(g)^{-1} \text{char}_{K_0(\mathfrak{c}) x K_0(\mathfrak{c})}(g). \tag{E.2.3}$$

We will now compute $K_0(\mathfrak{c}) x K_0(\mathfrak{c})^\chi \cdot h$ for $h \in \mathcal{A}^\mathfrak{c}$. For notational simplicity, temporarily set $K_0 := K_0(\mathfrak{c})$. For $h \in \mathcal{A}^\mathfrak{c}$, we have:

$$K_0(\mathfrak{c}) x K_0(\mathfrak{c})^\chi \cdot h(\alpha)$$

$$:= \int_{G(\mathbb{A}_f)} K_0(\mathfrak{c}) x K_0(\mathfrak{c})^\chi(g) h(\alpha g) dg$$

$$= \left(\int_{K_0} dg \right)^{-1} \sum_{x_i K_0 \in K_0 x K_0/K_0} \int_{x_i K_0} \chi(g)^{-1} h(\alpha g) dg$$

$$= \left(\int_{K_0} dg \right)^{-1} \sum_{x_i K_0 \in K_0 x K_0/K_0} \int_{K_0} \chi(x_i g)^{-1} h(\alpha x_i g) dg$$

$$= \left(\int_{K_0} dg \right)^{-1} \sum_{x_i K_0 \in K_0 x K_0/K_0} \chi(x_i)^{-1} h(\alpha x_i) \int_{K_0} dg \text{ (by definition of } \mathcal{A}^\mathfrak{c})$$

$$= \sum_i \chi(x_i)^{-1} h(\alpha x_i)$$

$$= (h|K_0(\mathfrak{c}) x K_0(\mathfrak{c}))(\alpha).$$

For any $t = \sum_j a_j K_0(\mathfrak{c}) x_j K_0(\mathfrak{c}) \in \mathbb{T}_\mathfrak{c}$, we set

$$t^\chi := \sum_j a_j K_0(\mathfrak{c}) x_j K_0(\mathfrak{c})^\chi. \tag{E.2.4}$$

(We used this notation for $t = T_{\mathfrak{c}}(\mathfrak{p})$ in Section E.1 above.) To summarize, the natural inclusion

$$S_\kappa(K_0(\mathfrak{c}), \chi) \longrightarrow \mathcal{A}_0(G, \chi) \tag{E.2.5}$$

is compatible with the map

$$\begin{aligned}\mathbb{T}_{\mathfrak{c}} &\longrightarrow \mathcal{H}(G(\mathbb{A}_f)) \\ t &\longmapsto t^\chi.\end{aligned} \tag{E.2.6}$$

This map is actually an injective algebra morphism, intertwining the usual product in $\mathbb{T}_{\mathfrak{c}}$ with convolution of functions in $\mathcal{H}(G(\mathbb{A}_f))$. The injectivity is clear, and the fact that the map respects the product structure on both sides is well known (see [Cart]).

If G' is a locally compact totally disconnected group and $K \subseteq G'$ is a compact open subgroup, denote by

$$\mathcal{H}(G'//K) \subseteq \mathcal{H}(G')$$

the subgroup consisting of those functions that are K-biinvariant. That is, $f \in \mathcal{H}(G')$ is an element of $\mathcal{H}(G'//K)$ if and only if $f(kg) = f(gk) = f(g)$ for all $g \in G'$ and $k \in K$. Then it is easy to check that the image of (E.2.6) is contained in $\mathcal{H}(G(\mathbb{A}_f)//K_0^\vee(\mathfrak{c}))$, where

$$K_0^\vee(\mathfrak{c}) := \left\{ \begin{pmatrix} a & b \\ c & d \end{pmatrix} \in K_0(\mathfrak{c}) : d - 1 \in \mathfrak{c} \right\}.$$

E.3 Agreement of *L*-functions

The purpose of this section is to prove the following theorem.

Theorem E.4. *Let $h \in S_\kappa^{\text{new}}(K_0(\mathfrak{c}), \chi)$ be a newform, and $\pi(h)$ its associated automorphic representation as in Theorem E.1. Then for every prime $\mathfrak{p}$ with associated finite place $v(\mathfrak{p})$, we have*

$$L_{\mathfrak{p}}(h, s) = L(\pi(h)_{v(\mathfrak{p})}, s)$$

where the local Euler factor on the left is that given in (5.9.6) and the Euler factor on the right is the "standard" Euler factor (see [JaLan]).

Proof. For ease of notation, write $\pi := \pi(h)$. By the classification of local admissible representations of GL_2, we have three possibilities for $\pi_{v(\mathfrak{p})}$; it is either principal, special (also known as Steinberg), or supercuspidal (see Section E.6 and [Gel, Section 4.2]). If $\pi_{v(\mathfrak{p})}$ is principal or special, the local Euler factors $L(\pi_{v(\mathfrak{p})}, s)$ are given by the table in Section E.6. If $\pi_{v(\mathfrak{p})}$ is supercuspidal, then $L(\pi_{v(\mathfrak{p})}, s) = 1$ [Gel, Theorem 6.15]. In any case, $L(\pi_{v(\mathfrak{p})})$ is a polynomial in $N_{L/\mathbb{Q}}(\mathfrak{p})^{-s}$ of degree

0, 1, or 2. It is clear that the coefficients of $(\mathrm{N}_{L/\mathbb{Q}}(\mathfrak{p})^{-s})^0$ and $(\mathrm{N}_{L/\mathbb{Q}}(\mathfrak{p})^{-s})^2$ in $L(\pi_{v(\mathfrak{p})}, s)$ and $L_\mathfrak{p}(h, s)$ agree. We are left with checking that the coefficients of $\mathrm{N}_{L/\mathbb{Q}}(\mathfrak{p})^{-s}$ agree.

The space $V(\pi)_{\kappa,\chi}^{K_0(\mathfrak{f}(\pi))}$ of (E.1.4) is one-dimensional as explained in the proof of Theorem E.1, and it is clear that $T_\mathfrak{c}(\mathfrak{p})^\chi$ both preserves this space and has image contained in it. It therefore acts as a scalar on it, say λ. By the discussion above (E.2.4) and the fact that we may regard h as an element of $V(\pi) \subset \mathcal{A}_0(G, \chi)$, we see that $\lambda = \lambda_h(\mathfrak{p})$. It therefore suffices, by the definition of $L_\mathfrak{p}(h, s)$ in (5.9.6), to prove that $\lambda \mathrm{N}_{L/\mathbb{Q}}(\mathfrak{p})^{-1/2}$ is the coefficient of $\mathrm{N}_{L/\mathbb{Q}}(\mathfrak{p})^{-s}$ in $L(\pi_{v(\mathfrak{p})}, s)$.

In order to do this, one calculates the eigenvalue of $T_\mathfrak{c}(\mathfrak{p})^\chi$ on $V(\pi)$. Note that $T_\mathfrak{c}(\mathfrak{p}) = K_0(\mathfrak{c}) \begin{pmatrix} 1 & 0 \\ 0 & \varpi_\mathfrak{p} \end{pmatrix} K_0(\mathfrak{c})$ where $\varpi_\mathfrak{p}$ is an idèle that is a uniformizer for the maximal ideal of $\mathcal{O}_{L,v(\mathfrak{p})}$ at $v(\mathfrak{p})$ and is trivial at all other places. Then, for any pure tensor $w = \otimes_v' w \in V(\pi)$, we have

$$T_\mathfrak{c}(\mathfrak{p})^\chi w := \left(\int_{K_0(\mathfrak{c})} dg \right)^{-1} \int_{K_0(\mathfrak{c}) \begin{pmatrix} 1 & 0 \\ 0 & \varpi_\mathfrak{p} \end{pmatrix} K_0(\mathfrak{c})} \chi(g)^{-1} \pi(g) w \, dg$$

$$= \left(\left(\int_{K_\mathfrak{p}} dg_{v(\mathfrak{p})} \right)^{-1} \int_{K_\mathfrak{p} \begin{pmatrix} 1 & 0 \\ 0 & \varpi_\mathfrak{p} \end{pmatrix} K_\mathfrak{p}} \pi(g_{v(\mathfrak{p})}) w_{v(\mathfrak{p})} \right) \otimes \left(\otimes_{v \neq v(\mathfrak{p})} w_v \right)$$

$$= \mathrm{char}_{K_\mathfrak{p} \begin{pmatrix} 1 & 0 \\ 0 & \varpi_\mathfrak{p} \end{pmatrix} K_\mathfrak{p}} \cdot w_{v(\mathfrak{p})} \otimes \left(\otimes_{v \neq v(\mathfrak{p})} w_v \right).$$

Here, at the last line, we regard $\varpi_\mathfrak{p}$ as uniformizer for the maximal ideal of $\mathcal{O}_{L,v(\mathfrak{p})}$, and $K_\mathfrak{p} = K_0(\mathfrak{f}(\pi_{v(\mathfrak{p})}))$. We conclude that λ is also the eigenvalue of $\mathrm{char}_{K_\mathfrak{p} \begin{pmatrix} 1 & 0 \\ 0 & \varpi_\mathfrak{p} \end{pmatrix} K_\mathfrak{p}} \in \mathcal{H}(G(L_{v(\mathfrak{p})}))$ acting on $\pi_{v(\mathfrak{p})}$. If $\pi_{v(\mathfrak{p})}$ is unramified, Proposition 4.6.6 of [Bu] computes this eigenvalue, and it turns out to be precisely $(\chi_1(\mathfrak{p}) + \chi_2(\mathfrak{p})) \mathrm{N}_{L/\mathbb{Q}}(\mathfrak{p})^{1/2}$.

In the cases where $\pi_{v(\mathfrak{p})}$ is ramified, Hida states the computation of λ as a theorem in [Hid4, Section 2] as a rephrasing of the results of [Cas]. A nice proof, following the proof of Proposition 4.6.6 of [Bu], is given in Proposition 2 of [Pop]. □

E.4 Langlands functoriality

In this section we indicate briefly how the base change map that we have used throughout this work fits into the general (conjectural) framework of Langlands functoriality. Our primary motivation is to make precise several statements made in the introduction and in Chapter 8 about how various maps on Hecke algebras are "induced by a map of L-groups."

Let H be a connected reductive F-group, where F is a number field. Langlands has conjectured that there is a partition of the set of all equivalence classes of automorphic representations of $H(\mathbb{A}_F)$ (resp. irreducible admissible representations of $H(F_v)$) into disjoint sets called L-packets enjoying certain functorial

E.4. Langlands functoriality

properties. Here v is any place of F. An L-packet of automorphic representations is called a global L-packet, and an L-packet of irreducible admissible representations is called a local L-packet. Describing what is known about these L-packets is beyond the scope of this work. However, in the case when $H = \mathrm{Res}_{M/F}\mathrm{GL}_n$ for some finite degree field extension M/F and some integer n the packets are singletons; that is, each irreducible representation defines an L-packet with isomorphism class of the representation itself as the sole member.

One reason Langlands introduced L-packets is to formulate the notion of a functorial transfer. More precisely, let H and G be connected reductive F-groups (resp. F_v-groups) and let

$$^LH \longrightarrow {}^LG$$

be an L-map (see [Bo79]). Langlands has conjectured that to each such map one has a functorial transfer associating an L-packet of representations of G to each L-packet of representations of H. The transfer of global L-packets is supposed to be compatible with the transfer of local L-packets. Again, describing what is known about this conjecture is beyond the scope of this work. The Paris book project organized by M. Harris [Harr] would be a place for the interested reader to begin, as well as forthcoming work of J. Arthur. The structure of L-packets in the archimedean case is due to Langlands and Shelstad (see [LanS] and the references therein). One should also consult the various papers of C. Moeglin et. al. for the local non-archimedean case. Also, though much more is known now than in 1979, the Corvallis proceedings [BoC] remain an invaluable resource.

In this book we only require a case of Langlands functoriality that was historically among the first to be proven [Lan], namely prime-degree base change for GL_2. Let L/E be a prime-degree Galois extension of number fields, let $H = \mathrm{GL}_n$ and $G = \mathrm{Res}_{L/E}\mathrm{GL}_n$. We will use the "finite form" of the L-group in what follows [Bo79, Section 2.4(2)]. One can find proofs and/or references for the statements we make in [Bo79].

For every place v of E write $L_v := L \otimes_E E_v$. We have

$$^LH = \mathrm{GL}_n(\mathbb{C}) \times \mathrm{Gal}(L/E)$$
$$^LG = \mathrm{GL}_n(\mathbb{C})^{\mathrm{Gal}(L/E)} \rtimes \mathrm{Gal}(L/E)$$
$$^LH_{E_v} = \begin{cases} \mathrm{GL}_n(\mathbb{C}) \times \mathrm{Gal}(L_v/E_v) & \text{if } v \text{ is inert or ramified} \\ \mathrm{GL}_n(\mathbb{C}) \times \langle 1 \rangle & \text{if } v \text{ is split} \end{cases}$$
$$^LG_{E_v} = \begin{cases} \mathrm{GL}_n(\mathbb{C})^{\mathrm{Gal}(L/E)} \rtimes \mathrm{Gal}(L_v/E_v) & \text{if } v \text{ is inert or ramified} \\ \mathrm{GL}_n(\mathbb{C})^{\mathrm{Gal}(L/E)} \times \langle 1 \rangle & \text{if } v \text{ is split.} \end{cases}$$

Here in the first (resp. second) semidirect product $\mathrm{Gal}(L/E)$ (resp. $\mathrm{Gal}(L_v/E_v) \hookrightarrow \mathrm{Gal}(L/E)$) acts by permuting the factors in the natural manner. The symbol $\langle 1 \rangle$ just means the trivial group with one element. For every place v of E there are natural embeddings $^LH_{E_v} \longrightarrow {}^LH$ and $^LG_{E_v} \to {}^LG$ given by the identity on the first factor and the natural embedding $\mathrm{Gal}(L_v/E_v) \to \mathrm{Gal}(L/E)$ (resp. $\langle 1 \rangle \to \mathrm{Gal}(L/E)$) on the second factor. The L-map that induces base change from GL_n to $\mathrm{Res}_{L/E} \mathrm{GL}_n$ is the map

$$b : \mathrm{GL}_n(\mathbb{C}) \times \mathrm{Gal}(L/E) \longrightarrow \mathrm{GL}_n(\mathbb{C})^{\mathrm{Gal}(L/E)} \rtimes \mathrm{Gal}(L/E) \qquad (\mathrm{E.4.1})$$

given by the diagonal embedding on the first factor and the identity on the second. Using the embeddings of the local L-groups mentioned above, b induces a map

$$b : {}^L\mathrm{GL}_{nE_v} \longrightarrow {}^L\mathrm{Res}_{L/E} \mathrm{GL}_{nE_v}$$

for every place v of F.

We now explain what we mean when we say that (E.4.1) induces base change. Choose a place v of E and assume that it is finite and unramified in L/E. Let $T \subseteq {}^L\mathrm{GL}^\circ_{nE_v}$ be a maximal torus, let $\sigma = 1$ if v is split in L/E and let σ be a generator of $\mathrm{Gal}(L \otimes_v E_v/E_v)$ if v is inert in L/E. Notice that $T^{\mathrm{Gal}(L/E)} \subseteq \mathrm{GL}_n^{\mathrm{Gal}(L/E)} = {}^L\mathrm{GL}^\circ_{nE_v}$ is also a maximal torus, and the restriction

$$b : T \times \sigma \longrightarrow T^{\mathrm{Gal}(L/E)} \rtimes \sigma$$

is just the natural diagonal embedding on the first factor and the identity on the second factor. Let W be the Weyl group of T in ${}^L\mathrm{GL}_{nE_v}$. Let R be a $\mathbb{C}$-algebra. Since the action of W on T commutes with the (trivial) action of σ on T, the map $T(R) \longrightarrow T(R) \times \sigma$ given by $t \mapsto t \times \sigma$ induces an isomorphism

$$T/W \longrightarrow T \times \sigma/W \qquad (\mathrm{E.4.2})$$

of affine schemes over $\mathbb{C}$. Define W_E to be the subgroup of the Weyl group of $T^{\mathrm{Gal}(L/E)}$ in ${}^L\mathrm{Res}_{L/E}\mathrm{GL}^\circ_{nE_v}$ that is fixed by σ. Let N be the normalizer of $T^{\mathrm{Gal}(L/E)}$ in ${}^L\mathrm{Res}_{L/E}\mathrm{GL}^\circ_{nE_v}$, and let N_E be the inverse image of W_E in N. Then b induces a morphism

$$b : T/W \longrightarrow T \times \sigma/W \longrightarrow (T^{\mathrm{Gal}(L/E)} \rtimes \sigma)/N_E$$

of affine schemes over $\mathbb{C}$. This is equivalent to the statement that b induces a $\mathbb{C}$-algebra homomorphism

$$b : \mathbb{C}[T^{\mathrm{Gal}(L/E)} \rtimes \sigma]^{N_E} \longrightarrow \mathbb{C}[T]^W. \qquad (\mathrm{E.4.3})$$

There are algebra isomorphisms

$$\mathbb{C}[T^{\mathrm{Gal}(L/E)} \rtimes \sigma]^{N_E} \cong \begin{cases} \mathbb{C}[t_1^{\pm 1}, \ldots, t_n^{\pm 1}]^{S_n} & \text{if } v \text{ is inert} \\ (\mathbb{C}[t_1^{\pm 1}, \ldots, t_n^{\pm 1}]^{S_n})^{[L:E]} & \text{if } v \text{ is split.} \end{cases}$$

$$\mathbb{C}[T]^W \cong \mathbb{C}[t_1^{\pm 1}, \ldots, t_n^{\pm 1}]^{S_n}$$

E.4. Langlands functoriality

where S_n acts by permuting the indices [ArtC, Chapter 1, Sections 4.1 and 4.2]. Moreover, one can arrange these isomorphisms so that the base change map of (E.4.3) is given by

$$b(f)(t_1,\ldots,t_n) = f(t_1^{[L:E]},\ldots,t_n^{[L:E]}) \text{ if } v \text{ is inert} \quad (E.4.4)$$
$$b(f_1,\ldots,f_{[L:E]}) = f_1 \cdots f_{[L:E]} \text{ if } v \text{ is split}$$

(the statement in the inert case is given in [ArtC, Chapter 1, Section 4.2], the statement in the split case is obvious).

Write $K_v := \mathrm{GL}_n(\mathcal{O}_{E_v})$ and $K_{L_v} := \mathrm{GL}_n(\mathcal{O}_{L_v})$. We let

$$b : \mathcal{H}(\mathrm{GL}_n(L_v)//K_{L_v}) \longrightarrow \mathcal{H}(\mathrm{GL}_n(E_v)//K_v) \quad (E.4.5)$$

be the algebra morphism defined by stipulating that the diagram

$$\begin{array}{ccc}
\mathbb{C}[T^{\mathrm{Gal}(L/E)} \rtimes \sigma]^{N_E} & \xrightarrow{\ b\ } & \mathbb{C}[T]^W \\
{\scriptstyle f \mapsto f^\vee} \uparrow & & \uparrow {\scriptstyle \phi \mapsto \phi^\vee} \\
\mathcal{H}(\mathrm{GL}_n(L_v)//K_{L_v}) & \xrightarrow{\ b\ } & \mathcal{H}(\mathrm{GL}_n(E_v)//K_v)
\end{array}$$

commutes, where the vertical arrows are given by the Satake isomorphism. Here we are using the normalization of the Satake isomorphism used in [ArtC, Chapter 1, Sections 4.1 and 4.2] (see also [Kot, Section 5]).

Let π_v be an irreducible admissible representation of $\mathrm{GL}_n(E_v)$ that is unramified (i.e., admits a K_v-fixed vector). Then $\mathcal{H}(\mathrm{GL}_n(E_v)//K_v)$ acts on the space of π_v via a character λ_{π_v}. An irreducible admissible representation Π_v of $\mathrm{GL}_n(L_v)$ is the **base change** of π_v if Π_v is unramified (i.e., contains a K_{L_v}-fixed vector) and $\mathcal{H}(\mathrm{GL}_n(L_v)//K_{L_v})$ acts on the space of Π_v through a character λ_{Π_v} satisfying

$$\lambda_{\Pi_v}(t) = \lambda_{\pi_v}(b(t)) \quad (E.4.6)$$

for all $t \in \mathcal{H}(\mathrm{GL}_n(L_v)//\mathrm{GL}_n(\mathcal{O}_{L_v}))$ (compare [ArtC, Chapter 3, Section 1]). The well-known, crucial observation is that the relation (E.4.6) uniquely determines Π_v. Globally, one says that an automorphic representation Π of $GL_n(\mathbb{A}_L)$ is a **weak base change** of an automorphic representation π of $\mathrm{GL}_n(\mathbb{A}_F)$ if Π_v is the base change of π_v for almost all places v. This makes sense because π and Π are necessarily unramified at almost all places. Moreover, by strong multiplicity one for $\mathrm{GL}_n(\mathbb{A}_L)$, if such a Π exists and is cuspidal then it is unique. We note that we have omitted a definition of the local base change for ramified representations; this involves a character identity [ArtC, Chapter 1, Definition 6.1] due to Shintani that we will not discuss here. We also note, though this is not quite obvious, that the definition of the unramified base change and global base change given here for general n coincides with that given in Section E.5 for $n=1$ and Section E.6 and Section E.8 for $n=2$.

Let $\mathfrak{c} \subset \mathcal{O}_L$ be an ideal, let $\mathfrak{c}_E = \mathfrak{c} \cap \mathcal{O}_E$, and let $\mathcal{D}_{L/E}$ (resp. $d_{L/E}$) denote the different (resp. discriminant) of L/E. We have an algebra morphism

$$b : \mathcal{H}(\mathrm{GL}_2(\mathbb{A}_L^{\mathfrak{c}\mathcal{D}_{L/E}})//K^{\mathfrak{c}\mathcal{D}_{L/E}}) \longrightarrow \mathcal{H}(\mathrm{GL}_2(\mathbb{A}_E^{\mathfrak{c}_E d_{L/E}})//K^{\mathfrak{c}_E d_{L/E}})$$

given by (E.4.5) at each place $v \nmid \mathfrak{c}_E d_{L/E}$. Here the superscripts indicate the subgroup of the finite adèles trivial at the places dividing the given ideal and

$$K^{\mathfrak{c}\mathcal{D}_{L/E}} := \prod_{v \nmid \infty \mathfrak{c} \mathcal{D}_{L/E}} K_{L_v}$$

$$K^{\mathfrak{c}_E d_{L/E}} := \prod_{v \nmid \infty \mathfrak{c}_E d_{L/E}} K_v.$$

Let

$$\mathbb{T}^{\mathfrak{c}_E d_{L/E}} := \mathbb{Z}[\{K_0(\mathfrak{c}_E) x K_0(\mathfrak{c}_E) \in \mathbb{T}_{\mathfrak{c}_E} : x_{v(\mathfrak{p})} = \begin{pmatrix} 1 & 0 \\ 0 & 1 \end{pmatrix} \text{ for } \mathfrak{p} | \mathfrak{c}_E d_{L/E}\}]$$

$$\mathbb{T}^{\mathfrak{c}\mathcal{D}_{L/E}} := \mathbb{Z}[\{K_0(\mathfrak{c}) x K_0(\mathfrak{c}) \in \mathbb{T}_{\mathfrak{c}} : x_{v(\mathfrak{p})} = \begin{pmatrix} 1 & 0 \\ 0 & 1 \end{pmatrix} \text{ for } \mathfrak{P} | \mathfrak{c}\mathcal{D}_{L/E}\}].$$

as in (8.2). Let $\kappa = (k, m) \in \mathcal{X}(E)$ be a weight, and fix a quasi-character $\chi_E : E^\times \backslash \mathbb{A}_E \to \mathbb{C}^\times$ of conductor dividing $\mathfrak{c}_E$ satisfying $\chi_\infty(\alpha_\infty) = \alpha^{-k-2m}$ for $\alpha_\infty \in \mathbb{A}_{L,\infty}^\times$. Let $\chi = \chi_E \circ \mathrm{N}_{L/E}$. The construction of Section E.2 provides us with injections

$$\chi_E : \mathbb{T}^{\mathfrak{c}_E d_{L/E}} \hookrightarrow \mathcal{H}(\mathrm{GL}_2(\mathbb{A}_E^{\mathfrak{c}_E d_{L/E}})//K^{\mathfrak{c}_E d_{L/E}})$$

$$\chi : \mathbb{T}^{\mathfrak{c}\mathcal{D}_{L/E}} \hookrightarrow \mathcal{H}(\mathrm{GL}_2(\mathbb{A}_L^{\mathfrak{c}\mathcal{D}_{L/E}}//K^{\mathfrak{c}\mathcal{D}_{L/E}})$$

(see (E.2.6)).

Lemma E.5. *Assume L/E is a quadratic extension. View $\mathbb{T}^{\mathfrak{c}_E d_{L/E}}$ as a subalgebra of*

$$\mathcal{H}(\mathrm{GL}_2(\mathbb{A}_E^{\mathfrak{c}_E d_{L/E}})//K^{\mathfrak{c}_E d_{L/E}})$$

and $\mathbb{T}^{\mathfrak{c}\mathcal{D}_{L/E}}$ as a subalgebra of

$$\mathcal{H}(\mathrm{GL}_2(\mathbb{A}_L^{\mathfrak{c}\mathcal{D}_{L/E}}//K^{\mathfrak{c}\mathcal{D}_{L/E}})$$

using (E.2.6). Let b be the base change map given in (E.4.5). We have

$$b(\mathbb{T}^{\mathfrak{c}\mathcal{D}_{L/E}}) \subset \mathbb{T}^{\mathfrak{c}_E d_{L/E}}.$$

The induced map

$$b : \mathbb{T}^{\mathfrak{c}\mathcal{D}_{L/E}} \longrightarrow \mathbb{T}^{\mathfrak{c}_E d_{L/E}} \tag{E.4.7}$$

is that given in Section 8.2 above.

E.4. Langlands functoriality

Proof. The proof proceeds by computing the effect of the base change map b on some elements of $\mathcal{H}(\mathrm{GL}_2(L_v)//K_{L_v})$. The normalization of the Satake isomorphism used in [ArtC, Chapter 1, Sections 4.1 and 4.2] is given explicitly in [Kot, Section 5]. This is the same as the normalization of the Satake isomorphism given in [Lau, Section 4.1]. Let ϖ_v be a uniformizer at the finite place v of E and let q be the order of the residue field at v. By [Lau, Claim 4.1.18] and [Lau, Corollary 4.1.19], we have

$$\mathrm{ch}^{\vee}_{K_v \begin{pmatrix}\varpi_v & 0\\ 0 & 1\end{pmatrix} K_v} = q^{1/2}(t_1+t_2)$$

$$\mathrm{ch}^{\vee}_{K_v \begin{pmatrix}\varpi_v & 0\\ 0 & \varpi_v\end{pmatrix} K_v} = t_1 t_2$$

If v splits in L/E, then a choice of isomorphism $L_v \cong E_v^{[L:E]}$ induces an isomorphism $\mathcal{H}(\mathrm{GL}_2(L_v)//K_{L_v}) \cong \mathcal{H}(\mathrm{GL}_2(E_v)//K_{E_v})^{[L:E]}$ and the Satake isomorphism

$$^{\vee} : \mathcal{H}(\mathrm{GL}_2(L_v)//K_{L_v}) \longrightarrow \mathbb{C}[T^{\mathrm{Gal}(L/E)} \rtimes \sigma]^{N_E}$$

is given by the Satake isomorphism for $\mathcal{H}(\mathrm{GL}_2(E_v)//K_{E_v})^{[L:E]}$ on each factor. In this case the base change morphism is given by

$$b: \mathbb{C}[t_1, t_1^{-1}, t_2, t_2^{-1}]^{S_2} \oplus \mathbb{C}[t_1, t_1^{-1}, t_2, t_2^{-1}]^{S_2} \longrightarrow \mathbb{C}[t_1, t_1^{-1}, t_2, t_2^{-1}]^{S_2}$$
$$(p_1, p_2) \longmapsto p_1 p_2.$$

We conclude that

$$b(\mathrm{ch}_{K_v\begin{pmatrix}\varpi_v & 0\\0 & 1\end{pmatrix}K_v}, \mathrm{ch}_{K_v}) = b(\mathrm{ch}_{K_v}, \mathrm{ch}_{K_v\begin{pmatrix}\varpi_v & 0\\0 & 1\end{pmatrix}K_v}) = \mathrm{ch}_{K_v\begin{pmatrix}\varpi_v & 0\\0 & 1\end{pmatrix}K_v} \quad (\mathrm{E.4.8})$$

$$b(\mathrm{ch}_{K_v\begin{pmatrix}\varpi_v & 0\\0 & \varpi_v\end{pmatrix}K_v}, \mathrm{ch}_{K_v}) = b(\mathrm{ch}_{K_v}, \mathrm{ch}_{K_v\begin{pmatrix}\varpi_v & 0\\0 & \varpi_v\end{pmatrix}K_v}) = \mathrm{ch}_{K_v\begin{pmatrix}\varpi_v & 0\\0 & \varpi_v\end{pmatrix}K_v}.$$

Now suppose v is inert in L/E. Let $\widetilde{\varpi}_v$ denote a uniformizer for L_v. By [Lau, Claim 4.1.18] and [Lau, Corollary 4.1.19] we have

$$\mathrm{ch}^{\vee}_{K_{L_v}\begin{pmatrix}\widetilde{\varpi}_v & 0\\0 & 1\end{pmatrix}K_{L_v}} = q^{[L:E]/2}(t_1+t_2)$$

$$\mathrm{ch}^{\vee}_{K_{L_v}\begin{pmatrix}\widetilde{\varpi}_v & 0\\0 & \widetilde{\varpi}_v\end{pmatrix}K_{L_v}} = t_1 t_2.$$

The base change morphism (E.4.1) is given by

$$b: \mathbb{C}[t_1, t_1^{-1}, t_2, t_2^{-1}]^{S_2} \longrightarrow \mathbb{C}[t_1, t_1^{-1}, t_2, t_2^{-1}]^{S_2}$$
$$p(t_1, t_2) \longmapsto p(t_1^2, t_2^2)$$

[ArtC, Section 4.2]. It follows that

$$b(\mathrm{ch}_{K_{L_v}\begin{pmatrix}\widetilde{\varpi}_v & 0\\0 & 1\end{pmatrix}K_{L_v}}) = \mathrm{ch}_{K_v\begin{pmatrix}\varpi_v & 0\\0 & 1\end{pmatrix}K_v} * \mathrm{ch}_{K_v\begin{pmatrix}\varpi_v & 0\\0 & 1\end{pmatrix}K_v} - 2q\mathrm{ch}_{K_v\begin{pmatrix}\varpi_v & 0\\0 & \varpi_v\end{pmatrix}K_v}$$
$$= \mathrm{ch}_{K_v\begin{pmatrix}\varpi_v^2 & 0\\0 & 1\end{pmatrix}K_v} - q\mathrm{ch}_{K_v\begin{pmatrix}\varpi_v & 0\\0 & \varpi_v\end{pmatrix}K_v} \quad (\mathrm{E.4.9})$$

$$b(\mathrm{ch}_{K_{L_v}}\begin{pmatrix}\tilde{\varpi}_v & 0 \\ 0 & \tilde{\varpi}_v\end{pmatrix}K_{L_v}) = \mathrm{ch}_{K_v}\begin{pmatrix}\varpi_v & 0 \\ 0 & \varpi_v\end{pmatrix}K_v * \mathrm{ch}_{K_v}\begin{pmatrix}\varpi_v & 0 \\ 0 & \varpi_v\end{pmatrix}K_v$$
$$= \mathrm{ch}_{K_v}\begin{pmatrix}\varpi_v^2 & 0 \\ 0 & \varpi_v^2\end{pmatrix}K_v$$

(compare [Bu, Proposition 4.6.4]). The lemma follows from (E.4.8) and (E.4.9). □

E.5 Prime degree base change for GL_1

Let K/F be a finite, prime-degree Galois extension of non-archimedean local fields, and let $\mathrm{N}_{K/F}$ be the norm map. An admissible representation of $\mathrm{GL}_1(F)$ is just a continuous quasicharacter $\chi : F^\times \to \mathbb{C}^\times$. Its base change to $\mathrm{GL}_1(K)$ is simply $\chi \circ \mathrm{N}_{K/F}$.

The Galois group $\mathrm{Gal}(K/F)$ acts on the set of admissible representations of $\mathrm{GL}_1(K)$ by
$$\chi^\sigma(x) := \chi(x^\sigma) \text{ for } \sigma \in \mathrm{Gal}(K/F).$$
Here $x \in K^\times$. On the other hand, we have an identification of the dual $\mathrm{Gal}(K/F)^\wedge$ with the set of characters $\eta : F^\times \to \mathbb{C}^\times$ which are trivial on $\mathrm{N}_{K/F}K^\times$. This gives us an action of $\mathrm{Gal}(K/F)^\wedge$ on the set of admissible representations of $\mathrm{GL}_1(F)$ by the rule sending an $\eta \in \mathrm{Gal}(K/F)^\wedge$ to
$$\chi \longmapsto \eta\chi.$$
The following proposition uses these two actions to give a description of the kernel and image of prime degree base change for GL_1:

Proposition E.6 (Prime-degree base change for GL_1). *Let K/F be a finite prime-degree Galois extension of non-archimedean local fields. Then $\chi \longmapsto \chi \circ \mathrm{N}_{K/F}$ gives a bijection between the $\mathrm{Gal}(K/F)^\wedge$-orbits of admissible representations of $\mathrm{GL}_1(F)$ and the set of $\mathrm{Gal}(K/F)$-fixed admissible representations of $\mathrm{GL}_1(K)$. Moreover,*
$$L(\chi \circ \mathrm{N}_{K/F}, s) = \prod_{\eta \in \mathrm{Gal}(K/F)^\wedge} L(\eta\chi, s).$$

This result is given as Proposition 1 in [Ger] and is a consequence of Frobenius reciprocity. Here, as usual, for non-archimedean local fields F, if $\mathfrak{p}$ is a uniformizer for the maximal ideal of $\mathcal{O}_F$, the $\mathfrak{p}$-adic valuation $|\cdot|$ on K is normalized so that $|\mathfrak{p}|^{-1} = |\mathcal{O}_F/\mathfrak{p}|$, and if $\chi : F^\times \to \mathbb{C}^\times$ is a quasicharacter one sets
$$L(\chi, s) := \begin{cases} (1 - \chi(\mathfrak{p})|\mathfrak{p}|^s)^{-1} & \text{if } \chi \text{ is unramified} \\ 1 & \text{otherwise.} \end{cases}$$

We now give a formula relating the conductor of a quasicharacter to its base change with respect to an unramified extension. Write $U_K^{(n)}$ (resp. $U_F^{(n)}$) for the higher unit groups of K (resp. F). As above, denote by $a(\star)$ the exponent of the conductor of $\star$. We have the following lemma:

Lemma E.7. *If K/F is an unramified prime-degree Galois extension of local fields and $\chi : F^\times \to \mathbb{C}^\times$ is a quasicharacter, then $a(\chi) = a(\chi \circ \mathrm{N}_{K/F})$.*

Proof. Since K/F is unramified, we have $\mathrm{N}_{K/F} U_K^{(n)} = U_F^{(n)}$ for all $m \geq 0$ (see [Neu, Section V.1 Corollary 1.2]). Thus χ is trivial on $U_F^{(m)}$ if and only if $\chi \circ \mathrm{N}_{K/F}$ is trivial on $U_K^{(m)}$, and the lemma follows. □

E.6 Conductors of admissible representations of GL_2

As in Section E.5, let K/F be a prime-degree Galois extension of non-archimedean local fields. In this paragraph, we recall the notion of a base change lifting of an admissible representation of $\mathrm{GL}_{2/F}$. The local base change lifting can be made explicit (at least in the "non-exceptional" case) and we will recall the explicit description in the interest of proving Proposition E.9, which relates the conductor of an admissible representation to the conductor of its base change.

We begin by listing the admissible representations of GL_2 over a characteristic zero non-archimedean local field F, together with their conductors. For the definitions of these representations, refer to [Schm] or [Ger]. Let χ_1, χ_2 be quasicharacters of $F^\times$ with $\chi_1 \chi_2^{-1} \neq |\cdot|^{\pm 1}$. Moreover, let E/F be a quadratic extension and $\xi : E^\times \to \mathbb{C}$ a quasicharacter non-trivial on $\mathrm{N}_{E/F} E^\times$.

Admissible Representations of GL_2		
Representation	Notation	Conductor
Principal series	$\pi(\chi_1, \chi_2)$	$\mathfrak{f}(\chi_1)\mathfrak{f}(\chi_2)$
Weil Representation	ω_ξ	$\mathfrak{p}_F^{f(E/F)a(\xi)} d_{E/F}$
Steinberg representations	$\rho(\chi_1|\cdot|^{1/2}, \chi_1|\cdot|^{-1/2})$	$\mathfrak{p}_F^{2a(\chi_1)} \cap \mathfrak{p}_F$
Exceptional representations		$\mathfrak{f}(\pi) \subseteq \mathfrak{p}_F^2$

Here $\mathfrak{p}_F$ is the maximal ideal of $\mathcal{O}_F$ (the ring of integers of F) and $f(E/F)$ is the degree of the extension of residue fields associated to E/F. The conductor calculations in this table are all given in [Schm], with the exception of the last line on exceptional representations, for this, see the proof of Theorem 1 of [Cas], especially the remark above the lemma on p. 303.

Remark. The exceptional representations can only occur in residue characteristic 2 (see [Ger, Appendix B]). Under the local Langlands correspondence, they correspond to Galois representations $\rho : \mathrm{Gal}(\overline{F}/F) \to \mathrm{GL}_2(\mathbb{C})$ such that the image of the inertia group in $\mathrm{PGL}_2(\mathbb{C})$ is A_4 or S_4.

We now define local base change liftings, following [Ger]:

Definition E.6.1. Let K/F be a prime-degree Galois extension with
$$\langle \sigma \rangle = \mathrm{Gal}(K/F).$$
We say that an admissible representation $\widehat{\pi}$ of $\mathrm{GL}_2(K)$ is a **base change lifting** of an admissible representation π of $\mathrm{GL}_2(F)$ if one of the following holds:

(1) We have $\widehat{\pi} \cong \pi(\chi_1 \circ \mathrm{N}_{K/F}, \chi_2 \circ \mathrm{N}_{K/F})$ and $\pi \cong \pi(\chi_1, \chi_2)$ for two quasicharacters $\chi_i : F^\times \to \mathbb{C}$.
(2) There is an extension $\widehat{\pi}'$ of $\widehat{\pi}$ to $\mathrm{Gal}(K/F) \ltimes \mathrm{GL}_2(K)$ such that $\mathrm{Tr}(\widehat{\pi}'(\sigma \times z)) = \mathrm{Tr}(\pi(x))$ whenever $\prod_{i=1}^{[K:F]} z^{\sigma^i}$ is conjugate in $\mathrm{GL}_2(K)$ to a regular semi-simple element $x \in \mathrm{GL}_2(F)$.

Remark. The isomorphism class of $\widehat{\pi}$ is independent of the choice of generator for $\mathrm{Gal}(K/F)$ (see [Ger, Theorem 1]).

We now wish to recall the analogue of Proposition E.6 for admissible representations of GL_2. For its statement, notice that for a prime degree Galois extension K/F, there is a natural action of $\mathrm{Gal}(K/F)$ on the set of admissible representations of $\mathrm{GL}_2(K)$ given by $\pi^\sigma(x) := \pi(x^\sigma)$. We then have the following:

Theorem E.8 (Local prime degree base change for GL_2). *Let π be an admissible representation of $\mathrm{GL}_2(F)$, where F is a characteristic zero non-archimedean local field. Then there is a unique base change lifting of π to an admissible representation $\widehat{\pi}$ of $\mathrm{GL}_2(K)$. The association $\pi \mapsto \widehat{\pi}$ enjoys the following properties:*

(1) *If $\widehat{\pi} \cong \widehat{\pi}_0$ then $\pi \cong \pi_0 \otimes \eta$ for some character $\eta : F^\times \to \mathbb{C}^\times$ trivial on $\mathrm{N}_{K/F}(K^\times)$ or $\pi = \pi(\chi_1, \chi_2)$ and $\pi_0 = \pi(\chi_1', \chi_2')$ where $\chi_1 \chi_1'^{-1}$ and $\chi_2 \chi_2'^{-1}$ are trivial on $\mathrm{N}_{K/F} K^\times$.*
(2) *We have*
$$L(\widehat{\pi}, s) = \prod_{\eta \in \mathrm{Gal}(K/F)^\wedge} L(\pi \otimes \eta, s)$$
where we use local class field theory to identify $\mathrm{Gal}(K/F)^\wedge$ with the group of characters of $F^\times$ trivial on $\mathrm{N}_{K/F} K^\times$.

This theorem is stated as Theorem 1 in [Ger]. See [Lan] for a proof. For the reader's convenience, we write down explicitly the L-functions of the various isomorphism classes of admissible representations of $\mathrm{GL}_2(F)$:

Representation	L-function
$\pi(\chi_1, \chi_2)$	$L(\chi_1, s) L(\chi_2, s)$
ω_ξ	$L(\xi, s)$
$\rho(\chi_1 \vert \cdot \vert^{1/2}, \chi_1 \vert \cdot \vert^{-1/2})$	$L(\chi_1 \vert \cdot \vert^{1/2}, s)$
Exceptional representations	1

E.6. Conductors of admissible representations of GL_2

See [Ger, Appendix B] for the first and the second statements, [JaLan, Proposition 3.6] for the third, and the paragraph after [Kutz, Theorem 1] for the fourth.

Let K/F be a finite prime-degree Galois extension. It is easy to write down the effect of the corresponding base change on the first three types of representation listed in the previous table:

Representation	Base Change
$\pi(\chi_1, \chi_2)$	$\pi(\chi_1 \circ N_{K/F}, \chi_2 \circ N_{K/F})$
ω_ξ	$\begin{cases} \omega_{\xi \circ N_{K/F}} & \text{if } E \not\subseteq K \\ \pi(\xi, \xi^{\sigma'}) & \text{if } E = K \end{cases}$
$\rho(\chi_1 \cdot \lvert\cdot\rvert^{1/2}, \chi_1 \cdot \lvert\cdot\rvert^{-1/2})$	$\rho(\chi_1 \circ N_{K/F} \cdot \lvert\cdot\rvert^{1/2}, \chi_1 \circ N_{K/F} \cdot \lvert\cdot\rvert^{-1/2})$

Here $\langle \sigma' \rangle = \mathrm{Gal}(E/F)$, and we view $\lvert\cdot\rvert$ as the normalized valuation of F on the left and $\lvert\cdot\rvert$ as the normalized valuation of K on the right.

We now give a proposition relating the conductor of an admissible representation to the conductor of its base change with respect to an unramified prime-degree Galois extension K/F.

Proposition E.9. *Let K/F be an unramified prime-degree Galois extension, let π be an admissible representation of $GL_2(F)$ and $\widehat{\pi}$ its base change to $GL_2(K)$. If the residue characteristic of F is not 2, or π is not exceptional, then $a(\pi) = a(\widehat{\pi})$. If the residue characteristic is 2, then $a(\pi) \leq 1$ implies $a(\widehat{\pi}) = a(\pi)$, and $a(\pi) = 0$ if and only if $a(\widehat{\pi}) = 0$.*

Here $a(\star)$ denotes the exponent of the conductor of $\star$.

Proof of Proposition E.9. If π is either a principal series representation or is a Steinberg representation, then the proposition follows immediately from Lemma E.7 combined with the first and the last table of this section.

Now assume that $\pi \cong \omega_\xi$ is the Weil representation attached to a quasicharacter $\xi : E^\times \to \mathbb{C}^\times$, where E/F is a quadratic extension. If $E = K$, then $\widehat{\omega_\xi}$ is isomorphic to the principal series representation $\pi(\xi, \xi^\sigma)$, where $\langle \sigma \rangle = \mathrm{Gal}(E/F)$. The exponent of the conductor of ω_ξ is $f(E/F)a(\xi) + a(E/F)$, where $f(E/F)$ is the degree of the extension of residue fields associated to E/F and the conductor of $\pi(\xi, \xi^\sigma)$ is $2a(\xi)$ (see [Schm]). We have that E/F is unramified, so we must show that

$$f(E/F)a(\xi) + a(E/F) = 2a(\xi),$$

which follows from Lemma E.7.

We now assume that $E \not\subseteq K$; in this case $\widehat{\omega_\xi} = \omega_{\xi \circ N_{EK/E}}$. Thus we must show that

$$f(E/F)a(\xi) + a(E/F) = f(EK/K)a(\xi \circ N_{EK/E}) + a(EK/K).$$

We have that $f(E/F) = f(EK/K)$ and $a(EK/K) = a(E/F)$. Moreover, EK/E is unramified because K/F is unramified, so another application of Lemma E.7 completes the proof in this case.

Assuming that the residue characteristic of F is not 2, we have exhausted all possibilities for π. We now assume that the residue characteristic is 2. If $a(\pi) \leq 1$, then by the conductor table in Section E.6, we conclude π is not exceptional, so the arguments above prove the desired result. Thus we are reduced to showing that $a(\widehat{\pi}) = 0$ implies that $a(\pi) = 0$. Again, arguments above prove this when π is in the principal series, is special, or is dihedral. If π is exceptional, then it is ramified, so $a(\pi) \neq 0$. We claim that in this case, $\widehat{\pi}$ must also be ramified (i.e., $a(\widehat{\pi}) \neq 0$). To see this, fix an algebraic closure $\overline{F}$ of F and an embedding $K \hookrightarrow \overline{F}$. Let

$$\rho(\pi) : \mathrm{Gal}(\overline{F}/F) \to \mathrm{GL}_2(\mathbb{C})$$
$$\rho(\widehat{\pi}) = \rho(\pi)|_{\mathrm{Gal}(\overline{F}/K)} : \mathrm{Gal}(\overline{F}/K) \to \mathrm{GL}_2(\mathbb{C})$$

be the Galois representations associate to π (resp. $\widehat{\pi}$) by the local Langlands correspondence for GL_2. Since the image of the inertia group under $\rho(\pi)$ composed with the canonical projection $P : \mathrm{GL}_2(\mathbb{C}) \to \mathrm{PGL}_2(\mathbb{C})$ is isomorphic to A_4 or S_4 and the extension K/F is abelian, it follows that the image of the inertia group under $P \circ \rho(\pi)|_{\mathrm{Gal}(\overline{F}/K)}$ is not trivial. Thus $\rho(\pi)|_{\mathrm{Gal}(\overline{F}/K)}$ is ramified, which implies $\widehat{\pi}$ is ramified. □

E.7 The archimedean places

Let F be a totally real number field and v a finite place of F. In the definition of an automorphic representation, the analogue at the archimedean places of an admissible representation of $\mathrm{GL}_2(F_v)$ is not a representation of $\mathrm{GL}_2(\mathbb{R})$ per ce, but rather an *admissible* $(\mathfrak{gl}_2(\mathbb{R}), O_2(\mathbb{R}))$-*module*. For the definition, see [Bu, Section 2.4].

We now recall the classification of the infinite-dimensional irreducible admissible $(\mathfrak{gl}_2(\mathbb{R}), O_2(\mathbb{R}))$-modules. Let $\chi_1, \chi_2 : \mathbb{R}^\times \to \mathbb{C}^\times$ be two quasicharacters such that $\chi_1 \chi_2^{-1}$ is not a quasicharacter of the form $y \mapsto \mathrm{sgn}(y)^\epsilon |y|^{\lambda-1}$, where $\lambda \in \mathbb{Z}$, $\epsilon \in \{0, 1\}$ and $\lambda \equiv \epsilon \pmod{2}$. The infinite-dimensional irreducible admissible $(\mathfrak{gl}_2(\mathbb{R}), O_2(\mathbb{R}))$-modules are given, up to isomorphism, in the following list:

(1) Principal series representations, denoted $\pi(\chi_1, \chi_2)$.
(2) Discrete series representations $\mathcal{D}_\mu(\lambda)$, where $\lambda \in \mathbb{Z}_{>1}$ and $\mu \in \mathbb{C}$.
(3) Limits of discrete series representations $\mathcal{D}_\mu(1)$, where $\mu \in \mathbb{C}$.

For a proof of this classification, see [Bu, Theorem 2.5.5]. As they do not play as much of a role in this work, we will not comment on the principal series representations. However, we will recall some basic properties of the discrete series representations that we used in the proof of Theorem E.1.

E.7. The archimedean places

We first set notation for the standard generators of $\mathfrak{sl}_2(\mathbb{R}) \otimes \mathbb{C}$:

$$X := \frac{1}{2}\begin{pmatrix} 1 & i \\ i & -1 \end{pmatrix} \quad Y := \frac{1}{2}\begin{pmatrix} 1 & -i \\ -i & -1 \end{pmatrix} \quad H := \begin{pmatrix} 0 & -i \\ i & 0 \end{pmatrix}.$$

Then the Casimir operator[1] for $\mathfrak{sl}_2(\mathbb{R})$, as in Section 5.4, is

$$C := XY + YX + \frac{H^2}{2}. \tag{E.7.1}$$

We will also require the element $I = \begin{pmatrix} 1 & 0 \\ 0 & 1 \end{pmatrix} \in Z(\mathfrak{gl}_2(\mathbb{R}))$.

As a means of distinguishing the discrete series and principal series representations, we have the following proposition:

Proposition E.10. *Let V be an irreducible, infinite-dimensional admissible $(\mathfrak{gl}_2(\mathbb{R}), O_2(\mathbb{R}))$-module. Then V is in the principal series if and only if the eigenvalue of C is not of the form $\left(\frac{\lambda^2}{2} - \lambda\right)$ where λ is an integer.*

See [Bu, Section 2.5] for a proof.

The discrete series representation $\mathcal{D}_\mu(\lambda)$ has the properties that

$$C \cdot w = \left(\frac{\lambda^2}{2} - \lambda\right) w \quad \text{and} \quad I \cdot w = \mu w \tag{E.7.2}$$

for all $w \in \mathcal{D}_\mu(\lambda)$. These two properties distinguish $\mathcal{D}_\mu(\lambda)$ among the infinite-dimensional $(\mathfrak{gl}_2, O_2(\mathbb{R}))$-modules that are not in the principal series. Fix $\mu \in \mathbb{C}$ and $\lambda \in \mathbb{Z}_{\geq 1}$. We then have a decomposition

$$\mathcal{D}_\mu(\lambda) = \bigoplus_{\substack{|\beta| \geq \lambda \\ \beta \equiv \lambda \pmod{2}}} \mathbb{C}w_\beta, \tag{E.7.3}$$

where

$$\begin{pmatrix} \cos(\theta) & \sin(\theta) \\ -\sin(\theta) & \cos(\theta) \end{pmatrix} \cdot w_\beta = e^{i\beta\theta} w_\beta \quad \text{and} \quad \begin{pmatrix} 1 & 0 \\ 0 & -1 \end{pmatrix} \cdot w_\beta = w_{-\beta}. \tag{E.7.4}$$

The action of $X, Y,$ and H on the w_β is given by

$$X \cdot w_\beta = \frac{1}{2}(\lambda + \beta) w_{\beta+2} \quad Y \cdot w_\beta = \frac{1}{2}(\lambda - \beta) w_{\beta-2} \quad H \cdot w_\beta = \beta w_\beta.$$

We see that this gives a complete description of $\mathcal{D}_\mu(\lambda)$ as a $(\mathfrak{gl}_2(\mathbb{R}), O_2(\mathbb{R}))$-module. For proofs of all of this, see [Bu, Section 2.5], and for a quick summary, see [Ku2, Section 3].

---

[1] In the notation of [Bu, Sections 2.2 and 2.3], $-2\Delta = C$. We note that it does not matter whether we define C in terms of the generators of $\mathfrak{sl}_2(\mathbb{R})$ or $\mathfrak{sl}_2(\mathbb{R}) \otimes \mathbb{C}$ because $C \in Z(\mathfrak{gl}_2)$.

E.8 Global base change

Let L/E be a prime-degree Galois extension of totally real number fields. For any place v of E, let $L_v := L \otimes_E E_v$. If v is a finite place of E and π_v is an admissible representation of $\mathrm{GL}_2(E)$, we let $BC(\pi_v)$ be the admissible representation of $\otimes_{w|v} \mathrm{GL}_2(L_w)$ that is given by the base change $\widehat{\pi}$ of π to L_w for every place $w|v$. If v is an infinite place, then $L_v \cong \bigoplus_{w|v} \mathbb{R}$. If π_v is an infinite-dimensional irreducible admissible $(\mathfrak{gl}_2, O_2(\mathbb{R}))$-module, define $BC(\pi_v)$ to be the irreducible admissible $\prod_{w|v}(\mathfrak{gl}_2, O_2(\mathbb{R}))$-module given by the product $\otimes_{w|v} \pi_v$.

Now let π be an automorphic representation of $\mathrm{GL}_2(\mathbb{A}_E)$. An automorphic representation $\widetilde{\pi}$ of $\mathrm{GL}_2(\mathbb{A}_L)$ is a *base change lifting* of π if for all places v of E we have that $\prod_{w|v} \widetilde{\pi}_w \cong BC(\pi_v)$. If $\widetilde{\pi}$ is any base change lifting of π we write $\widehat{\pi} := \widetilde{\pi}$.

In analogy with the case of automorphic representations of GL_1, we have an action of $\mathrm{Gal}(L/E)$ on the set of admissible representations of $\mathrm{GL}_2(\mathbb{A}_L)$ given by $\pi^\sigma(x) := \pi(x^\sigma)$. Moreover, we have an action of $\{\eta\} = \mathrm{Gal}(L/E)^\wedge$ on the set of admissible representations of $\mathrm{GL}_2(\mathbb{A}_E)$ given by $\pi \mapsto \pi \otimes \eta$, where we identify $\mathrm{Gal}(L/E)^\wedge$ with the group of Hecke characters trivial on $\mathrm{N}_{L/E} \mathbb{A}_L^\times$. With this in mind, we state the main theorem of base change for GL_2.

Theorem E.11 (Base change for GL_2). *Let L/E be a prime-degree Galois extension of totally real number fields. Every automorphic representation π of $\mathrm{GL}_2(\mathbb{A}_E)$ has a unique base change lifting to an automorphic representation $\widehat{\pi}$ of $\mathrm{GL}_2(\mathbb{A}_L)$. Moreover, any cuspidal automorphic representation $\widetilde{\pi}$ of $\mathrm{GL}_2(\mathbb{A}_L)$ that is isomorphic to its conjugates under the action of $\mathrm{Gal}(L/E)$ is of the form $\widetilde{\pi} = \widehat{\pi}$ for a cuspidal automorphic representation π of $\mathrm{GL}_2(\mathbb{A}_E)$. The association $\pi \mapsto \widehat{\pi}$ enjoys the following properties:*

(1) *If π and $\widehat{\pi}$ are both cuspidal, then any automorphic representation π' such that $\widehat{\pi'} = \widehat{\pi}$ satisfies $\pi \otimes \eta \cong \pi'$ for some $\eta \in \mathrm{Gal}(L/E)^\wedge$.*

(2) *If π and $\widehat{\pi}$ are both cuspidal, we have*

$$L_v(\widehat{\pi}, s) = \prod_{\eta \in \mathrm{Gal}(L/E)^\wedge} L_v(\pi \otimes \eta, s)$$

for every place v.

For a proof, see [Lan]. This formulation of the theorem is an adaptation of that given in [Ger, Theorem 2]. We are using Langlands' notation for the L-functions (see [Lan]).

Remarks.

(1) This theorem holds for L/E an arbitrary prime-degree Galois extension, though to state it precisely we would have to deal with infinite places v such that L_v or E_v is isomorphic to $\mathbb{C}$.

E.8. Global base change

(2) We could have also incorporated non-cuspidal automorphic representations into the theorem at the cost of introducing additional notation. See [Lan] or [Ger] for details. Philosophically, we used the theory of base change for Eisenstein series in Chapter 11, but since base change liftings of Eisenstein series on GL_2 can be defined entirely in terms of quasicharacters $\chi : E^\times \backslash \mathbb{A}_E^\times \to \mathbb{C}$, there was no need to invoke the machinery of an analogue of Theorem E.11.

(3) The statement of base change for GL_2 given in [Ger] uses admissible $\mathrm{GL}_2(\mathbb{R})$-modules as opposed to admissible $(\mathfrak{gl}_2(\mathbb{R}), O_2(\mathbb{R}))$-modules, thus some "translation" has to be done to obtain the precise statement of Theorem E.11.

As above, let L/E be a prime-degree Galois extension of totally real number fields. Fix a weight $\kappa = (k, m) \in \mathcal{X}(E)$ and let χ_E be a character of $E^\times \backslash \mathbb{A}_E^\times$ satisfying $\chi_{E\infty}(a_\infty) = a_\infty^{-k-2m}$. Moreover, set $\chi = \chi_E \circ \mathrm{N}_{L/E}$ and define $\widehat{\kappa}$ as in the introduction to Chapter 8. The following is a corollary of of Theorem E.11:

Corollary E.12. *Suppose that $\widehat{h} \in S_{\widehat{\kappa}}^{\mathrm{new}}(K_0(\mathfrak{c}), \chi)$ is a simultaneous eigenform for all Hecke operators for some $\mathfrak{c} \subset \mathcal{O}_L$. Assume that for almost all primes $\mathfrak{P} \subset \mathcal{O}_L$ we have $\lambda_{\widehat{h}}(\mathfrak{P}) = \lambda_{\widehat{h}}(\mathfrak{P}^\sigma)$ for all $\sigma \in \mathrm{Gal}(L/E)$. Then there is an ideal $\mathcal{N}'(\mathfrak{c}) \subseteq \mathcal{O}_E$ and a newform $h \in S_\kappa^{\mathrm{new}}(K_0(\mathcal{N}'(\mathfrak{c})), \chi_E)$ such that $\widehat{\pi(h)} = \pi(\widehat{h})$; in particular*

$$\prod_{\mathfrak{P}|\mathfrak{p}} L_\mathfrak{P}(\widehat{h}, s) = \prod_{\eta \in \mathrm{Gal}(L/E)^\wedge} L_\mathfrak{p}(h \otimes \eta, s) \tag{E.8.1}$$

for all primes $\mathfrak{p}$. Moreover the ideal $\mathcal{N}'(\mathfrak{c}) \subset \mathcal{O}_E$ satisfies

$$\mathcal{N}'(\mathfrak{c}) \supseteq \mathfrak{m}_2 \mathfrak{b}_{L/E}(\mathfrak{c} \cap \mathcal{O}_E)$$

where $\mathfrak{m}_2 \subset \mathcal{O}_E$ is an ideal divisible only by dyadic primes which we may take to be $\mathcal{O}_E$ if $\mathfrak{c} + 2\mathcal{O}_L = \mathcal{O}_L$ and $\mathfrak{b}_{L/E}$ is an ideal divisible only by those primes dividing $d_{L/E}$.

Conversely, if $h \in S_\kappa^{\mathrm{new}}(K_0(\mathfrak{c}_E), \chi_E)$ is a newform, then there is a unique newform $\widehat{h} \in S_{\widehat{\kappa}}^{\mathrm{new}}(K_0(\mathfrak{c}), \chi)$ (for some ideal $\mathfrak{c} \subset \mathcal{O}_L$) such that $\pi(\widehat{h}) = \widehat{\pi(h)}$. In particular,

(1) *For all $\mathfrak{p}$ the equality (E.8.1) holds.*
(2) *For all $\mathfrak{p} \nmid \mathfrak{c}$ we have $\lambda_{\widehat{h}}(\mathfrak{P}) = \lambda_{\widehat{h}}(\mathfrak{P}^\sigma)$ for all $\sigma \in \mathrm{Gal}(L/E)$.*

Moreover, in this situation we have:

$$\lambda_{\widehat{h}}(t) = \lambda_h(b(t)) \tag{E.8.2}$$

for all $t \in \mathbb{T}^{\mathfrak{c}\mathcal{D}_{L/E}}$, where $b : \mathbb{T}^{\mathfrak{c}\mathcal{D}_{L/E}} \longrightarrow \mathbb{T}^{\mathfrak{c}_E d_{L/E}}$ is the base change mapping of (E.4.7) and of Section 8.2.

Remark. The η of Corollary E.12 are trivial on $\mathbb{A}_{E\infty}^\times \subset \mathrm{N}_{L/E}\mathbb{A}_L^\times$, and hence $h \otimes \eta$ has the same weight as h.

Proof. Let S be the finite (possibly empty) set of primes of $\mathcal{O}_L$ where $\lambda_{\widehat{h}}(\mathfrak{P}) \neq \lambda_{\widehat{h}}(\mathfrak{P}^\sigma)$ for some $\sigma \in \mathrm{Gal}(L/E)$ together with the primes ramifying in L/E and the primes where $\pi(\widehat{h})$ is ramified. Here $\pi(\widehat{h})$ is the automorphic representation generated by $\widehat{h}$. Let $\mathfrak{P} \subset \mathcal{O}_L$, $\mathfrak{P} \notin S$, be a prime such that $\mathfrak{p} = \mathrm{N}_{L/E}(\mathfrak{P})$ is prime. We claim that $\pi(\widehat{h})_{v(\mathfrak{P})}^\sigma \cong \pi(\widehat{h})_{v(\mathfrak{P}^\sigma)}$ as irreducible admissible representations of $\mathrm{GL}_2(L_{v(\mathfrak{P})}) \cong \mathrm{GL}_2(L_{v(\mathfrak{P})^\sigma}) \cong \mathrm{GL}_2(E_{v(\mathfrak{p})})$ for all $\sigma \in \mathrm{Gal}(L/E)$.

To see this, note that our assumptions that $\lambda_{\widehat{h}}(\mathfrak{P}) = \lambda_{\widehat{h}}(\mathfrak{P}^\sigma)$ and $\chi = \chi_E \circ \mathrm{N}_{L/E}$ imply that $L_{\mathfrak{P}}(\widehat{h}, s) = L_{\mathfrak{P}^\sigma}(\widehat{h}, s)$ for all $\mathfrak{P} \notin S$. This implies, by Theorem E.4, that

$$L(\pi(\widehat{h})_{v(\mathfrak{P})}, s) = L(\pi(\widehat{h})_{v(\mathfrak{P}^\sigma)}, s). \tag{E.8.3}$$

Since $\pi(\widehat{h})_{v(\mathfrak{P})}$ and $\pi(\widehat{h})_{v(\mathfrak{P}^\sigma)}$ are both principal series representations, it makes sense to speak of their Satake parameters as representations of $\mathrm{GL}_2(L_{v(\mathfrak{P})}) \cong \mathrm{GL}_2(L_{v(\mathfrak{P})^\sigma}) \cong \mathrm{GL}_2(E_{v(\mathfrak{p})})$. Equation (E.8.3) implies that these Satake parameters are equal, hence $\pi(\widehat{h})_{v(\mathfrak{P})} \cong \pi(\widehat{h})_{v(\mathfrak{P}^\sigma)}$ (see, e.g., [Ku2, Section 3]).

Now let $\rho(\pi(\widehat{h}))$ be the (ℓ-adic) Galois representation attached to $\widehat{h}$ by the work of Taylor and Blasius-Rogawski (see [Tay1, pp. 265–266] and [BlaR]). Since the set of primes $\mathfrak{P} \subset \mathcal{O}_L$ not in S such that $\mathrm{N}_{L/E}(\mathfrak{P})$ is a prime of $\mathcal{O}_E$ has Dirichlet density 1 as a set of places of L [Ser, Section 1.2.2], a standard argument using the Chebatarev density theorem implies that $\rho(\pi(\widehat{h}))^\sigma \cong \rho(\pi(\widehat{h}))$ for $\sigma \in \mathrm{Gal}(L/E)$. This implies by the local Langlands correspondence[2] that $\pi(\widehat{h})_{v^\sigma} \cong \pi(\widehat{h})_v^\sigma$ for almost all finite places v of L. Thus, in the terminology of [Lan, Section 2], $\pi(\widehat{h})$ is a "quasi-lifting" of some automorphic representation π of $\mathrm{GL}_2(\mathbb{A}_E)$, which implies, again as stated in [Lan, Section 2], that $\pi(\widehat{h})$ is a lifting. The proof of the first statement now follows from Theorem E.11 combined with Theorem E.1, Theorem E.4, and Proposition E.9.

For the converse statement, we first claim that $\widehat{\pi(h)}$ is cuspidal. If $\widehat{\pi(h)}$ is not cuspidal, then L/E must be quadratic, and $\pi(h)$ must be induced from a quasicharacter of $L^\times \backslash \mathbb{A}_L^\times$ (see Appendix C and Theorem 2(b) of [Ger]). This implies that $\pi(h)$ is in the principal series at every infinite place, which contradicts part (3) of Theorem E.1.

Thus the base change $\widehat{\pi(h)}$ is cuspidal. By Theorem E.1, we have $\widehat{\pi(h)} = \pi(\widetilde{h})$ for a unique normalized newform $\widetilde{h} \in S^{\mathrm{new}}(K_0(\mathfrak{c}), \chi)$ for some ideal $\mathfrak{c} \subset \mathcal{O}_L$. We claim that we may take $\widehat{h} = \widetilde{h}$. Indeed, Theorem E.4 implies (1). Moreover, for any prime $\mathfrak{P} \subset \mathcal{O}_L$, the fact that $\left(\widehat{\pi(h)}_{v(\mathfrak{P})}\right)^\sigma \cong \widehat{\pi(h)}_{v(\mathfrak{P}^\sigma)}$ implies that these two irreducible admissible representations are isomorphic as Hecke modules, and thus the translation between the "automorphic" Hecke algebra and the classical Hecke algebra outlined in Section E.2 implies that $\lambda_{\widetilde{h}}(\mathfrak{P}) = \lambda_{\widetilde{h}}(\mathfrak{P}^\sigma)$ for all primes $\mathfrak{P} \subset \mathcal{O}_L$ and $\sigma \in \mathrm{Gal}(L/E)$. $\square$

---

[2] See [Tay2] for a discussion of the compatibility of $\rho(\pi(\widehat{h}))$ and the local Langlands correspondence.

Bibliography

[Adem] A. Adem, J. Leida and Y. Ruan, **Orbifolds and Stringy Topology**, Cambridge University Press, Cambridge UK, 2007. [187, 197]

[Andr] G.E. Andrews, R. Askey and R. Roy, **Special Functions**, Cambridge University Press, Cambridge UK, 2000. [211]

[Art] J. Arthur, D. Ellwood, R. Kottwitz (ed.), **Harmonic Analysis, the Trace Formula, and Shimura Varieties**, Clay Mathematics Proceedings **4**, Amer. Math. Soc., Providence R.I., 2005. [240]

[ArtC] J. Arthur and L. Clozel, **Simple Algebras, Base Change, and the Advanced Theory of the Trace Formula**, Annals of Math. Studies **120**, Princeton University Press, 1989. [225, 227]

[AshG] A. Ash and D. Ginzburg, p-adic L-functions for $GL(2n)$, Invent. Math. **116** (1994) 27–74. [13]

[AshM] A. Ash, D. Mumford, M. Rapoport, and Y.-S. Tai, **Smooth Compactifications of Locally Symmetric Varieties**, Math. Sci. Press, Brookline MA, 1975. [42]

[Atk] A.O. Atkin and J. Lehner, Hecke operators on $\Gamma_0(m)$, Math. Ann. **185** (1970), 134–160. [71]

[Bai] W. Baily, The decomposition theorem for V-manifolds, Amer. J. Math **78** (1956), 862–888. [187]

[BaiB] W. Baily and A. Borel, Compactification of arithmetic quotients of bounded symmetric domains, Ann. of Math. **84** (1966), pp. 442–528. [32, 42, 43]

[Bei] A. Beilinson, J. Bernstein and P. Deligne, **Faisceax Pervers**, Astérisque **100**, Soc. Math. de France, Paris, 1982. [31]

[Bie] E. Bierstone, The structure of orbit spaces and the singularities of equivariant mappings. Monografias de matemática **35**, Instituto de Matemática Pure e Aplicada, Rio de Janeiro, 1980. [194]

[Bis] E. Bishop, Conditions for the analyticity of certain sets, I. Mich. Math. Jour. 11(1964) 289–304. [55]

[Bla]   D. Blasius, *Hilbert modular forms and the Ramanujan conjecture*, **Noncommutative geometry and number theory**, **Aspects Math. E37** 35–56 (2006). [177]

[BlaR] D. Blasius and J. Rogawski, Motives for Hilbert modular forms, Invent. Math. 114 (1994) no. 1, 55–87. [236]

[Borc] R. Borcherds, The Gross-Kohnen-Zagier theorem in higher dimensions, Duke Math. J. **97** (1999), 219–233. [3]

[Bo69] A. Borel, **Introduction aux groupes arithmétiques**, Hermann, Paris, 1969. [47]

[Bo79] A. Borel, Automorphic L-functions, in **Automorphic Forms, Representations, and L-Functions** Proc. Symp. in Pure Math. **33.2** (1979) 27–61. [223]

[Bo84] A. Borel, Sheaf theoretic intersection cohomology. **Seminar on Intersection Cohomology**, Progress in Mathematics **50**, Birkhäuser Boston, 1984. [25, 30, 240]

[BoC]   **Automorphic Forms, Representations, and L-Functions**, A. Borel and W. Casselman, eds., Proc. Symp. in Pure Math. **33.2** (1979). [223]

[BoHC] A. Borel and Harish-Chandra, Arithmetic subgroups of algebraic groups, Ann. Math. **75** (1962), 485–535. [42, 54]

[BoS]   A. Borel and J.P. Serre, Corners and arithmetic groups, Comment. Math. Helv. **48** (1973), pp. 436–491. [41, 49]

[BoW]   A. Borel and N. Wallach, **Continuous Cohomology, Discrete Subgroups, and Representations of Reductive Groups** (second ed.), Math. Surv. **67**, Amer. Math. Soc., Providence, R. I., 2000. [13, 107]

[Bre]   G. Bredon, **Sheaf Theory**, Graduate Texts in Mathematics **170**, second edition, Springer Verlag, New York, 1997. [25]

[Bru]   J.H. Bruinier, J. Burgos Gil, and U. Kühn, Borcherds products and arithmetic intersection theory on Hilbert modular surfaces, Duke Math. J. **139** (2007) 1–88. [3]

[Bry]   J.L. Brylinski and J.P. Labesse, Cohomologie d'intersection et fonctions L de certaines variétés de Shimura, Ann. Sci. de l'E. N. S. **17** (1984), 361–412. [3, 15]

[Bu]   D. Bump, **Automorphic Forms and Representations**, Cambridge Studies in Advanced Mathematics **55**, Cambridge University Press, Cambridge, 1997. [137, 138, 182, 208, 209, 210, 214, 215, 216, 219, 222, 228, 232, 233]

[Cart] P. Cartier, Representations of p-adic groups: a survey, **Automorphic Forms, Representations, and L-functions**, Proc. Symp. in Pure Math. **33**, AMS, 1979. [140, 217, 218, 219, 221]

[Cas]   W. Casselman, On some results of Atkin and Lehner, Math. Ann. **201** (1973), 301–314. [72, 216, 222, 229]

Bibliography

[Ch1] J. Cheeger, On the Hodge theory of Riemannian pseudomanifolds, in **Geometry of the Laplace Operator**, Proc. Symp. Pure Math. **XXXVI**, Amer. Math. Soc., Providence, R. I., 1980. [53]

[Ch2] J. Cheeger, Spectral geometry of singular Riemannian spaces, J. Diff. Geom. **18** (1983), 575–657. [53]

[ChR] W. Chen and Y. Ruan, A new cohomology theory of orbifolds, Comm. Math. Physics **248** (2004), 1–31. arxiv.math.AG/0004129 v3, 15 March, 2001. [187]

[Cog] J.W. Cogdell, Arithmetic cycles on Picard modular surfaces and modular forms of Nebentypus, J. reine angew. Math. **357** (1984) 115–137. [3]

[Cu] R. Cushman and R. Sjamaar, On singular reduction of Hamiltonian spaces, Symplectic Geometry and Mathematical Physics (Aix-en-Provence, 1990) (P. Donato et al., eds.), Progress in Mathematics, **99**, Birkhäuser, Boston, 1991. [194]

[De] P. Deligne, Travaux de Shimura, in Séminaire Bourbaki, 23ème année (1970/71), Exp. No. 389, Lecture Notes in Mathematics **244** 1971. [58]

[Di] F. Diamond and J. Shurman, **A First Course in Modular Forms**, Graduate Texts in Mathematics **228**, Springer Verlag N.Y., 2005. [72]

[Fed1] H. Federer, Some Theorems on integral currents, Trans. A. M. S. 117 (1965) 43–67. [55]

[Fed2] H. Federer, **Geometric Measure Theory**, Springer Verlag, Berlin, 1969. [55]

[Fed3] H. Federer, Colloquium lectures on geometric measure theory, Bull. Amer. Math. Soc. 84 (1978), 291–338. [55]

[Fl] Y. Flicker, On distinguished representations, Jour. für die reine und angewandte Mathematik 418 (1991), 139–172 [14]

[Fr] E. Freitag, **Hilbert Modular Forms**, Springer Verlag, Berlin, 1990. [14]

[Fu] W. Fulton, **Young Tableaux**, London Math. Soc. Student Texts **35**, Cambridge University Press, Cambridge UK, 1997. [100]

[FuH] W. Fulton and J. Harris, **Representation Theory**, Graduate Texts in Mathematics **129**, Springer Verlag, New York, 1991. [100]

[Ga] P. Garrett, **Holomorphic Hilbert Modular forms**, Wadsworth & Brooks, 1990, Belmong CA. [15, 205]

[Ge] G. van der Geer, **Hilbert Modular Surfaces**, Ergebnisse Math. **16**, Springer Verlag, Berlin, 1988. [14, 16, 17]

[Gel] S. Gelbart, **Automorphic Forms on Adele Groups**, Annals of Mathematics Studies **83** (1975), Princeton University Press. [14, 205, 221]

[GelJ] S. Gelbart and H. Jacquet, A relation between automorphic representations of GL(2) and GL(3), Ann. scient. de l'É.N.S. 4e série **11** 4 (1978) 471–542. [87, 88]

[GelM]   S. Gelfand and Y. Manin, **Algebra V: Homological Algebra**, Encyclopedia of Mathematical Sciences **38** (1994), Springer Verlag, New York. [23, 45]

[Ger]   P. Gérardine, and J.P. Labesse, The solution of a base change problem for GL(2) (following Langlands, Saito, Shintani), **Automorphic Forms, Representations, and L-functions**, Proc. Symp. in Pure Math. **33**, Amer. Math. Soc., Providence R. I., 1979. [144, 228, 229, 230, 231, 234, 235, 236]

[Gh]   E. Ghate, Adjoint L-values and primes of congruence for Hilbert modular forms, Compositio Math. **132** (2002), 243–281. [61, 84, 112, 133]

[Go1]   B. Gordon, Intersections of higher weight cycles over quaternionic modular surfaces and modular forms of Nebentypus, Bull. Amer. Math. Soc. **14** (1986), 293–298. [3]

[Go2]   B. Gordon, Intersections of higher weight cycles and modular forms, Comp. Math. **89** (1993), 1–44. [3]

[Goren] E. Goren, **Lectures on Hilbert Modular Varieties and Modular Forms**, CRM Monograph Series **14**, Amer. Math. Soc., Providence, R.I., 2002. [15]

[Gre1]   M. Goresky, Triangulation of stratified objects, Proc. Amer. Math. Soc. **72** (1978), 193–200. [23, 194]

[Grc2]   M. Goresky, Compactifications and cohomology of modular varieties, in [Art], 551–582. [41, 49, 50]

[Gre3]   M. Goresky, G. Harder, and R. MacPherson, Weighted cohomology, Inv. Math. 116 (1993), 139–213. [41, 50]

[Gre4]   M. Goresky and R. MacPherson, Intersection homology theory, Topology 19 (1980), 135–162. [29, 36, 37]

[Gre5]   M. Goresky and R. MacPherson, Intersection homology II. Inv. Math. **71** (1983), 77–129. [23, 29, 30, 31, 32, 33, 34, 36, 37]

[Gre6]   M. Goresky and R. MacPherson, The topological trace formula, in **The Zeta Function of Picard Modular Surfaces**, R. Langlands and D. Ramakrishnan, eds. Publ. C.R.M., Univ. of Montreal Press (1992) 465–478. [29, 41]

[Gre7]   M. Goresky and W. Pardon, Chern classes of automorphic vector bundles, Inv. Math. **147** (2002), 561–612. [95]

[Gre8]   M. Goresky and P. Siegel, Linking pairings on singular spaces, Comment. Math. Helv. 59 (1983), 96–110. [134]

[Gro]   B. Gross and K. Keating, On the intersection of modular correspondences, Invent. Math. **112** (1993), 225–245. [3]

[Hab]   N. Habegger, From PL to sheaf theory, in [Bo84]. [25, 30]

[Hae]   A. Haefliger, Groupoïdes d'holonomie et classifiants, Astérisque **116**, Soc. Math. de France (1984), 70–79. [198]

Bibliography

[Har] G. Harder, Eisenstein cohomology of arithmetic groups. The case GL_2. Invent. Math. **89** (1987), 33–118. [115, 150, 180]

[HarL] G. Harder, R. Langlands, and M. Rapoport, Algebraische Zyklen auf Hilbert-Blumenthal Flächen, Jour. für die reine und angewandte Mathematik **366** (1986), 53–120. [13, 86]

[Hardt1] R. Hardt, Slicing and intersection theory for chains associated with real analytic varieties. Acta Math. **129** (1972), 75–136. [23, 55]

[Hardt2] R. Hardt, Slicing and intersection theory for chains modulo ν associated with real analytic varieties. Trans. Amer. Math. Soc. **183** (1973), 327–340. [23, 55]

[Hardt3] R. Hardt, Homology and images of semianalytic sets. Bull. Amer. Math. Soc. **80** (1974), 675–678. [23]

[Hardt4] R. Hardt, Homology theory for real analytic and semianalytic sets. Ann. Scuola Norm. Sup. Pisa Cl. Sci. (4) 2 (1975), no. 1, 107–148. [23]

[Hardt5] R. Hardt, Topological properties of subanalytic sets. Trans. Amer. Math. Soc. **211** (1975), 57–70. [23]

[Hardt6] R. Hardt, Stratification of real analytic mappings and images. Invent. Math. **28** (1975), 193–208. [23]

[Hardt7] R. Hardt, Triangulation of subanalytic sets and proper light subanalytic maps. Invent. Math. **38** (1976/77), no. 3, 207–217. [23]

[Harish] Harish-Chandra, Automorphic forms on semi-simple Lie groups, Notes by J.G.M. Mars, Lecture Notes in Mathematics **68** Springer-Verlag (1968). [70]

[Harr] M. Harris et al., **Stabilization of the trace formula, Shimura varieties, and arithmetic applications**, http://people.math.jussieu.fr/ harris/. [223]

[Hid1] H. Hida, On the abelian varieties with complex multiplication as factors of the abelian varieties attached to Hilbert modular forms, Jap. J. Math. New Ser. **5** (1979), 157–208. [160]

[Hid2] H. Hida, On abelian varieties with complex multiplication as factors of the Jacobian of Shimura curves, Amer. J. Math. **103** (1981), 727–776. [160]

[Hid3] H. Hida, On p-adic Hecke algebras for GL_2 over totally real fields. Annals of Math. **128** (1988) 295–384. [57, 66, 89, 91, 112, 116, 119]

[Hid4] H. Hida, Nearly ordinary Hecke algebras and Galois representations of several variables. **Proceedings of the JAMI Inaugural Conference**, Johns Hopkins University Press (1989) 115–134. [222]

[Hid5] H. Hida, On p-adic L-functions of $GL(2) \times GL(2)$ over totally real fields. Annales de l'institut Fourier, **41** no. 2 (1991), 311–391. [15, 57, 70, 74, 76, 77, 84, 89, 116, 121, 173, 174, 175, 180]

[Hid6]   H. Hida, **Elementary Theory of Eisenstein Series and L-functions** London Mathematical Society Student texts **26**, Cambridge University Press (1993). [122, 175, 180]

[Hid7]   H. Hida, On the critical values of L-functions of GL(2) and GL(2) × GL(2), Duke Math. Journal. **74** (1994), 431–529. [5, 15, 57, 61, 64, 65, 74, 89, 112, 115, 150, 212]

[Hid8]   Hida, H., Non-critical values of adjoint L-functions for SL(2), Proc. Symp. Pure Math. 66 (1999) Part I, 123–175. [86, 167, 170, 173, 175]

[Hid9]   H. Hida, **Hilbert Modular Forms and Iwasawa Theory** Oxford University Press (2006). [64]

[HidT]   Hida, H. and Tilouine, J., Anti-cyclotomic Katz p-adic L-functions and congruence modules, Annales scient. de l'È.N.S. 4 série, tome 26, no. 2 (1993) 189–259. [83, 84, 88]

[Hir1]   H. Hironaka, Introduction to real-analytic sets and real-analytic maps. Quaderni dei Gruppi di Ricerca Matematica del Consiglio Nazionale delle Ricerche. Istituto Matematico "L. Tonelli" dell'Università di Pisa, Pisa, 1973. [23]

[Hir2]   H. Hironaka, Subanalytic sets. Number theory, algebraic geometry and commutative algebra, in honor of Yasuo Akizuki, pp. 453–493. Kinokuniya, Tokyo, 1973. [23]

[Hir3]   H. Hironaka, Introduction to the theory of infinitely near singular points. Memorias de Matematica del Instituto "Jorge Juan", No. 28 Consejo Superior de Investigaciones Científicas, Madrid, 1974. [23]

[Hir4]   H. Hironaka, Triangulations of algebraic sets. Algebraic geometry (Proc. Sympos. Pure Math., **29**, Humboldt State Univ., Arcata, Calif., 1974), pp. 165–185. Amer. Math. Soc., Providence, R.I., 1975. [23]

[Hir5]   H. Hironaka, Stratification and flatness. Real and complex singularities (Proc. Ninth Nordic Summer School/NAVF Sympos. Math., Oslo, 1976), pp. 199–265. Sijthoff and Noordhoff, Alphen aan den Rijn, 1977. [23]

[Hirz]   F. Hirzebruch, and D. Zagier, Intersection numbers of curves on Hilbert modular surfaces and modular forms of Nebentypus, Invent. Math., **36** (1976), 57–114. [1, 3, 16, 135]

[Hud]   J. Hudson, **Piecewise Linear Topology**, W.A. Benjamin Inc., New York, 1969. [21, 184]

[Iv]   B. Iverson, **Cohomology of Sheaves**, Universitext, Springer Verlag, Berlin, 1986. [23, 45]

[Iwan]   H. Iwaniec and E. Kowalski, **Analytic Number Theory**, American Mathematical Society, Providence R.I., 2004. [81]

[JaLai] H. Jacquet, K.F. Lai, and S. Rallis, A trace formula for symmetric spaces, Duke Math. J. **70** No. 2 (1993) 305–372. [14]

Bibliography

[JaLan] H. Jacquet and R. Langlands, **Automorphic forms on** GL(2), Lecture Notes in Mathematics **114**, Springer Verlag, N.Y., 1970. [88, 217, 221, 231]

[JaS1] Jacquet, H. and Shalika, J.A., On Euler products and the classification of automorphic representations I, American Journal of Mathematics, **103** 3 (1981) 499–558. [83]

[JaS2] Jacquet, H. and Shalika, J.A., On Euler products and the classification of automorphic representations II, American Journal of Mathematics, **103** 4 (1981) 777–815. [83, 176, 177]

[Joh] F. Johnson, On the triangulation of stratified sets and singular varieties, Mathematika **29** (1982), 137–170. [23]

[Kaw] T. Kawasaki, The signature theorem for V-manifolds, Topology **17** (1978), 75–83. [187, 190, 191]

[KnL] A. Knightly and C. Li, **Traces of Hecke Operators**, Mathematical Surveys and Monographs **133**, American Mathematical Society, Providence, R.I., 2006. [203]

[Koba] S. Kobayashi and K. Nomizu, **Foundations of Differential Geometry**, John Wiley & Sons, New York, 1963, 1991. [188, 189]

[Kobl] N. Koblitz, **Introduction to Elliptic Curves and Modular Forms**, Graduate Texts in Mathematics **97**, Springer Verlag, N.Y., 1984. [76]

[Kot] R.E. Kottwitz, Orbital integrals on GL_3, Amer. Jour. Math. **102** No. 2 (1980) 327–384. [225, 227]

[Ku1] S. Kudla, Algebraic cycles on Shimura varieties of orthogonal type, Duke Math. J. **86** (1997), 39–78. [3]

[Ku2] S. Kudla, From modular forms to automorphic representations, **An Introduction to the Langlands Program**, Birkhäuser, 2004, 133–151. [214, 215, 233, 236]

[Ku3] S. Kudla, Special cycles and derivatives of Eisenstein series, **Heegner Points and Rankin L-series**, MSRI Publications **49**, Cambridge University Press, 2004, 243–270. [3, 6]

[KuM1] S. Kudla and J. Millson, The theta correspondence and harmonic forms I, Math. Ann. **274** (1986), 353–378. [3, 8]

[KuM2] S. Kudla and J. Millson, The theta correspondence and harmonic forms II, Math. Ann. **277** (1987), 267–314. [3, 8]

[KuM3] S. Kudla and J. Millson, Intersection numbers of cycles on locally symmetric spaces and Fourier coefficients of holomorphic modular forms in several complex variables, Publ. Math. de l'IHÉS, **71** 1990, 121–172. [3, 6, 15]

[KuR] S. Kudla, M. Rapoport, and T. Yang, **Modular Forms and Special Cycles on Shimura Curves**. Annals of Mathematics Studies **161** (2006). [3]

[Kutz] P. Kutzko, The Langlands conjecture for GL_2 of a local field, Annals of Math. **112** 2, (1980) 381–412. [231]

[La] S. Lang, **Introduction to Modular Forms**, Grundlehren der mathematischen Wissenschaften **222**, Springer, 2001. [72]

[Lan] R. Langlands, **Base Change for $GL(2)$**, Annals of Mathematics Studies **96**, Princeton University Press, 1980. [87, 136, 176, 223, 230, 234, 235, 236]

[LanS] R. Langlands and D. Shelsted, Descent for transfer factors, in **The Grothendieck Festschrift**, vol. II, Birkhäuser, 1990. [223]

[Lau] G. Laumon, **Cohomology of Drinfeld Modular Varieties, Part I: Geometry, counting of points and local harmonic analysis**, Cambridge studies in adv. math. **41** Cambridge University Press 1996. [227]

[Li] W.-C. Li, Newforms and functional equations, Math. Ann. **212** (1975), 285–315. [71]

[Lo1] S. Lojasiewicz, Triangulations of semi-analytic sets. Ann. Scu. Norm. Pisa **18** (1964), 449–474. [23]

[Lo2] S. Lojasiewicz, Ensembles semi-analytiques, Lectures Notes (1965), I.H.E.S., Bures-sur-Yvette, France. [23]

[Loo] E. Looijenga, L^2 cohomology of locally symmetric varieties. Comp. Math. **67** (1988), pp. 3–20. [44]

[Ma1] J. Mather, Notes on topological stability, unpublished lecture notes, Harvard University, 1970.
http://www.math.princeton.edu/facultypapers/mather/notes_on_topological_stability.pdf [22]

[Ma2] J. Mather, Differentiable invariants, Topology **16**, 1977, 145–155. [194]

[Mi] J. Millson, Intersection numbers of cycles and Fourier coefficients of holomorphic modular forms in several complex variables, Proc. Symp. Pure Math **41** (1989), Part 2, 129–142. [3]

[Mil] J. Milne, Introduction to Shimura varieties, in **Harmonic Analysis, the Trace Formula, and Shimura Varieties**, Clay Mathematical Proceedings **4**, American Mathematical Society, Providence R.I., 2005 [42, 59, 104]

[Miy] T. Miyake, On automorphic forms on GL_2 and Hecke operators, Ann. of Math. **94** (1971) 174–189. [72]

[Mo1] I. Moerdijk and D.A. Pronk, Simplicial cohomology of orbifolds, Indag. Math. **10** (1999), 269–293. [190, 194, 197, 198]

[Mo2] I. Moerdijk. Orbifolds as Groupoids: an Introduction. **Orbifolds in mathematics and physics** (Madison, WI, 2001), Contemp. Math. **310**, Amer. Math. Soc., Providence, RI, 2002. pp. 205–222. [187, 197]

[MP] I. Moerdijk and D.A. Pronk, Orbifolds, Sheaves and Groupoids, K-Theory **12** (1997), 3–21. [187, 188, 195, 197, 198]

[Mu] W. Müller, Signature defects of cusps of Hilbert modular varieties and values of L-series at $s = 1$, Jour. Diff. Geom. **20** (1984), 55–119. [55]

[Mun] J. Munkres, **Elements of Algebraic Topology**, Addison-Wesley, New York, 1984. [185]

[MurR] R. Murty and D. Ranakrishnan, Period relations and the Tate conjecture for Hilbert modular surfaces, Inv. Math. **89** (1987), 319–345. [77, 151, 160]

[Neu] J. Neukirch, **Algebraic Number Theory**, Grundlehren der mathematischen Wissenschaften **332**, Springer-Verlag, 1999. [229]

[No] K. Nomizu, Invariant affine connections on homogeneous spaces, Amer. J. Math **76** (1954), 33–65. [95]

[Oda1] T. Oda, On modular forms associated with indefinite quadratic forms with signature $(2, n - 2)$. Math. Ann. **231** (1977) 97–144. [3]

[Oda2] T. Oda **Periods of Hilbert Modular Surfaces**, Progress in Math. **19** Birkhäuser, 1982. [3, 14, 15]

[Oda3] T. Oda, A note on a geometric version of the Siegel formula for quadratic forms of signature $(2, 2k)$, Science Reports of Niigata University, Series A **20** (1984) 13–24. [3, 16]

[Oda4] T. Oda, The Riemann-Hodge period relation for Hilbert modular forms of weight 2, in **Cohomology of Arithmetic Groups and Automorphic Forms**, Lecture Notes in Mathematics **1447**, Springer Verlag, 1990. [16]

[PlaR] V. Platonov and A. Rapinchuk, **Algebraic Groups and Number Theory**, Academic Press, N.Y., 1994. [4, 42]

[Pop] A. Popa Whittaker newforms for local representations of GL(2), J. of Number Theory **128/6** (2008) 1637–1645. [222]

[Pro] C. Procesi and G. Schwarz, Inequalities defining orbit spaces, Invent. Math. **81** (1985), 539–554. [194]

[Rag] A. Raghuram and F. Shahidi, Functoriality and special values of L-functions, in **Eisenstein Series and Applications**, Progress in Mathematics **258**, Birkhäuser 2008. [13]

[Rap] M. Rapoport, On the shape of the contribution of a fixed point on the boundary, in **The Zeta Functions of Picard Modular Surfaces**, R. Langlands and D. Ramakrishanan, eds., Publ. CRM, Univ. of Montréal, 1992, 479–488. [11, 51]

[Ram1] Ramakrishnan, D., Modularity of Rankin-Selberg L-series, and multiplicity one for SL_2, Annals of Math. **152** (2000) 45–111. [86]

[Ram2] Ramakrishnan, D., Modularity of solvable Artin representations of GO(4)-type, IMRN **1** (2002) 1–54. [85, 86, 175]

[Ram3] D. Ramakrishnan, Algebraic cycles on Hilbert modular fourfolds and poles of L-functions, in **Algebraic groups and arithmetic**, S.G. Dani and

G. Prasad, eds., Tata Inst. Fund. Res., Mumbai, 2004, pp. 221–274. [77, 151, 160]

[RamV] D. Ramakrishnan and R. Valenza, **Fourier Analysis on Number Fields**, Graduate Texts in Mathematics **186**, Springer Verlag, New York, 1999. [203, 205]

[Rib] K. Ribet, Twists of modular forms and endomorphisms of abelian varieties, Math. Ann. **253** (1980), 43–62. [160]

[Rog] J.D. Rogawski, **Automorphic Representations of the Unitary group in Three Variables**, Annals of Math. Studies **123** 1990. [14]

[Rou] C. Rourke and B. Sanders, Homology stratifications and intersection homology, **Proceedings of the Kirbyfest** Geom. Topol. Monogr., 2, Geom. Topol. Publ., Coventry, 1999. [32]

[Sa1] L. Saper, $\mathcal{L}$-modules and micro-support, Ann. Math., to appear. [11, 50, 51]

[Sa2] L. Saper, L_2-cohomology of locally symmetric spaces. I, Pure and Applied Mathematics Quarterly, **1** (2005), 889–937. [41]

[SaS1] L. Saper and M. Stern, L_2-cohomology of arithmetic varieties, Ann. of Math. **132** (1990), pp. 1–69. [44]

[SaS2] L. Saper and M. Stern, Appendix, in **The Zeta Functions of Picard Modular Surfaces**, R. Langlands and D. Ramakrishanan, eds., Publ. CRM, Univ. of Montréal, 1992, 489–491. [11, 51]

[Sat] I. Satake, The Gauss-Bonnet theorem for V-manifolds. J. Math. Soc. Japan **9** (1957), 464–492. [42, 43, 187, 190, 194, 197]

[Schm] R. Schmidt, Some remarks on local newforms for GL(2), J. Ramanujan Math. Soc. 17 (2002) 115–147. [229, 231]

[Schwa] G. Schwarz, Smooth functions invariant under the action of a compact Lie group, Topology **14** (1975), 63–68. [194]

[Schwe] J. Schwermer, Cohomology of arithmetic groups, automorphic forms and L-functions, in **Cohomology of Arithmetic Groups and Automorphic Forms**, Lecture Notes in Mathematics **1447**, Springer Verlag, 1990. [46]

[Ser] J-P. Serre, Abelian ℓ-adic Representations, W.A. Benjamin, 1968. [236]

[Sha] Shahidi, F., Automorphic L-functions: A survey in **Automorphic Forms, Shimura Varieties, and L-functions, Volume I**, Perspectives in Mathematics **11** (1990), 415–437. [177]

[She] Shemanske, T. and Walling, L., Twists of Hilbert modular forms, TAMS, **338** (1993) 375–403. [83, 177]

[Shim1] G. Shimura, **Introduction to the Arithmetic Theory of Automorphic Functions**, Princeton University Press, 1971. [47, 66, 122, 127, 139, 182]

[Shim2] G. Shimura, On elliptic curves with complex multiplication as factors of the Jacobians of the modular function fields. Nagoya Math. J. **43** (1971), 199–208. [160]

Bibliography

[Shim3] G. Shimura, On the factors of the Jacobian variety of a modular function field. J. Math. Soc. Japan **25** (1973), 523–544. [160]

[Shim4] G. Shimura, The special values of zeta functions associated with Hilbert modular forms, Duke. Math. J. **45** (1978) 637–679. [68, 116, 121, 180]

[Sja] R. Sjamaar and E. Lerman, Stratified symplectic spaces and reduction, Ann. of Math. (2) **134** (1991), 375–422. [194]

[Sto] W. Stoll, The growth of the area of a transcendental analytic set, I and II, Math. Ann. 156 (1964) 47–78 and 144–170. [55]

[Stz] G. Stolzenberg, **Volumes Limits, and Extensions of Analytic Varieties**, Lecture Notes in Mathematics **19** Springer Verlag, NY 1966. [55]

[Tay1] R. Taylor, On Galois representations associated to Hilbert modular forms, Invent. math. **98** (1989), 265–280. [236]

[Tay2] R. Taylor, Representations of Galois groups associated to Hilbert modular forms, in **Automorphic Forms, Shimura Varieties, and L-functions, Volume II**, Perspectives in Mathematics **11** (1990), 323–336. [236]

[Th] R. Thom, Ensembles et morphismes stratifiés, Bull. Amer. Math. Soc. **75** (1969), 240–284. [22]

[Tol] D. Toledo and Y.L. Tong, Duality and intersection theory in complex manifolds 1, Math. Ann. **237** (1978), 41–77. [15]

[Ton] Y.L. Tong, Weighted intersection numbers on Hilbert modular surfaces. Comp. Math. **38** (1979) 299–310. [3, 15, 147]

[TonW1] Y.L. Tong and S.P. Wang, Harmonic forms dual to geodesic cycles in quotients of $SU(p,1)$, Math. Ann. **258** (1982), 189–318. [3]

[TonW2] Y.L. Tong and S.P. Wang, Construction of cohomology of discrete groups, Trans. Amer. Math. Soc. **306** (1988), 735–763. [3]

[Weib] C. Weibel, **An Introduction to Homological Algebra**, Cambridge studies in advanced mathematics **38**, Cambridge University Press, Cambridge, 1994. [195]

[Weil] A. Weil, **Dirichlet Series and Automorphic Forms**, Lecture Notes in Mathematics **189**, Springer-Verlag 1971. [74]

[Ya] C.T. Yang, The triangulability of the orbit space of a differentiable transformation group, Bull. Amer. Math. Soc. **69** (1963), 405–408. [194]

[Za] D. Zagier, Modular forms whose Fourier coefficients involve zeta functions of quadratic fields, **Modular Functions in One Variable VI**, Lecture Notes in Mathematics **627** Springer-Verlag (1976) 105–169. [9, 16]

[Zu1] S. Zucker, L_2 cohomology of warped products and arithmetic groups. Inv. Math **70** (1982), 169–218. [44, 50, 55]

[Zu2] S. Zucker, Satake compactifications, Comm. Math. Helv. **58** (1983), 312–343. [50]

Index of Notation

$|\cdot|_{\mathfrak{c}}$ restricted absolute value, 203
$|\xi|$ support of ξ, 27
$\langle \cdot, \cdot \rangle_{IH}$ intersection pairing, 36
$\langle \cdot, \cdot \rangle_K$ Kronecker pairing, 36
$(\cdot, \cdot)_P$ Petersson product, 11, 70, 125
$[x_0]$ fractional ideal, 4, 200
$[\cdot, \cdot]_{IH}$ adjusted product, 125
$[k+2m]$ $k_\sigma + 2m_\sigma$, 82, 116, 125, 152
$\{k\}$ $\sum_\sigma k_\sigma$, 125
$|\cdot|_{\mathbb{A}}$ normalized absolute value, 3, 201
$\langle p, p' \rangle$ inner product, 123
$\gg 0$ totally positive, 4
α_z, 64
$\gamma_{\chi_E}(\mathfrak{m})$ Hecke translate, 8, 142
Γ_c sections with compact support, 45
Γ_0^{ifr} torsion free subgroup, 133
$\Gamma_j(\mathfrak{c}), \Gamma_j^0(\mathfrak{c})$ arithmetic group, 59, 60
Γ_i^{qu} quotient group, 133
ϵ augmentation, 36
$\epsilon(z)$ sign of Im(z), 157
ι main involution, 66, 111
θ Cartan involution, 49
Θ bilinear form, 155
κ weight, 61, 63
$\widehat{\kappa}$ weight, 136, 179
$\widehat{\kappa} - w$, 160
$\lambda_f(\mathfrak{m})$ Hecke eigenvalue, 8, 68
$\pi_\mathfrak{a}(f)$, 75
$\widehat{\pi}$ base change of π, 140, 230
σ real embedding, 58
$\sigma'_{\mathfrak{c}_E, \vartheta}$, 180
$\Sigma(L)$ infinite places of L, 3, 58
$\tau_{\leq k}$ truncation, 31
Υ fractional ideal for twisting, 77, 160
$\phi_\mathfrak{c}$ restricted Hecke character, 203
χ_E quasicharacter on $\mathbb{A}_E^\times$, 136
$\chi_0, \chi_0^\vee$ character, 63, 66, 113, 152
ψ Hecke character, 119
$\omega_f, \omega_J(f^{-\iota})$
   differential form, 9, 108, 110, 114
$[\omega_f]$ cohomology class, 108, 110
$\Omega^r(X, E)$ differential forms, 92, 107
$\Omega_2^\bullet$ L^2 differential forms, 43
$\Omega_{\text{inv}}^\bullet$ invariant differential forms, 46

$\lambda^\vee$ representation, 101
1 the weight $(1, \ldots, 1)$, 61
A subalgebra of $\mathbb{C}$, 132
$\mathcal{A}_0(G, \chi)$ cusp forms, 215
$\mathbb{A}$ adèles of $\mathbb{Q}$, 3
$\mathbb{A}_L$ adèles of L, 3, 199
A_G torus, 41
$\mathcal{A}(G, \chi)$ automorphic forms, 214
$a(\mathfrak{m}, f)$
   mth Fourier coefficient, 5, 73, 206
$a(\xi y \mathcal{D}, f)$ Fourier coefficient, 73
A'_P A_P/A_G, 49
b_0 non-archimedean part, 3, 199
b_∞ archimedean part, 3, 199
B' upper triangular subgroup, 172
BG classifying space, 194
C, C_σ Casimir, 62
$\mathfrak{c}$ ideal, 4
$\mathfrak{c}_E = \mathfrak{c} \cap \mathcal{O}_E$, 7, 136
C_{BM}^i Borel-Moore chains, 24
$\widehat{C}_r^K(Y)$
   cellular Borel-Moore chains, 27
CLC
   cohomologically loc. const., 23, 32
$\mathbf{C}^\bullet(X, \mathbf{E})$ sheaf of chains, 24
D symmetric space, 58
$d\mu_K, d\mu_\mathfrak{c}$ canonical measure, 68
$D_c^b(X)$ derived category, 23
$\det_\infty$ determinant, 103
$\det$ determinant, 61
$\mathcal{D}_{L/E}$ different, 4, 136
$d_{L/E}$ discriminant, 4, 136
$\mathcal{D}_\mu(k)$ discrete series, 216
$\mathbb{D}_X^\bullet$ dualizing sheaf, 25
E totally real field, 4
$\mathbf{E}, \mathcal{E}(\chi_0)$
   local coefficient system, 15, 24, 48
$\mathcal{E}(\chi_0)$ local coefficient system, 103
$\mathcal{E}'_{\mathfrak{c}_E, \vartheta}$ Eisenstein series, 179
EG universal space, 194
e_L standard additive character, 3, 200

249

Index of Notation

f_c complex conjugate, 10, 116, 121
$f \otimes \eta$ twist of f, 76
$\widehat{f}$ base change of f, 10, 140, 230
$f^{-\iota}$ main involution, 66, 111
$\mathfrak{f}(\psi)$ conductor, 82, 202
f^ς Galois twist of f, 86

$G(\eta)$ Gauss sum, 78
^0G, 41
$\mathrm{Gal}(L/E)^\wedge$ Galois characters, 4
G, G_L
  algebraic group, 4, 58, 135, 147, 152
$G(\mathbb{Q})^+$ $G(\mathbb{Q}) \cap G(\mathbb{R})^0$, 59

$\mathfrak{h}$ upper half-plane, 59
$\mathcal{H}$ convolution (Hecke) algebra, 218
$\mathfrak{h}^\pm$ $\mathbb{C} - \mathbb{R}$, 101
H^{BM} Borel-Moore homology, 25
H^i_c compact support cohomology, 24
$H^j(\mathfrak{g}_\mathbb{R}, K^1_\infty)$
  Lie algebra cohomology, 46, 180
H^j_{inv} invariant cohomology, 46
$\widehat{H}^K_r$ cellular homology, 27
$h(K_0(\mathfrak{c}))$
  narrow class number, 59, 60, 133

$\mathbf{I}^\mathfrak{p}\mathbf{C}$ intersection chain complex, 30
$I(f,s)$ Rankin-Selberg integral, 170
$I^\mathfrak{p} H^{BM}_i$ closed support, 30
IH^E isotypical part, 143
$I^\mathfrak{m} H^i_{\mathrm{inv}}$ invariant cohomology, 115, 180
IH^{new}, 11
$IH(X_0(\mathfrak{c}))(f)$ isotypical part, 8
$\mathcal{I}_L$ fractional ideals, 73
$\mathbf{I}^\mathfrak{p}\mathbf{S}$ intersection cochain complex, 30

$j(\alpha, z)$ automorphy factor, 101
j_x inclusion of x, 24
$j^!_x$ extraordinary pullback, 24

$K_0(\mathfrak{c})$ Hecke type subgroup, 4, 60
$K_{11}(\mathfrak{c})$ congruence group, 89
$K_1(\mathfrak{c})$
  Hecke congruence group, 153, 156
$K^\vee_1(\mathfrak{c})$ Hecke congruence group, 156
K^1_∞ maximal
  compact subgroup, 41, 58, 101
$|K|$ geometric realization, 21
K^{fr}
  torsion-free subgroup, 60, 104, 132

K_∞, 4, 41, 58
$K_{L,\infty}$, 58
(k, m) weight, 61, 136, 179
ℓ_x link at x, 30
L totally real field, 4, 136
L/E totally real extension, 136
$L_\mathfrak{p}(f \otimes \eta, s)$ local factor, 77, 82
$L(f \otimes \eta, s)$ standard L function, 77, 82
$L^c(s), L_\mathfrak{c}(s)$, 81
$L(\mathrm{Ad}(f) \otimes \phi, s)$ adjoint L-function, 84
$L(\mathrm{As}(f \otimes \phi), s)$ Asai L-function, 85
$L(f \times g, s)$
  Rankin-Selberg L-function, 83
$\mathcal{L}(\kappa, \chi_0)$ local system, 62, 103, 113
$\mathcal{L}^\vee(\kappa, \chi_0)$ local system, 113
$L(\kappa)$ representation of GL_2, 61, 102

M large number field, 132
m lower middle perversity, 29, 50
n upper middle perversity, 29, 50
$\mathcal{M}(f), \mathcal{M}^E$ isotypical component, 141
$M_\kappa(K_0(\mathfrak{c}))$ Hilbert modular forms, 64
$M^{\mathrm{coh}}_\kappa(K_0(\mathfrak{c}))$
  Hilbert modular forms, 65
$M(\kappa)$ representation of GL_2, 100, 102

$\mathcal{N}(\mathfrak{c})$ upper bound, 7, 142, 235
N norm map, 8, 136, 138, 228

$\mathcal{O}_L$ integers in L, 3
$\widehat{\mathcal{O}}_L$, 3, 70, 77, 200
$\mathcal{O}^\times_{L,+}$ positive units, 60
$\mathcal{O}_v$ valuation ring, 199
$\mathcal{O}_Y$ orientation sheaf, 26

$\mathcal{P}$ Poincaré duality, 31, 36
p perversity, 29
$P^{(k)}$ section of Sym^k, 102
P_{new}, 11, 151, 168
$\mathfrak{p}_v$ prime
  ideal corresponding to v, 63, 199
P_z section, 102

$q_\kappa(x, y)$ Whittaker function, 5, 73, 205

R coefficient ring, 23
$R(\mathfrak{c})$ semigroup, 66, 129
$R^\vee_\mathfrak{c}$ opposite semigroup, 129, 131
$\mathrm{Res}_{L/\mathbb{Q}}$ restriction of scalars, 4, 58
RHom derived Hom, 24

$S_\kappa^{\mathrm{coh}}(K_0(\mathfrak{c}),\chi)$ cusp forms, 65
$S_\kappa(K_0(\mathfrak{c}),\chi)$ cusp forms, 5, 63
$S_\kappa^{\mathrm{new}}(K_0(\mathfrak{c}),\chi)$ newforms, 9, 71, 144
S_κ^+ plus space, 7, 143
Stab Stabilizer, 95

$T_\mathfrak{c}(\mathfrak{n}), T_\mathfrak{c}(\mathfrak{n},\mathfrak{n})$
   Hecke operator, 6, 67, 131
$t_j(\mathfrak{c})$, 59, 60
$\widehat{T}(\mathfrak{m})$ section of b, 8, 137, 138
$\mathbb{T}_\mathfrak{c}$ Hecke algebra, 6, 67, 218

$\mathcal{U}_P$ unipotent radical, 49

V_{χ_E}, 181

W_m archimedean
   Whittaker function, 205
$\mathcal{W}_\pi$ Whittaker model, 207
$W_\mathfrak{c}$ Atkin-Lehner matrix, 118
$W_\mathfrak{c}^*$ Atkin-Lehner
   operator, 10, 17, 119, 125
$W(f)$ eigenvalue of $W_\mathfrak{c}^*$, 10, 121
w_σ, 158

$X_0(\mathfrak{c})$ Baily-Borel
   compactification, 4, 59, 60
$\overline{X}^{\mathrm{RBS}}$ reductive Borel-Serre, 49
x^ι main involution, 66
$\mathcal{X}(L)$ set of weights, 61, 63, 103
$\widehat{X}_P$ stratum, 50

$Y_0, Y_0(\mathfrak{c})$
   Hilbert modular variety, 4, 58–60
Y_0^{fr} torsion-free quotient, 61, 104, 133

$\mathcal{Z}$ Zucker map, 44
$Z_1(\mathfrak{c})$ Hilbert modular variety, 153
$\mathbb{Z}[\chi]$ subalgebra of $\mathbb{C}$, 137
$Z(\mathfrak{g})$-finite, 214
Z_θ twisted cycle, 160, 165

Index of Terminology

absolute value, 3, 201
adèles, 3, 199
adjunction, 39
admissible module, 232
admissible representation, 219
algebraic cycle, 28
allowability condition, 30
arithmetic group, 42
Artin map, 203
Asai L-function, 85
Atkin-Lehner
    operator, 10, 111, 117–119, 125
atlas, R-orbifold, 190
augmentation, 36
automorphic form, 105, 214
automorphic representation, 214, 215
automorphy factor, 64, 97, 101

Baily-Borel
    compactification, 4, 6, 42, 59
base change,
    6, 8, 9, 86, 136, 140, 213, 228, 230
basic neighborhood, 22
Borel-Moore chains, 24, 27
boundary component, 42

Cartan involution, 49
Casimir, 62, 233
category, derived, 23
cell complex, 21
cell, convex, 21
central character, 5, 63
chains, 21
    Borel-Moore, 24, 27
    elementary, 183
    intersection, 30
    sheaf of, 24
    support, 27
character
    central, 5, 63
    Hecke, 4, 202
    standard additive, 3, 200
chart, orbifold, 190
class group, 60, 201
class number, narrow, 60

classifying space, 194
cochain, 21
    intersection, 30
coefficient system, 91
cohomologically constructible, 23
cohomologically locally constant, 23, 32
cohomology
    cuspidal, 114
    invariant, 115, 180
    isotypical part, 143
    Lie algebra, 46
    local, 24
    sheaf, 23
    stalk, 24
compactification
    Baily-Borel, 59, 60
    reductive Borel-Serre, 42, 49
complex
    dualizing, 37
    simplicial, 21
complex conjugate, 111, 116, 121
component group, 109
conductor, 63, 202, 216, 229, 231
conjecture, Zucker, 43
connection
    flat, 28
connection, flat, 92, 95
constructible, 23
contragredient, 61, 154
convex cell, 21
correspondence, 38
    Hecke, 6, 47, 126
    twisting, 160
cosheaf
    definition of, 183
    homology of, 183
cuspidal
    cohomology, 114
    function, 63
    representation, 215
cycle
    algebraic, 28
cycle, modular, 51

degree (of Hecke operator), 182

253

derived category, 23
diagonal embedding, 16
different, 4, 136, 200
differential form, 107, 196
  L^2, 44
  closed, 108
  invariant, 46
Dirichlet unit theorem, 65
discrete series, 216, 233
discrete structure group, 92
discriminant, 4, 136
dual representation, 61, 154
duality
  Poincaré, 26, 31, 37
  Verdier, 45
dualizing sheaf, 25

effective action, 187
eigenvalue, Hecke, 68
Eisenstein series, 64, 179
elementary chain, 183
elliptic modular form, 2
embedding, diagonal, 16
equivariant sheaf, 194
Euler factor, 221
Euler product, 82

face (of a cell), 21
flat connection, 28, 92, 95
form, differential, 108
Fourier
  coefficient, 5, 10, 74, 144, 146
  expansion, 5, 205
  series, 72
fractional ideal, 4, 200
function
  cuspidal, 63
  Whittaker, 5, 72, 205, 207
fundamental class, 26, 51

Gauss sum, 78
geodesic action, 49
groupoid, 197

Haar measure, 203
Hecke
  algebra, 6, 67, 136, 218
  character, 4, 202
  correspondence, 6, 47, 126
  eigenvalue, 8, 68

  operator, 6, 66, 67
    degree of, 182
  subgroup, 4, 60
Hermitian symmetric space, 42
highest weight, 100
Hilbert modular
  form, 62
  variety, 4, 58–60
homogeneous vector bundle, 95
homological stratification, 32
homology manifold, 25, 191
homology sheaf, 25

idèle class group, 201
inner product, 123
intersection chains, 30
intersection pairing, 35
invariant cohomology, 115, 180
invariant differential form, 46
invariant metric, 41
involution
  Cartan, 49
  complex conjugation, 111
  main, 66, 111

L-function, 221
  adjoint, 84
  Asai, 85
  Rankin-Selberg, 83
  standard, 82
Langlands decomposition, 49
leading coefficient, 74
level, 5
lifting, 230
link, 22, 30, 54
local coefficient
  system, 62, 91, 103, 113, 133, 196
  simplicial, 183
local homology, 25
locally symmetric space, 42

main involution, 66, 111
measure
  canonical, 68
  Haar, 203
metric, invariant, 41
middle perversity, 29, 50
minimal representation, 84
model, Whittaker, 205, 207

Index of Terminology

modular cycle, 51
modular form
  complex conjugate, 116
  elliptic, 2
  Hilbert, 62
modulus function, 203
multiplicity one, 72, 145, 216

narrow class group, 60, 133, 201
nebentypus, 5, 63
new subspace, 11
newform, 8, 9, 16, 71, 86, 217
  normalized, 216
norm, 8, 136, 138
normal operator, 72
normalized
  absolute value, 3, 201
  cusp form, 74
  newform, 216

old subspace, 11
operator
  Atkin-Lehner, 10, 111
  Hecke, 66
orbifold, 187
  atlas, 190
  chart, 190
  local system, 95, 96, 100
  stratification, 32
  structure, 193
orientation, 21, 36

pairing, 93
  dual, 37
  intersection, 35, 36
  Kronecker, 36
partition, 100
perversity, 29
  middle, 29, 50
Petersson product, 11, 68, 203
Poincaré duality, 26, 31, 37
polynomials, space of, 113
Pontrjagin dual, 200
principal series, 229
product
  intersection homology, 35, 36
  Kronecker, 36
  Petersson, 11, 68, 203
projection formula, 38

pseudo cell decomposition, 21
pseudomanifold, 21
  oriented, 23

quasi-isomorphism, 23
quasicharacter, 4, 202

ramified character, 202
Rankin-Selberg method, 170
representation
  admissible, 219, 229
  automorphic, 214, 215
  cuspidal, 215
  discrete series, 233
  dual, 61, 154
  principal series, 229
  Steinberg, 229
  Weil, 229, 231
restricted Hecke character, 203
restriction of scalars, 4, 58
de Rham theorem, 196
R-orbifold
  atlas, 190
  chart, 190
  structure, 193

Saper's theorem, 50
Satake topology, 43
scalars, restriction of, 4, 58
Schur module, 100
semi-algebraic set, 22
series, Fourier, 72
set
  semi-algebraic, 22
  semi-analytic, 22
  subanalytic, 22
sheaf
  cohomology, 23
  complex of, 23
  dualizing, 25
  equivariant, 194
  local homology, 25
  of chains, 24
  on an orbifold, 195
  orientation, 26
  simplicial, 184
Shimura variety, 58
simplicial local system, 183
simplicial sheaf, 184

simultaneous eigenform, 74
standard additive character, 3, 200
Steinberg representation, 229
stellar subdivison, 183
stratification, 193
    homological, 32
    orbifold, 32
    Whitney, 22
stratum, 50
strong approximation, 59
structure group, 92
subanalytic set, 22
subdivision, stellar, 183
support, 27
symmetric space, 41, 58

topological triviality, 22
torsion-free subgroup, 60, 133
totally positive, 4
triangulation, 22, 27
twisted cycle, 165
twisting, 76
    correspondence, 160

unimodular, 203
unipotent radical, 49
unit theorem, Dirichlet, 65
universal space, 194
unramified character, 202

valuation ring, 199
variety
    Hilbert modular, 4, 58–60
    Shimura, 58
vector bundle, 91, 95, 104
Verdier dual, 45
volume, 42

weakly increasing, 64
weights, 61
    lifted from E, 136
Weil representation, 229, 231
Whitney conditions, 22
Whittaker function, 5, 72, 205, 207
Whittaker model, 205, 207

Young diagram, 100

Zucker conjecture, 43
Zucker map, 44

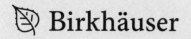

Progress in Mathematics (PM)

Edited by
Hyman Bass, University of Michigan, USA
Joseph Oesterlé, Institut Henri Poincaré, Université Paris VI, France
Alan Weinstein, University of California, Berkeley, USA
Yuri Tschinkel, Courant Institute of Mathematical Sciences, New York, USA

Progress in Mathematics is a series of books intended for professional mathematicians and scientists, encompassing all areas of pure mathematics. This distinguished series, which began in 1979, includes research level monographs, polished notes arising from seminars or lecture series, graduate level textbooks, and proceedings of focused and refereed conferences. It is designed as a vehicle for reporting ongoing research as well as expositions of particular subject areas.

PM 296: Itenberg, I.; Jöricke, B.; Passare, M. (Eds.)
Perspectives in Analysis, Geometry, and Topology. On the Occasion of the 60th Birthday of Oleg Viro (2012).
ISBN 978-0-8176-8276-7

PM 295: Joseph, A.; Melnikov, A.; Penkov, I. (Eds.)
Highlights in Lie Algebraic Methods (2012).
ISBN 978-0-8176-8273-6

PM 294: Barreira, L.
Thermodynamic Formalism and Applications to Dimension Theory (2011).
ISBN 978-3-0348-0205-5

PM 293: Mazzucchelli, M.
Critical Point Theory for Lagrangian Systems (2011).
ISBN 978-3-0348-0162-1

PM 292: van den Ban, E. P.; Kolk, J.A.C. (Eds.)
Geometric Aspects of Analysis and Mechanics. In Honor of the 65th Birthday of Hans Duistermaat (2011).
ISBN 978-0-8176-8243-9

PM 291: Greene, R.E.; Kim, K.-T.; Krantz, S.G.
The Geometry of Complex Domains (2011).
ISBN 978-0-8176-4139-9

PM 290: Mantegazza, C.
Lecture Notes on Mean Curvature Flow (2011).
ISBN 978-3-0348-0144-7

PM 289: Colombo, F.; Sabadini, I.; Struppa, D. C.
Noncommutative Functional Calculus (2011).
ISBN 978-3-0348-0109-6

PM 288: Neeb, K.-H.; Pianzola, A. (Eds.)
Developments and Trends in Infinite-Dimensional Lie Theory (2011).
ISBN 978-0-8176-4740-7

PM 287: Cattaneo, A.S.; Giaquinto, A.; Xu, P. (Eds.)
Higher Structures in Geometry and Physics (2011).
ISBN 978-0-8176-4734-6

PM 286: Abbes, A.
Éléments de Géométrie Rigide.
Volume 1: Construction et Étude Géométrique des Espaces Rigides (2011).
ISBN 978-3-0348-0011-2

PM 285: Soifer, A.
Ramsey Theory. Yesterday, Today, and Tomorrow (2010).
ISBN 978-0-8176-8091-6

PM 284: Gyoja, A.; Nakajima, H.; Shinoda, K.-I.; Shoji, T.; Tanisaki, T. (Eds.)
Representation Theory of Algebraic Groups and Quantum Groups (2010).
ISBN 978-0-8176-4696-7

PM 283: El Zein, F.; Suciu, A.I.; Tosun, M.; Uludag, M.; Yuzvinsky, S. (Eds.)
Arrangements, Local Systems and Singularities (2010). ISBN 978-3-0346-0208-2

PM 282: Bogomolov, F.; Tschinkel, Y. (Eds.)
Cohomological and Geometric Approaches to Rationality Problems (2010).
ISBN 978-3-7643-8798-3